国外优秀数学著作
原 版 系 列

U0139743

Affine Elastic Curves in R² and R³ —Concepts and Methods

R²和R³中的仿射弹性曲线——概念与方法

● 黄荣培 著

（英文）

哈尔滨工业大学出版社
HARBIN INSTITUTE OF TECHNOLOGY PRESS

黑版贸审字 08 - 2019 - 180 号

图书在版编目(CIP)数据

$\mathbf{R}^2$ 和 $\mathbf{R}^3$ 中的仿射弹性曲线:概念与方法 = Affine Elastic Curves in $\mathbf{R}^2$and $\mathbf{R}^3$:Concepts and Methods:英文/黄荣培著. —哈尔滨:哈尔滨工业大学出版社,2022.8
ISBN 978-7-5767-0279-8
Ⅰ.①R… Ⅱ.①黄… Ⅲ.①仿射几何-英文 Ⅳ.①O186.13

中国版本图书馆 CIP 数据核字(2022)第 121756 号

$\mathbf{R}^2$ HE $\mathbf{R}^3$ ZHONG DE FANGSHE TANXING QUXIAN:GAINIAN YU FANGFA

策划编辑 刘培杰 杜莹雪
责任编辑 张永芹 邵长玲
封面设计 孙茵艾
出版发行 哈尔滨工业大学出版社
社　　址 哈尔滨市南岗区复华四道街 10 号　邮编 150006
传　　真 0451 - 86414749
网　　址 http://hitpress.hit.edu.cn
印　　刷 黑龙江艺德印刷有限责任公司
开　　本 886 mm×1 230 mm　1/32　印张 8.875　字数 176 千字
版　　次 2022 年 8 月第 1 版　2022 年 8 月第 1 次印刷
书　　号 ISBN 978-7-5767-0279-8
定　　价 38.00 元

Contents

Preface

This booklet is devoted to the study of the variational properties of affine curves invariant under centroaffine transformations or equiaffine transformations in $\mathbf{R}^2$ and $\mathbf{R}^3$. It can be considered as a counterpart of the study of the classical Euclidean elastic curves. This work is based on my thesis for degree of Doctor of Philosophy at Case Western Reserve University and my further study on this subject.

I would like to express my appreciation for the guidance, support and kindness of my advisor Professor David Singer, who brings me into this field. From him, I realized how to find some research idea from the "simple" discrete objects. This changes the way I used to study mathematics. I would also like to thank Professor Joel Langer for his assistance and extensive technique comments.

I would like to thank Professor Chunli Shen for his long time encouragement and helpful combined academic and life guidance.

Finally, I should like to thank Hannah Olsen for inviting me to prepare this booklet with Lambert Academic Publishing.

Rongpei Huang
At East China Normal University

Chapter 1

Introduction

1.1 Background of elastica theory

The formal investigation of the elastica began 300 years ago with a challenge posed and later solved by James Bernoulli. That investigation was motivated by a physical problem of minimizing the bending energy of a thin inextensible wire. One might wish to determine the results of applying loads to a beam, either horizontal or vertical, supported and secured at one end. A practical application might be to determine the maximum load that could be supported by a beam of specified size and composition. There was a great deal of experimental data dating back to the construction of the medieval cathedrals([23]). In the seventeenth century this so called problem of fracture had been formalized and progress was made by Galileo Galilei, Edme Mariotte, Gottfried Leibniz, and Pierre Varignon([12]).

One observes the shape of the beam under a load, where it is assumed that the beam will recover its size and shape when the load is removed. In 1691 James Bernoulli posed this problem for the special case of the rectangular elastica of unit excursion. This elastic beam is vertical and secured at the bottom and perpendicular to the floor. A load is attached to the top that acts vertically and bends the beam until the top is parallel to the floor. Three years later Bernoulli had received no answers to his challenge and so published his solution([39]). Christiaan Huygens criticized Bernoulli for not showing all of the possible cases that could have resulted from his theory and sketched several of the elastica not included in Bernoulli's analysis. A year before posing the problem of the elastica, James Bernoulli had posed the problem of catenary. The challenge was to find the equation of the shape taken by a flexible but inextensible cord fixed at two points. By the following year, Leibniz, Huygens and James' brother John Bernoulli had published solutions showing that the solution is $y = c\cosh(x/c)$.

The problems of the catenary and the elastica were two examples that led Bernoulli to invent the calculus of variations and each could be posed in terms of a variational principle. The catenary is the curve for which the center of gravity is as low as possible and the elastica is the curve such that the center of gravity of the included area is farthest from the straight line connecting two end points. In 1742 James' nephew Daniel Bernoulli wrote a letter to Leonhard Euler proposing the total squared curvature as the correct quantity to minimize([14], [31]). Euler finished his book on the Calculus of Variations in 1744 and applied his theory to the elastic curves using Daniel's suggestion. As has happened with the circumstances surrounding James Bernoulli's posing of the problem, the history of Euler's analysis of elastic curves has been overshadowed by other achievements. Pierre-Louis Maupertuis had published a paper on the Principle of Least Action earlier in 1744. For four years prior to this, Euler had corresponded for four years with Maupertuis about theology and the Principle of the Least Action. Euler believed that the nature behaves so as to minimize or maximize some functions.

The Euler-Bernoulli treatment of the elastica transformed a physics problem into one in mathematics. Examining the bending energy of a physical rod is replaced by investigating the total squared curvature of a regular curve. Much of the development in the theory of elastic rods is based on a discovery of Gustav Kirchhoff in 1850's, known as the kinetic analogue of the elastic problem, that the equations that describe the thin elastic rod in equilibrium are mathematically identical to those used to describe the dynamics of the heavy top which was solved exactly by Joseph-Louis Lagrange in 1788. The geometric significance of minimizing the total squared curvature functional was recognized by Wilhelm Blaschke under the name of Radon's problem since Johann Radon studied the Euclidean elastic curve and got the complete solutions.

In 1970's Hidenori Hasimoto([17]) discovered the relation of the localized induction equation (LIE): $\gamma_t = \gamma_s \times \gamma_{ss} = \kappa B$ to soliton theory by showing that it induces an evolution on the complex curvature $\psi = \kappa \exp(\int^s \tau du)$ governed by the nonlinear Schrödinger equation (NLS): $\psi_t = i(\psi_{ss} + \frac{1}{2}|\psi|^2\psi)$. NLS is a well-known example of completely integrable systems with infinitely many conserved quantities and soliton solutions([24],[34]). LIE is also known as the Betchov Da Rios equation. It is a completely integrable Hamiltonian system and possesses infinite many conserved integrals involving the curvature and torsion of the curve. Furthermore the elastica is a congruence solution of the LIE. Since Hasimoto's discovery, during recent three decades, the Euler-Bernoulli model has been reconsidered for numerous reasons([27], [30], [39]). The total squared curvature functional has emerged as a useful quantity in the study of geodesics and the closed thin elastic rod is often used as a model for the DNA molecule([24]). There has been a recent resurgence of interest in applying the theory for the Kirchhoff elastic rod to the phenomenon of DNA super-coiling([37]). Joel Langer and David Singer started the research in a series of papers dealing with closed elastic curves in spaces of constant sectional curvature([27]), particularly concerning the existence of closed elastica on the sphere $\mathbf{S}^2$, and furthermore study the convergence of the negative gradient flow of the total squared curvature([28]). Robert Bryant and Phillip Griffiths exploit the natural symmetries of the problem to obtain interesting results concerning the existence of closed, free elastica in the hyperbolic plane $\mathbf{H}^2$([7]).

1.2 The goal of this book

This book deals with the affine starlike curves and non-degenerate affine curves in $\mathbf{R}^2$ and $\mathbf{R}^3$. This study is stimulated by the elastic curve and the vortex filament problem which describes a curve $\gamma_t(s) = \gamma(s,t)$ evolving in three-dimensional space $\mathbf{R}^3$. The basic idea here is to replace the Euclidean space with an affine space. We try to develop some properties about the elastic curve which is invariant under some specific subgroups of the affine groups $A(2)$ and $A(3)$. For the case of affine plane $\mathbf{R}^2$, we study the non-degenerate affine curve $X(t)$ for which $X'(t)$ and $X''(t)$ are linearly independent. The non-degenerate affine curve can always be reparametrized by an affine arclength parameter s which is defined by the condition that the area of the parallelogram spanned by $X'(s)$ and $X''(s)$ is equal to 1 everywhere. This kind of curve has no inflection point, that means all points of the curve lie on one side of the tangent line locally. The affine curvature $k(s)$ is defined by the (signed) area of the parallelogram spanned by $X''(s)$ and $X^{(3)}(s)$([33], [38]). The affine arclength parameter and affine curvature are invariant under any equiaffine transformation of the affine plane. It turns out that such a curve is unique up to an equiaffine transformation of the plane for a given smooth function $k(s)$. We study the critical point of the affine curvature energy functional $F(X) = \int_X p(k)ds$ on those non-degenerate curves with fixed length satisfying given boundary conditions. Here $p(k)$ is a smooth function of k. The Euler-Lagrange equation of this critical point(affine elastica) is a fourth order ordinary differential equation. We find a special case which is integrable by quadratures. For this special case, we develop a theory on the starlike affine curves which can be considered as an independent integrable geometric variational problem. We study the starlike affine curve $X(t)$ for which $X(t)$ and $X'(t)$ are linearly independent. The starlike affine curve can be reparametrized by the parameter s which is defined by the condition that the area of the parallelogram spanned by $X(s)$ and $X'(s)$ is equal to 1 on the entire curve. We call this parameter s centroaffine arclength parameter. We define the centroaffine curvature by the (signed) area of the parallelogram spanned by $X'(s)$ and $X''(s)$. The centroaffine arclength parameter and centroaffine curvature are invariant under the action of the unimodular group $SL(2, \mathbf{R})$ which is a subgroup of the equiaffine transformation of the affine plane. We study the critical point of the centroaffine curvature energy functional and call it centroaffine elastica. The Euler-Lagrange equation of the centroaffine elastica is an integrable second order ordinary differential equation. We solve it by quadratures and reduce the second order structure equations(centroaffine Frenet equations) to a first order linear system by using Killing field and the classification of the conjugate class of the linear Lie group $SL(2, \mathbf{R})$. Therefore we solve the centroaffine elastica in the affine plane $\mathbf{R}^2$ completely by quadratures. The non-degenerate affine curve and the starlike curve are closely related to each other. If s is an affine arclength parameter of the non-degenerate curve X, then s is a centroaffine arclength parameter of the starlike curve X', the tangent image of the curve X. Furthermore the affine curvature of X becomes the centroaffine curvature of X' at the corresponding point. From this point of view, we can partially solve the affine elastica problem by using the solution of the centroaffine case. Along the similar line, we study the case of the affine space $\mathbf{R}^3$. We define the invariants: affine curvature and affine torsion for the non-degenerate curve and develop its corresponding concepts in centroaffine state. This will deal with the relation between the curvature and torsion. It would be more complicated. Since the non-degenerate curve in the affine plane $\mathbf{R}^2$ is a degenerate curve in the affine space $\mathbf{R}^3$, the three-dimensional case is distinct from the two dimensional case.

The rest of this chapter will be a brief review of the classical elastica in 3-dimensional Euclidean space $\mathbf{E}^3$ and a simple introduction to elliptic functions and elliptic equations. At last, we provide some basic concepts of the affine space which we will use later.

Chapter 2 deals with the centroaffine elastica in the affine plane $\mathbf{R}^2$. We derive the Euler-Lagrange equation and its first integral for arbitrary curvature energy functionals. We express the centroaffine curvature function $k(s)$ of the centroaffine elastica by the Jacobi Elliptic functions, work out one Killing field along the centroaffine elastica and solve the structure equation by using that Killing field and the classification of the conjugate class of $sl(2, \mathbf{R})$(Theorem 2.2). We discuss the closed centroaffine elastica(Theorem 2.3) with non-constant centroaffine curvature by using two routes: one is from the expression of the solution and the other is using the Lamé's equation. The result(Theorem 2.6) from the second method could support the existence of the closed affine elastica with non-constant affine curvature in chapter 4(Theorem 4.1). We will discuss the stability of the closed centroaffine elastica with constant centroaffine curvature via the second variation formula and get the Jacobi form result(Theorem 2.7).

Chapter 3 treats the critical curves about the energy functionals with the centroaffine elastica and centroaffine torsion of starlike curves in the affine space $\mathbf{R}^3$. If the energy functional depends only on the centroaffine curvature, we call the critical curve the generalized centroaffine elastica. We find that the centroaffine torsion has a simple closed relation to the centroaffine curvature and can solve the centroaffine curvature by using the same method used in chapter 2. We solve the structure equation by using the Killing field and the classification of the conjugate class of $sl(3, \mathbf{R})$. From the definition of the centroaffine curvature and torsion of the generalized centroaffine elastica and their simple relation, we find that the Killing field corresponding to the conjugate class of $sl(3, \mathbf{R})$ is determined by the constants of the first integral of the Euler-Lagrange equation. By using the expression of the centroaffine elastica, for which the Lagrange is a degree 2 polynomial of the centroaffine curvature k, we find the closed centroaffine elastica is plane curve. We discuss the energy functional for linear function of the centroaffine torsion and find that the critical curve can be solved by quadratures. We study the infinite dimensional manifold of closed starlike curves and find a closed 2-form on it. By using this 2-form, some Hamiltonian flow related to the starlike curves was studied. At last we study the flow generated by the field $-k'X + kX'$ in the affine space $\mathbf{R}^3$. This vector field has the form of the Killing field along the centroaffine elastica. From this observation, centroaffine elastic curves evolve without changing form under this flow. We find that the centroaffine curvature evolves under the flow by a KdV equation, which is known to be a completely integrable partial differential equation. We obtain the formulae of the centroaffine curvature and torsion of the middle curve.

Chapter 4 deals with the affine elastica in the affine plane $\mathbf{R}^2$. As a side result, we find that the affine geodesic in $\mathbf{R}^2$ is a parabola and it obtains the maximum affine arclength. The Euler-Lagrange equation in this case does not appear to be completely integrable by quadratures, we only solve a special case. By using the result of chapter 2, we partially solve the affine elastica and show the existence of the closed non-constant affine curvature elastica(Theorem 4.2). We discuss the stability of the closed affine elastica with constant affine curvature via the second variation. At last, we study the critical curves with respect to the energy functional which is the integral of a linear function of the affine curvature and solve its Euler-Lagrange equation by Jacobi elliptic functions.

Chapter 5 deals with the affine elastica with respect to the affine curvature in the affine space $\mathbf{R}^3$. We find the affine geodesic in $\mathbf{R}^3$ obtains the maximum affine arclength and partially solved the affine elastica by using the result of chapter 3.

1.3 A brief review of the classical elastica in three-dimensional Euclidean space $\mathbf{E}^3$

After we construct a rectangular coordinate system $O_{x^1x^2x^3}$, the three-dimensional Euclidean space $\mathbf{E}^3$ is isomorphic to the 3-dimensional vector space $\mathbf{R}^3$ with the standard inner product

$$x \cdot y = x^1y^1 + x^2y^2 + x^3y^3,$$

for any vectors $x = (x^1, x^2, x^3)$ and $y = (y^1, y^2, y^3)$. The standard vector product of x and y is defined as

$$x \times y = (x^2y^3 - x^3y^2, x^3y^1 - x^1y^3, x^1y^2 - x^2y^1).$$

We can think of curves in $\mathbf{R}^3$ as paths of a point in motion. The rectangular coordinate (x^1, x^2, x^3) of a point can then be expressed as functions of a parameter t inside a certain closed interval:

$$x_1 = x_1(t), \quad x_2 = x_2(t), \quad x_3 = x_3(t), \qquad a \leq t \leq b. \tag{1.1}$$

For convenience, the equation of the curve takes the vector form

$$\gamma = \gamma(t) = (x_1(t), x_2(t), x_3(t)), \qquad a \leq t \leq b. \tag{1.2}$$

The curve γ is called a regular curve, if its tangent vector $\gamma'(t) = (x_1'(t), x_2'(t), x_3'(t)) \neq 0$, for any $t \in [a, b]$. This means a regular curve has a nonvanishing velocity vector. We will only consider regular curves in this book. We can express the arclength of a segment of the curve γ between the points $\gamma(t_0)$ and $\gamma(t)$ by means of the integral

$$s(t) = \int_{t_0}^{t} \sqrt{(\frac{dx^1}{dt})^2 + (\frac{dx^2}{dt})^2 + (\frac{dx^3}{dt})^2} dt = \int_{t_0}^{t} |\gamma'(t)| dt. \tag{1.3}$$

The direction with arc length s increasing is called the positive direction of the curve. This kind of curve is called an oriented curve. When we introduce s as a parameter of the curve γ instead of t, we have $|\gamma'(s)| = 1$. From this point of view, a parameter t with the property $|\gamma'(t)| = 1$ is called an arc length parameter of the curve γ. We usually denote s as the arclength parameter of the curve. The vector $\frac{d\gamma}{ds}$ is therefore a unit vector. We call the vector $T(s) = \frac{d\gamma}{ds}$ the unit tangent vector of the curve γ at $\gamma(s)$. When $\gamma''(s) \neq 0$, the direction of $\gamma''(s)$ is called the principal normal of the curve γ and we denote it by $N(s)$. By differentiating $T(s) \cdot T(s) = 1$, we have $T(s) \cdot \frac{dT(s)}{ds} = 0$. This shows that the curvature vector $\frac{dT(s)}{ds}$ is perpendicular to $T(s)$. The length of this vector, $k(s) = |\frac{dT(s)}{ds}|$, is called the curvature of the curve γ at $\gamma(s)$. Thus we have

$$\frac{dT(s)}{ds} = k(s)N(s). \tag{1.4}$$

The curvature measures the rate of change of the tangent when moving along the curve.

We define the unit binormal vector $B(s)$ by the formula

$$B(s) = T(s) \times N(s). \tag{1.5}$$

By differentiating $B(s) \cdot B(s) = 1$ and $T(s) \cdot B(s) = 0$, we have $B'(s) \cdot B(s) = 0$ and $T(s) \cdot B'(s) = 0$. This means the vector $\frac{dB(s)}{ds}$ is parallel to the principal normal $N(s) = B(s) \times T(s)$. We define $\tau(s) = -N(s) \cdot \frac{dB(s)}{ds}$ the torsion of the curve γ at $\gamma(s)$. Thus we have

$$\frac{dB(s)}{ds} = -\tau(s)N(s). \tag{1.6}$$

Differentiating the identity $N(s) = B(s) \times T(s)$ gives

$$\frac{dN(s)}{ds} = -k(s)T(s) + \tau(s)B(s). \tag{1.7}$$

The three vector formulas

$$\begin{cases} \frac{dT(s)}{ds} = & k(s)N(s), & \\ \frac{dN(s)}{ds} = -k(s)T(s) & & +\tau(s)B(s), \\ \frac{dB(s)}{ds} = & -\tau(s)N(s) & \end{cases} \tag{1.8}$$

together with $T(s) = \frac{d\gamma}{ds}$ describe the motion of the moving trihedron $\{\gamma(s); T(s), N(s), B(s)\}$ of the curve γ. They take the central rule in the theory of space curves and are known as the formulas of Frenet.

In 1691 James Bernoulli posed the problem of bent beam, elastic bar, or simply "elastica". In 1738 Daniel Bernoulli suggested that an elastic rod in equilibrium is a curve in the Euclidean space $\mathbf{R}^2$ or $\mathbf{R}^3$ which is critical for the total squared curvature defined on regular curves with a fixed length satisfying given first order boundary data. Radon got the solutions for this problem([4]).

In this section, we study the critical point of the curvature energy functional $F(\gamma) = \int_\gamma (k^2 + \lambda)ds$ on those immersed curves in $\mathbf{R}^3$ satisfying suitable boundary conditions for a constant λ(Lagrange multiplier). λ may be thought of as a length penalty. The critical point of this functional is called an elastica. When $\lambda = 0$, we call it a free elastica.

We consider a variation $\gamma = \gamma(w,t) : (-\varepsilon, \varepsilon) \times I \to \mathbf{R}^3$ with $\gamma(0,t) = \gamma(t)$ an immersed regular curve on $\mathbf{R}^3$. Associated with such a variation is the variation vector field $W = W(0,t) = (\partial\gamma/\partial w)(0,t)$ along the curve γ. We will also write $V = V(w,t) = (\partial\gamma/\partial t)(w,t)$,$v(w,t) = |V(w,t)|$, $T(w,t) = V(w,t)/v(w,t)$, $W(w,t) = (\partial\gamma/\partial w)(w,t)$ etc., with the obvious meaning. Let s denote the arclength parameter of the curve $\gamma_w(t) = \gamma(w,t)$, and we write $\gamma(s)$, $k(w,s)$, etc., for the corresponding reparametrizations. L is the arclength of γ. We may assume $t = s$ be the arclength parameter of γ and then $I = [0, L]$.

$$[V(w,t), W(w,t)] = [d\gamma(\frac{\partial}{\partial t}), d\gamma(\frac{\partial}{\partial w})] = d\gamma([\frac{\partial}{\partial t}, \frac{\partial}{\partial w}]) = 0.$$

By using the torsion free relation $\nabla_V W = \nabla_W V + [V, W] = \nabla_W V$,

$$\frac{\partial v^2}{\partial w} = 2 < \nabla_W V, V >= 2 < \nabla_V W, V >= 2v^2 < \nabla_T W, T > .$$

Thus we have

$$W(v) = -gv, \qquad \text{for } g = -<\nabla_T W, T>. \tag{1.9}$$

$$0 = [V, W] = [vT, W] = v[T, W] - W(v)T = v([T, W] + gT).$$

This means

$$[W, T] = gT.$$

By using $\nabla_X \nabla_Y Z - \nabla_Y \nabla_X Z - \nabla_{[X,Y]} Z = 0$, we can now compute

$$\begin{aligned} \tfrac{\partial k^2}{\partial w} &= 2 <\nabla_W \nabla_T T, \nabla_T T> \\ &= 2 <\nabla_T \nabla_W T + \nabla_{[W,T]} T, \nabla_T T> \\ &= 2 <\nabla_T \nabla_T W + \nabla_T [W, T] + g \nabla_T T, \nabla_T T> \\ &= 2 <\nabla_T^2 W + g \nabla_T T + g_s T, \nabla_T T> + 2gk^2 \\ &= 2 <\nabla_T^2 W, \nabla_T T> + 4gk^2. \end{aligned} \tag{1.10}$$

We consider the curvature energy functional defined on a class of regular curves in $\mathbf{R}^3$.

$$F(\gamma) = \int_0^{L(w)} (k^2 + \lambda) ds. \tag{1.11}$$

Here $L(w)$ is the arclength of $\gamma_w(t) = \gamma(w, t)$. When we confine on the curve $\gamma(t) = \gamma(0, t)$, since $s = t$, we will drop t and s. The first variation formula is

$$\begin{aligned} &\delta F(W) = \tfrac{d}{dw} \textstyle\int_I (k^2 + \lambda) v dt|_{w=0} \\ &= \textstyle\int_I [W(k^2) v + (k^2 + \lambda) \tfrac{\partial v}{\partial w}] dt|_{w=0} \\ &= \textstyle\int_0^L <\nabla_T \nabla_T W, 2\nabla_T T> + (3k^2 - \lambda)) g] ds \\ &= \textstyle\int_0^L <2\nabla_T^3 T + \nabla_T [(3k^2 - \lambda) T], W> ds \\ &\quad + [<\nabla_T W, 2\nabla_T T> - <W, (k^2 - \lambda) T + 2k' N + 2k\tau B>]_0^L. \end{aligned} \tag{1.12}$$

By the Frenet equations(1.8), we have

$$\begin{aligned} \delta F(W) = &[<\nabla_T W, 2\nabla_T T> - <W, 2\nabla_T^2 T + (3k^2 - \lambda) T>]_0^L \\ &+ \textstyle\int_0^L <[2k'' + k(k^2 - 2\tau^2 - \lambda)] N + 2(2k'\tau + k\tau') B, W> ds. \end{aligned} \tag{1.13}$$

We obtain the Euler-Lagrange equation.

$$E = [2k'' + k(k^2 - 2\tau^2 - \lambda)] N + 2(2k'\tau + k\tau') B = 0. \tag{1.14}$$

This equation is equivalent to the following pair of equations.

$$\begin{aligned} &2k'' + k(k^2 - 2\tau^2 - \lambda) = 0, \\ &2k'\tau + k\tau' = 0. \end{aligned} \tag{1.15}$$

For k is constant, we know τ is constant too and they satisfy

$$k(k^2 - 2\tau^2 - \lambda) = 0, \tag{1.16}$$

from the Euler-Lagrange equations (1.15). At this case, the Frenet equation is a linear system of ordinary differential equations with constant coefficients, we can give the formula directly. Now we assume k is not constant. From the second equation of (1.15), we know

$$k^2\tau = c_1. \tag{1.17}$$

Here c_1 is a constant. The integral of the first equation becomes:

$$(k')^2 + \frac{1}{4}k^4 - \frac{\lambda}{2}k^2 + \frac{c_1^2}{k^2} = c_2. \tag{1.18}$$

Here c_2 is a constant. Therefore we can express the curvature $k(s)$ by quadratures

$$\pm \int \frac{2k}{\sqrt{-(k)^6 + 2\lambda k^4 + 4c_2k^2 - 4c_1^2}} dk = \int ds. \tag{1.19}$$

This integral is not easy. We give another way to solve the equation(1.18). Making the change of variable $u = k^2$, we arrive at

$$(u')^2 + u^3 - 2\lambda u^2 - 4c_2u + 4c_1^2 = 0. \tag{1.20}$$

This equation is of the form $(u')^2 = P(u)$, P a third degree polynomial, it can be solved by standard techniques in term of elliptic functions. The cubical polynomial $P(u)$ satisfies $P(0) = -4c_1^2 \leq 0$ and $\lim_{u\to\pm\infty} P(u) = \mp\infty$. Then we have a non-positive root. If $u = k^2$ is a nonconstant solution of equation(1.20), it must take on the values at which $P(u) > 0$. It follows that $P(u)$ has three real roots $\alpha_1, \alpha_2, \alpha_3$ satisfying $\alpha_1 \leq 0 \leq \alpha_2 \leq \alpha_3$. The solution of equation(1.20) is given by

$$u(s) = \alpha_3(1 - q^2\mathrm{sn}^2(rs, p)). \tag{1.21}$$

where $sn(x, p)$ is the elliptic sine function(for information about the elliptic function, see next section), and

$$p^2 = \frac{\alpha_3 - \alpha_2}{\alpha_3 - \alpha_1}, \quad q^2 = \frac{\alpha_3 - \alpha_2}{\alpha_3}, \quad r = \sqrt{\frac{\alpha_3 q^2}{4p^2}} = \frac{\sqrt{\alpha_3 - \alpha_1}}{2}.$$

This is a stable solution.

If γ happens to be an extremal of $F(\gamma)$, then standard arguments imply that γ satisfies the Euler-Lagrange equation $E = 0$. In this situation, the first variation formula reduces to

$$\delta F(W) = [< \nabla_T W, 2\nabla_T T > - < W, J >]_0^L. \tag{1.22}$$

Here we have set

$$J = (k^2 - \lambda)T + 2k'N + 2k\tau B. \tag{1.23}$$

We consider the translational symmetries of a constant vector field W. For such W, the variation of $F(\gamma)$ is zero, so we have

$$\delta F(W) =< W(0), J(0) > - < W(L), J(L) >= 0.$$

This variation formula continues to hold when L is replaced with any intermediate value L', $0 < L' < L$. It follows that $< W, J >$ is constant on $[0, L]$. But W is an arbitrary constant vector field, so we obtain that J is a constant vector field along any elastic curve and $|J|^2 = \lambda^2 + 4c_2$.

Rotations of $\mathbf{R}^3$ also leave $F(\gamma)$ unchanged. A vector field on $\mathbf{R}^3$ generates a one-parameter family of rotations when it is of the form $\rho \times Z$, where ρ is the position vector in $\mathbf{R}^3$ and Z is a constant vector field. Restricting such vector fields to γ gives vector fields $\gamma \times Z$, for which the variation of $F(\gamma)$ is zero. So (1.22) gives

$$\delta F(\gamma \times Z) = -[< J \times \gamma + 2kB, Z >]_0^L.$$

Therefore $J \times \gamma + 2kB$ is a constant vector field along the critical curve. Denote $I = 2kB$, then we have

$$I = -J \times \gamma + V, \tag{1.24}$$

where V is a constant vector field along γ. That is I is the restriction to γ of a Killing field on $\mathbf{R}^3$, a sum of a rotation field and a translation field. If we translate the origin, then γ is replaced by $\gamma + V \times J/|J|^2$. Therefore we may choose a new coordinate system in which V is a multiple of J. Then we may replace (1.24) with

$$I = -J \times \gamma + \alpha J, \tag{1.25}$$

where $\alpha =< V, J > /|J|^2 =< I, J > /|J|^2 = 4c_1/(\lambda^2 + 4c_2)$ is a constant.

A Killing field along a curve is a vector field along the curve which is the restriction of an infinitesimal isometry of the ambient space. Along an elastica, the Killing fields J and I play an important role in the integration of the equations of the elastica and can be used to construct a system of cylindrical coordinates. Since the Killing field J is a constant vector field, it is a translation field. We can choose one coordinate field $\partial/\partial z = J/|J|$. From equation(1.25), we may choose a cylindrical coordinate system $\{r, \theta, z\}$ such that

$$I = -|J|\frac{\partial}{\partial\theta} + \alpha|J|\frac{\partial}{\partial z}. \tag{1.26}$$

That means I defines a rotation along z direction. $\frac{\partial}{\partial\theta} = (\alpha J - I)/|J|$ is a rotation field perpendicular to J. Then $\partial/\partial r$ is given by a cross product

$$\frac{\partial}{\partial r} = \frac{\frac{\partial}{\partial\theta} \times \frac{\partial}{\partial z}}{|\frac{\partial}{\partial\theta} \times \frac{\partial}{\partial z}|} = \frac{B \times J}{|B \times J|}. \tag{1.27}$$

The first fundamental form of the standard cylindrical coordinate system is $ds^2 = dz^2 + dr^2 + r^2 d\theta^2$. Then equation(1.26) gives

$$r^2 = \frac{|I|^2 - \alpha^2|J|^2}{|J|^2} = \frac{4k^2(\lambda^2 + 4c_2) - 16c_1^2}{(\lambda^2 + 4c_2)^2}. \tag{1.28}$$

In the cylindrical coordinate system (z, r, θ), we can write the unit tangent vector of the elastica as $T = r_s\frac{\partial}{\partial r} + \theta_s\frac{\partial}{\partial\theta} + z_s\frac{\partial}{\partial z}$. Taking the inner products with the above formulas for $\partial/\partial z$ and $\partial/\partial\theta$, we can obtain

$$z_s = T \cdot \frac{\partial}{\partial z} = \frac{k^2 - \lambda}{|J|} = \frac{k^2 - \lambda}{\sqrt{\lambda^2 + 4c_2}}, \tag{1.29}$$

$$\theta_s = \frac{T \cdot \frac{\partial}{\partial\theta}}{r^2} = \frac{c_1(k^2 - \lambda)\sqrt{\lambda^2 + 4c_2}}{k^2(\lambda^2 + 4c_2) - 4c_1^2}. \tag{1.30}$$

Therefore we have the theorem

Theorem 1 *Let (r, θ, z) be cylindrical coordinates given above, and $\gamma(s) = (r(s), \theta(s), z(s))$. Then, for elastica, we have*

$$r^2 = \frac{4k^2(\lambda^2 + 4c_2) - 16c_1^2}{(\lambda^2 + 4c_2)^2},$$

$$z = \int \frac{k^2 - \lambda}{\sqrt{\lambda^2 + 4c_2}} ds, \tag{1.31}$$

$$\theta = \int \frac{c_1(k^2 - \lambda)\sqrt{\lambda^2 + 4c_2}}{k^2(\lambda^2 + 4c_2) - 4c_1^2} ds.$$

From the formula of r^2, r has the same periodicity and critical points as k. Formula (1.21) means $\alpha_2 \leq k^2 \leq \alpha_3$. Then we have

$$\frac{4\alpha_2(\lambda^2 + 4c_2) - 16c_1^2}{(\lambda^2 + 4c_2)^2} \leq r^2 \leq \frac{4\alpha_3(\lambda^2 + 4c_2) - 16c_1^2}{(\lambda^2 + 4c_2)^2}.$$

Therefore the elastica lies between two concentric cylinders around the z-axis. The maxima of the curvature occur on the outer cylinder and the minima on the inner cylinder.

For more information about elastica in $\mathbf{R}^3$, see Refs.([29], [22], [10]).

1.4 A simple introduction to elliptic functions and elliptic equations

In this section, we introduce some basic concepts of the elliptic functions and elliptic equations.

For $0 \leq p \leq 1$, we call the integral

$$\int_0^x \frac{dt}{\sqrt{(1-t^2)(1-p^2t^2)}} \tag{1.32}$$

the first kind of Legendre elliptic integral. It depends on the parameter p, known as the modulus. The complimentary modulus p' is defined as $\sqrt{1-p^2}$.

When $p = 0$,

$$u(x) = \int_0^x \frac{dt}{\sqrt{(1-t^2)(1-p^2t^2)}} = \int_0^x \frac{dt}{\sqrt{1-t^2}} = \arcsin x. \tag{1.33}$$

Its inverse function is $x = \sin u$.

When $p \neq 0$, the inverse function of the elliptic integral

$$u(x) = \int_0^x \frac{dz}{\sqrt{(1-z^2)(1-p^2z^2)}} \tag{1.34}$$

is called a Jacobi elliptic sine function. We denote it by $x = \mathrm{sn}(u) = \mathrm{sn}(u,p)$. Setting an integral transformation $z = \sin\theta$ and denoting $\phi = \arcsin x$, the elliptic integral of the first kind becomes

$$F(\phi,p) = u(\sin\phi) = \int_0^{\phi} \frac{d\theta}{\sqrt{1-p^2\sin^2\theta}}. \tag{1.35}$$

Denoting the inverse function $u(\sin\phi)$ by $\phi = \mathrm{am}(u,p)$, the amplitude of u, then we have $\mathrm{sn}u = \mathrm{sn}(u,p) = \sin\phi$. We define the Jacobi elliptic cosine function and Jacobi elliptic delta function as

$$\mathrm{cn}u = \mathrm{cn}(u,p) = \cos\phi, \qquad \mathrm{dn}u = \mathrm{dn}(u,p) = \sqrt{1-p^2\sin^2\phi}. \tag{1.36}$$

The Jacobi elliptic functions have relations

$$\mathrm{sn}^2(u,p) + \mathrm{cn}^2(u,p) = 1 = p^2\mathrm{sn}^2(u,p) + \mathrm{dn}^2(u,p), \tag{1.37}$$

$$p^2\mathrm{cn}^2(u,p) + 1 - p^2 = \mathrm{dn}^2(u,p) = \mathrm{cn}^2(u,p) + (1-p^2)\mathrm{sn}^2(u,p). \tag{1.38}$$

The derivatives of the Jacobi elliptic functions are

$$\frac{d}{du}\mathrm{sn}(u,p) = \mathrm{cn}(u,p)\mathrm{dn}(u,p), \tag{1.39}$$

$$\frac{d}{du}\mathrm{cn}(u,p) = -\mathrm{sn}(u,p)\mathrm{dn}(u,p), \tag{1.40}$$

$$\frac{d}{du}\mathrm{dn}(u,p) = -p^2\mathrm{sn}(u,p)\mathrm{cn}(u,p). \tag{1.41}$$

The Jacobi elliptic functions have periodic relations

$$\mathrm{sn}(u+2K) = -\mathrm{sn}(u),\ \mathrm{cn}(u+2K) = -\mathrm{cn}(u),\ \mathrm{dn}(u+2K) = \mathrm{dn}(u),$$

where $K(p) = F(\pi/2,p) = \int_0^{\pi/2} \frac{d\theta}{\sqrt{1-p^2\sin^2\theta}}$ is the complete elliptic integral of first kind.

The degeneracy of the Elliptic functions is:

As $p \to 0$,

$$\mathrm{sn}(u,p) \to \sin u, \quad \mathrm{cn}(u,p) \to \cos u, \quad \mathrm{dn}(u,p) \to 1.$$

As $p \to 1$,

$$\mathrm{sn}(u,p) \to \tanh u, \quad \mathrm{cn}(u,p) \to \mathrm{sech}u, \quad \mathrm{dn}(u,p) \to \mathrm{sech}u.$$

We call the ordinary differential equation

$$(y')^2 = a_0 + a_1 y + a_2 y^2 + a_3 y^3 + a_4 y^4, \tag{1.42}$$

or

$$y'' = A_0 + A_1 y + A_2 y^2 + A_3 y^3 \tag{1.43}$$

the elliptic equation, here a_0, a_1, a_2, a_3, a_4, A_0, A_1, A_2, A_3 are constants.

Multiplying equation(1.43) by y' and integrating it, we can get the equation(1.42). Thus elliptic equation(1.42) and elliptic equation(1.43) are equivalent.

A direct calculation gives the following results:

(1) Functions $y = \mathrm{sn}(x,p)$, $y = \mathrm{cn}(x,p)$ and $y = \mathrm{dn}(x,p)$ satisfy the elliptic equations

$$(y')^2 = (1-y^2)(1-p^2y^2),$$
$$(y')^2 = (1-y^2)(1-p^2+p^2y^2),$$
$$(y')^2 = (1-y^2)(y^2-1+p^2),$$

respectively.

(2) Functions $y = A\mathrm{sn}(x,p)$, $y = A\mathrm{cn}(x,p)$ and $y = A\mathrm{dn}(x,p)$ satisfy the elliptic equations

$$(y')^2 = \frac{1}{A^2}(A^2-y^2)(A^2-p^2y^2),$$

$$(y')^2 = \frac{1}{A^2}(A^2-y^2)((1-p^2)A^2+p^2y^2),$$

$$(y')^2 = \frac{1}{A^2}(A^2-y^2)(y^2-(1-p^2)A^2),$$

respectively.

(3) Functions $y = \mathrm{sn}^2(x,p)$, $y = \mathrm{cn}^2(x,p)$ and $y = \mathrm{dn}^2(x,p)$ satisfy the elliptic equations

$$(y')^2 = 4y(1-y)(1-p^2y),$$
$$(y')^2 = 4y(1-y)(1-p^2+p^2y),$$
$$(y')^2 = 4y(1-y)(y-1+p^2),$$

respectively.

For more information about the Jacobi elliptic functions and Jacobi elliptic equations, see Refs.([8], [15]).

1.5 Some basic concepts of affine space

In this section, we recall some basic concepts of the affine space of arbitrary dimension n. An early systematic description of affine differential geometry can be found in Blaschke's book([5]). Affine geometry is, loosely speaking, Euclidean geometry stripped of its metric structure, measuring volumes in stead of measuring distances or angles. There is a close analogy between the development of affine differential geometry and classical differential geometry. A more rigorous explanation of non-metric or pure affine space is given below.

Definition 1 *Let V be an n-dimensional real vector space. A non-empty set Ω is said to be an affine space associated to V if there is a mapping*

$$\Omega \times \Omega \to V$$

denoted by

$$(p, q) \in \Omega \times \Omega \mapsto \overrightarrow{pq} \in V$$

satisfying the following conditions:

(1) for any $p, q, r \in \Omega$, we have $\overrightarrow{pq} + \overrightarrow{qr} = \overrightarrow{pr}$;

(2) for any $p \in \Omega$ and any vector $x \in V$, there is one and only one $q \in \Omega$ such that $x = \overrightarrow{pq}$.

The condition (1) implies that $\overrightarrow{pp} = 0$ and $\overrightarrow{pq} = -\overrightarrow{qp}$ for any $p, q \in \Omega$. For simplicity, we may write $x = \overrightarrow{pq}$ as $q = p + x$. The dimension of Ω is defined as that of V. We say that V is the associated vector space for the affine space Ω.

We may choose the n-dimensional real vector space V as the set Ω and for any $(p, q) \in \Omega \times \Omega$, define $x = \overrightarrow{pq}$ to be the vector $q - p \in V$. Then V becomes an n-dimensional affine space. In particular, we choose V be the standard real n-dimensional vector space $\mathbf{R}^n$. Then we regard it as an affine space and call it the standard n-dimensional affine space.

Definition 2 *Let $o \in \Omega$ be a fixed point and $\{e_1, \cdots, e_n\}$ a basis of V . For any point $p \in \Omega$, we can write*

$$\overrightarrow{op} = \sum_{i=1}^{n} x^i(p) e_i,$$

where $(x^1(p), \cdots, x^n(p))$ *is a uniquely determined n-tuple of real numbers, called the coordinate of* p. *The set of functions* $\{x^1, \cdots, x^n\}$ *is called an affine coordinate system of* Ω.

After we choose an affine coordinate system, the mapping $p \mapsto (x^1, \cdots, x^n)$ defines an isomorphism between Ω and the standard real affine space $\mathbf{R}^n$. From this point of view, we can only consider the standard n-dimensional affine space in this book. When we consider the standard n-dimensional affine space $\mathbf{R}^n$ as an n-dimensional differentiable manifold, for each point $p \in \mathbf{R}^n$, we may identify the tangent space $T_p\mathbf{R}^n$ with the vector space $\mathbf{R}^n$. We may consider each $x \in V$ as a vector field that assigns to each $p \in \Omega$ a tangent vector $\overrightarrow{pq}$ determined by x. Geometrically, all these vectors determined by x are parallel. From the construction of the affine coordinate system $\{x^1, \cdots, x^n\}$, the tangent vector $\partial/\partial x^i$ of the coordinate curve as a vector field corresponds to e_i, where $\{e_1, \cdots, e_n\}$ is a basis of $V = \mathbf{R}^n$ on which the affine coordinate system is based.

Let $Y = \sum_{i=1}^n Y^i(x)\partial/\partial x^i$ be a vector field on $\mathbf{R}^n$ and $Y = \sum_{i=1}^n a^i\partial/\partial x^i$ a vector, we define the covariant derivative $\nabla_X Y$ of Y along X by

$$\nabla_X Y = \sum_{i,j=1}^n a^j \frac{\partial Y^i}{\partial x^j}\frac{\partial}{\partial x^i}. \tag{1.44}$$

This defines a usual affine connection on $\mathbf{R}^n$. It is a torsion-free connection. That means the torsion tensor field

$$T(X,Y) = \nabla_X Y - \nabla_Y X - [X,Y] = 0. \tag{1.45}$$

The Lie bracket $[X,Y]$ of $X = \sum_{i=1}^n X^i(x)\partial/\partial x^i$ and $Y = \sum_{i=1}^n Y^i(x)\partial/\partial x^i$ is defined by

$$[X,Y] = \sum_{i,j=1}^n (X^j \frac{\partial Y^i}{\partial x^j} - Y^j \frac{\partial X^i}{\partial x^j})\frac{\partial}{\partial x^i}. \tag{1.46}$$

The affine connection we defined is a flat connection. This implies that the Riemannian curvature vector field

$$R(X,Y)Z = \nabla_X\nabla_Y Z - \nabla_Y\nabla_X Z - \nabla_{[X,Y]}Z = 0. \tag{1.47}$$

Definition 3 *Let* $f : \Omega \to \Omega$ *be a one-to-one mapping of* Ω *onto itself. For each* $p \in \Omega$, *we define a mapping* $F_p : V \to V$ *as follows. For each* $x \in V$, *let* $r \in \Omega$ *be a uniquely determined point in* Ω *such that* $\overrightarrow{pr} = x$. *Then we set*

$F_p(x) = \overrightarrow{f(p)f(r)}$. *We call f an affine transformation of Ω if, for a certain point $p \in \Omega$, the map F_p is a linear transformation of V onto itself. In this way, it follows that for any point $q \in \Omega$ the map F_q coincides with F_p. We call this map the associated linear transformation and denote it by F.*

Let us see why $F_p(x) = F_q(x)$ for any $p, q \in \Omega$ and $x \in V$. We set $y = \overrightarrow{pq}$ and $z = x + y$. There exist points $r, s \in \Omega$ such that $x = \overrightarrow{pr}$ and $z = \overrightarrow{ps}$. Then we have

$$x = z - y = \overrightarrow{ps} - \overrightarrow{pq} = \overrightarrow{qp} + \overrightarrow{ps} = \overrightarrow{qs}.$$

The linear property of the map F_p implies

$$F_p(x) = F_p(z) - F_p(y) = \overrightarrow{f(p)f(s)} - \overrightarrow{f(p)f(q)} = \overrightarrow{f(q)f(s)} = F_q(x).$$

Since f is a one-to-one map of Ω onto itself, F_p is a one-to-one linear map of V onto itself. This means F_p is nonsingular.

Let $\{x^1, \cdots, x^n\}$ be an affine coordinate system with origin o and based on a basis $\{e_1, \cdots, e_n\}$ and $\{y^1, \cdots, y^n\}$ be an affine coordinate system with origin $f(o)$ and based on a basis $\{F(e_1), \cdots, F(e_n)\}$. Then we have $y_i(f(p)) = x_i(p)$ for $p \in \Omega$. We can write the relationship between the coordinate systems $\{x^1, \cdots, x^n\}$ and $\{y^1, \cdots, y^n\}$ in the form

$$x^i = \sum_{j=1}^{n} a^i_j y^j + c^i, \qquad i = 1, \cdots, n, \tag{1.48}$$

where $A = (a^j_i)$ is a nonsingular $n \times n$ matrix and $c = (c^1, \cdots, c^n)^T$ is a column vector. Therefore relative to the one coordinate system $\{x^1, \cdots, x^n\}$, the coordinates $\bar{x}^i = x^i(f(p))$ of the image $f(p)$ can be expressed in terms of the coordinates of p in the form

$$\bar{x}^i = x^i(f(p)) = \sum_{j=1}^{n} a^i_j y^j(f(p)) + c^i = \sum_{j=1}^{n} a^i_j x^j(p) + c^i, \quad i = 1, \cdots, n, \tag{1.49}$$

This relation may be expressed by the equation $\bar{x} = Ax + c$, or in its expanded matrix form,

$$\begin{pmatrix} \bar{x} \\ 1 \end{pmatrix} = \begin{pmatrix} A & c \\ 0 & 1 \end{pmatrix} \begin{pmatrix} x \\ 1 \end{pmatrix}. \tag{1.50}$$

A volume element ω in the vector space $V = \mathbf{R}^n$ is a nonzero alternating n-form. Once an orientation of V is fixed, ω is determined up to a positive constant factor, that is, for any oriented basis $\{e_1, \cdots, e_n\}$, the value

$\omega(e_1, \cdots, e_n)$ can be assigned to be an arbitrary positive number c, which determines ω uniquely.

A volume element ω in V determines a volume element on the manifold Ω, that is, a non-vanishing differential n-form, denoted by the same letter ω, such that $\omega(\partial/\partial x^1, \cdots, \partial/\partial x^n) = c$, where the vector fields $\partial/\partial x^1, \cdots, \partial/\partial x^n$ correspond to $e_1, \cdots, e_n$ as explained above. It is obvious that $\omega(X_1, \cdots, X_n) = \omega(Y_1, \cdots, Y_n)$ if each $Y_i \in T_y\mathbf{R}^n$ is parallel to $X_i \in T_x\mathbf{R}^n$. Thus ω is said to be parallel.

When a parallel volume element is determined in the affine space Ω, an affine transformation f is said to be equiaffine (or unimodular) if it preserves the volume element, that is, the associated linear transformation F preserves the corresponding volume element in V. It can be verified that this is the case if and only if f is expressed by (1.49), where the matrix (a_i^j) satisfies $\det(a_i^j) = 1$. The set of all equiaffine transformations forms a subgroup of the group of all affine transformations. We will study the affine properties invariant under the equiaffine transformation.

We can think of curves in the affine space $\mathbf{R}^n$ as paths of a point in motion. Its coordinate functions $x^1(t), \cdots, x^n(t)$ are continuously $n+1$ times differentiable inside a certain interval. We call the direction with parameter t increasing the direction of the curve. Similar to the arclength parameter for the curve in Euclidean space, we will introduce a special parameter called affine arclength parameter. We can express the coordinate of a point as a column vector $x = (x^1(t), \cdots, x^n(t))^T$ and assume that

$$\det(\frac{dx}{dt}, \cdots, \frac{d^n x}{dt^n}) \neq 0$$

along the curve $x(t)$. Let $u = u(t)$ with $du/dt > 0$ be another parameter of the curve $x(t)$. The condition $du/dt > 0$ means that this parameter transformation preserves the direction of the curve. By the chain rule of the derivative, we have

$$\frac{dx}{dt} = \frac{dx}{du}\frac{du}{dt},$$

$$\frac{d^2x}{dt^2} = \frac{d^2x}{du^2}(\frac{du}{dt})^2 + \frac{dx}{du}\frac{d^2u}{dt^2},$$

$$\cdots\cdots,$$

$$\frac{d^n x}{dt^n} = \frac{d^n x}{du^n}(\frac{du}{dt})^n + (\text{terms including } \frac{dx}{du}, \frac{d^2x}{du^2}, \cdots, \frac{d^{n-1}x}{du^{n-1}}).$$

This implies

$$\det(\frac{dx}{dt},\cdots,\frac{d^nx}{dt^n})=\det(\frac{dx}{du},\cdots,\frac{d^nx}{du^n})(\frac{du}{dt})^{\frac{n(n+1)}{2}}.$$

Then we have

$$|\det(\frac{dx}{dt},\cdots,\frac{d^nx}{dt^n})|^{\frac{2}{n(n+1)}}=|\det(\frac{dx}{du},\cdots,\frac{d^nx}{du^n})|^{\frac{2}{n(n+1)}}.$$

The affine arclength of a segment of the curve $x(t)$ between $x(a)$ and $x(t)$ can be expressed as

$$s=s(a,t)=\int_a^t|\det(\frac{dx}{dt},\cdots,\frac{d^nx}{dt^n})|^{\frac{2}{n(n+1)}}dt. \tag{1.51}$$

For simplicity, we may assume $\det(\frac{dx}{dt},\cdots,\frac{d^nx}{dt^n})>0$. When we use the affine arclength as a parameter of the curve, we have

$$\det(\frac{dx}{ds},\cdots,\frac{d^nx}{ds^n})=1. \tag{1.52}$$

Usually a parameter t is called an affine arclength parameter of the curve $x(t)$, if $\det(\frac{dx}{dt},\cdots,\frac{d^nx}{dt^n})=1$. This means the volume of the n-dimensional parallelotope spanned by $\frac{dx}{dt},\cdots,\frac{d^nx}{dt^n}$ is 1. We will use s to express the affine arclength parameter. Set the frame fields along $x(s)$ as $e_1=\frac{dx}{ds},e_2=\frac{d^2x}{ds^2},\cdots,e_n=\frac{d^nx}{ds^n}$. Taking the derivative of equation(1.52), we obtain

$$\det(e_1,\cdots,e_{n-1},e_n^{\prime})=0.$$

This means the vector $e_n^{\prime}$ can be expressed as a linear combination of $e_1,\cdots,e_{n-1}$, that is

$$e_n^{\prime}=k_1e_1+k_2e_2+\cdots+k_{n-1}e_{n-1}. \tag{1.53}$$

and

$$k_i(s)=\det(\frac{dx}{ds},\cdots,\frac{d^{i-1}x}{ds^{i-1}},\frac{d^{n+1}x}{ds^{n+1}},\frac{d^{i+1}x}{ds^{i+1}},\cdots,\frac{d^nx}{ds^n}), \tag{1.54}$$
$$i=1,2,\cdots,n-1.$$

We call $k_i(s)$ the affine curvature of the curve $x(s)$. The Frenet equation can be written as

$$\begin{pmatrix} e_1^{\prime}\\ e_2^{\prime}\\ e_3^{\prime}\\ \vdots\\ e_{n-1}^{\prime}\\ e_n^{\prime}\end{pmatrix}=\begin{pmatrix} 0&1&0&\cdots&0&0\\ 0&0&1&\cdots&0&0\\ 0&0&0&\cdots&0&0\\ \vdots&\vdots&\vdots&\ddots&\vdots&\vdots\\ 0&0&0&\cdots&0&1\\ k_1&k_2&k_3&\cdots&k_{n-1}&0\end{pmatrix}\begin{pmatrix} e_1\\ e_2\\ e_3\\ \vdots\\ e_{n-1}\\ e_n\end{pmatrix}. \tag{1.55}$$

By using the existence and uniqueness theorem of the ordinary differential equations, we have the following theorem:

Theorem 2 *Let* $k_1, k_2, \cdots, k_{n-1} : [a, b] \to \mathbf{R}$ *be* $n-1$ *differentiable functions. There exists a unique curve* $x(s)$ *in* $\mathbf{R}^n$ *with* $k_1, k_2, \cdots, k_{n-1}$ *as affine curvatures up to an equiaffine transformation of* $\mathbf{R}^n$.

For more information about the affine spaces, see Refs.([16], [33], [38]).

Chapter 2

The Centroaffine elastica in the affine plane $\mathbf{R}^2$

In this chapter, we will study the variational properties of plane curves invariant under the group of area-preserving homogeneous affine transformations or centroaffine transformations in the punctured plane. The group is generated by the action of the unimodular linear group $SL(2,\mathbf{R})$. We may expect to characterize the centroaffine properties of critical plane curves with respect to certain energy functionals by the centroaffine arclength parameter and centroaffine curvature.

2.1 Definition and variation formula of the generalized centroaffine elastica in $\mathbf{R}^2$

We consider curves in the affine plane $\mathbf{R}^2$([33], [38]). Let (x, y) be a standard coordinate system of the affine plane $\mathbf{R}^2$, and $X : I \to \mathbf{R}^2$ be a smooth curve. We write it as a column vector

$$X(t) = (x(t), y(t))^T, \tag{2.1}$$

here $x(t), y(t)$ are smooth functions of t defined on a certain interval I. We identify the tangent space of $\mathbf{R}^2$ with $\mathbf{R}^2$ and write the tangent vector as $X'(t) = dX/dt = (x'(t), y'(t))^T$ or simply X', similarly for X'' and $(X(t), X'(t))$ be the 2×2 matrix

$$\begin{pmatrix} x(t) & x'(t) \\ y(t) & y'(t) \end{pmatrix}. \tag{2.2}$$

Definition 4 *A curve* $X : I \to \mathbf{R}^2$ *is called starlike if the tangent vector and its position vector are not parallel, that is*

$$|X(t), X'(t)| = \det(X(t), X'(t)) \neq 0, \tag{2.3}$$

for any $t \in I$. *A parameter* s *is called a centroaffine arclength parameter of the starlike curve* X *if*

$$(X(s), X'(s)) \in SL(2, \mathbf{R}), \tag{2.4}$$

that means $|X(s), X'(s)| = 1$, *for any* $s \in I$*(Kepler's first law).*

Definition 5 *Let* s *be a centroaffine arclength parameter of a curve* $X : I = [0, L] \to \mathbf{R}^2$, *here* L *is the centroaffine arclength of* X *defined as in (2.9). We define the centroaffine curvature of the curve* X *by*

$$k(s) = |X'(s), X''(s)|. \tag{2.5}$$

Differentiating $|X(s), X'(s)| = 1$, we obtain

$$|X(s), X''(s)| = 0.$$

This means that $X(s)$, $X''(s)$ are linearly dependent and from the centroaffine curvature definition(2.5), we have the structure equation.

$$X''(s) = -k(s)X(s). \tag{2.6}$$

Let t be an arbitrary parameter of the curve X.

$$X'(t) = X'(s)\frac{ds}{dt}. \tag{2.7}$$

We have the formula of the velocity of the curve X

$$|X(t), X'(t)| = |X(s), X'(s)|\frac{ds}{dt} = \frac{ds}{dt}. \tag{2.8}$$

Hence we obtain the centroaffine arclength formula of a segment of the curve X between $X(t_0)$ and $X(t)$ for arbitrary parameter t

$$s(t) = \int_{t_0}^{t} |X(t), X'(t)| dt. \tag{2.9}$$

If t is also a centroaffine arclength parameter of the curve, then $ds/dt = 1$. Therefore $s = t + c$, for a constant c. This shows that a starlike curve $X(t)$ can admit a centroaffine arclength parameter uniquely up to a constant, and the starlike property and the centroaffine curvature are invariant under the action of the group $SL(2, \mathbf{R})$ on $\mathbf{R}^2$.

The area of the parallelogram spanned by $X(t)$ and $X'(t)$ is

$$|X(t) \times X'(t)| = |x(t)y'(t) - x'(t)y(t)| = ||X(t), X'(t)||.$$

Thus a parameter t is a centroaffine arclength parameter of the starlike curve X if and only if the area of the parallelogram spanned by $X(t)$ and $X'(t)$ is 1. The centroaffine curvature is the (signed) area of the parallelogram spanned by $X'(s)$ and $X''(s)$.

We denote ∇ the usual affine connection of $\mathbf{R}^2$. This is a torsion-free and flat connection. The letter X will also denote a variation $X = X_w(t) = X(w, t) : (-\varepsilon, \varepsilon) \times I \to \mathbf{R}^2$ with $X(0, t) = X(t)$. Associated with such a variation is the variational vector field $W = W(t) = (\partial X/\partial w)(0, t) = dX(\partial/\partial w)(0, t)$ along the curve $X(t)$. We will also write $W = W(w, t) = dX(\partial/\partial w)(w, t)$, $X' = X'(w, t) = dX(\partial/\partial t)(w, t)$. We know

$$[X'(w,t), W(w,t)] = [dX(\frac{\partial}{\partial t}), dX(\frac{\partial}{\partial w})] = dX([\frac{\partial}{\partial t}, \frac{\partial}{\partial w}]) = 0. \tag{2.10}$$

We denote s the centroaffine arclength parameter of the curve $X_w(t)$ for a fixed w and write $X(s), k(w, s)$ for the corresponding reparametrizations and $s \in [0, L(w)]$, where $L(w)$ is the centroaffine arclength of $X_w(s)$. We may assume $t = s$ be the centroaffine arclength parameter of $X(t)$ and then $I = [0, L]$. Equation(2.8) gives

$$W(\frac{ds}{dt}) = W(|X(t), X'(t)|) = -g\frac{ds}{dt}, \tag{2.11}$$

here we write $g = -|X(t), X'(t)|^{-1} W(|X(t), X'(t)|)$.

$$\begin{aligned} W(|X(t), X'(t)|) &= |\tfrac{\partial X}{\partial w}(t), X'(t)| + |X(t), \tfrac{\partial^2 X}{\partial w \partial t}(t)| \\ &= |W(t), X'(t)| + |X(t), \tfrac{\partial}{\partial t} W(t)|. \end{aligned} \tag{2.12}$$

We simply write $X^{(n)}(s) = X^{(n)}(w, s)$, $X^{(n)}(t) = X^{(n)}(w, t)$. Since

$$0 = [W, X'(t)] = [W, \frac{ds}{dt} X'(s)] = \frac{ds}{dt}[W, X'(s)] + W(\frac{ds}{dt}) X'(s). \tag{2.13}$$

By using equation(2.11), we have

$$[W, X'(s)] = -\frac{dt}{ds}W(\frac{ds}{dt})X'(s) = gX'(s).$$

By using equations(1.43) and (1.45),

$$\begin{aligned}
\nabla_W X'(s) &= \nabla_{X'(s)}W + [W, X'(s)] + T(W, X'(s)) \\
&= \nabla_{X'(s)}W + gX'(s). \\
\nabla_W \nabla_{X'(s)} X'(s) &= \nabla_{X'(s)}\nabla_W X'(s) + \nabla_{[W,X'(s)]}X'(s) + R(W, X'(s))X'(s) \\
&= \nabla_{X'(s)}(\nabla_{X'(s)}W + gX'(s)) + gX''(s) \\
&= \nabla^2_{X'(s)}W + g_s X'(s) + 2gX''(s).
\end{aligned}$$

Thus inserting them to the centroaffine curvature formula, we have

$$\begin{aligned}
\nabla_W |X'(s), X''(s)| &= |\nabla_W X'(s), X''(s)| + |X'(s), \nabla_W X''(s)| \\
&= |\nabla_W X'(s), X''(s)| + |X'(s), \nabla_W \nabla_{X'(s)} X'(s)| \\
&= |\nabla_{X'(s)}W, X''(s)| + |X'(s), \nabla^2_{X'(s)}W| + 3gk.
\end{aligned} \tag{2.14}$$

We write these results into the following lemma.

Lemma 1 *Using the above notation, we have the following formulas:*

1. $[X'(t), W(t)] = 0.$

2. $W(\frac{ds}{dt}) = -g\frac{ds}{dt}$, *where* $g = -|X(t), X'(t)|^{-1}W(|X(t), X'(t)|).$

3. $[W, X'(s)] = gX'(s).$

4. $W(k) = |\nabla_{X'(s)}W, X''(s)| + |X'(s), \nabla^2_{X'(s)}W| + 3gk.$

We consider the curvature energy functional defined on a class of starlike curves in the affine plane $\mathbf{R}^2$.

$$\int_0^{L(w)} p(k)ds = \int_0^L p(|X'(w,s), X''(w,s)|)|X(w,t), X'(w,t)|dt. \tag{2.15}$$

Here $p(k)$ is a smooth function of k. When we confine on the curve $X(t) = X(0,t)$, since $s = t$, we will drop t, s in $X'(t), X'(s)$ respectively.

$$\begin{aligned}
&\tfrac{d}{dw}\textstyle\int_0^{L(w)} p(k)ds|_{w=0} \\
&= \textstyle\int_0^L [p'(k)W(k)\frac{ds}{dt} + p(k)W(\frac{ds}{dt})]dt \\
&= \textstyle\int_0^L [p'(k)(|\nabla_{X'}W, X''| + |X', \nabla^2_{X'}W| + 3gk) - p(k)g]ds \\
&= \textstyle\int_0^L [p'(k)(|\nabla_{X'}W, X''| + |X', \nabla^2_{X'}W|) + (3p'(k)k - p(k))g]ds.
\end{aligned} \tag{2.16}$$

Here we have used $p'(k)$ to denote the partial derivative of $p(k)$ with respect to k and the prime $(\ )'$ means the usual derivative with respect to s.

We give the variation $X(w,t)$ a boundary condition such that $W(0,0) = W(0,L) = 0$, $\nabla_{X'}W(0,0) = \nabla_{X'}W(0,L) = 0$. We calculate the integral part by part.

$$\begin{aligned}
&\int_0^L p'(k)(|\nabla_{X'}W, X''| + |X', \nabla^2_{X'}W|)ds \\
&= \int_0^L(|\nabla_{X'}W, p'(k)X''| + |p'(k)X', \nabla^2_{X'}W|)ds \\
&= \int_0^L(|\nabla_{X'}W, p'(k)X'' + (p'(k)X')'|)ds + |p'(k)X', \nabla_{X'}W|_0^L \\
&= -\int_0^L |W, (p'(k)X'')' + (p'(k)X')''|ds \\
&\quad +[|W, p'(k)X'' + (p'(k)X')'| + |p'(k)X', \nabla_{X'}W|]_0^L \\
&= \int_0^L |W, (3p''(k)k + 2p'(k))k'X - (p^{(3)}(k)(k')^2 + p''(k)k'' - 2p'(k)k)X'|ds \\
&\quad +[|W, p'(k)X'' + (p'(k)X')'| + |p'(k)X', \nabla_{X'}W|]_0^L.
\end{aligned}$$

$$\begin{aligned}
&\int_0^L(3p'(k)k - p(k))gds \\
&= -\int_0^L(3p'(k)k - p(k))(|W, X'| + |X, \nabla_{X'}W|)ds \\
&= -\int_0^L(|W, (3p'(k)k - p(k))X'| + |W, ((3p'(k)k - p(k))X)'|)ds \\
&\quad +|W, (3p'(k)k - p(k))X|_0^L \\
&= -\int_0^L(|W, (3p''(k)k'k + 2p'(k)k')X + 2(3p'(k)k - p(k))X'|ds \\
&\quad +|W, (3p'(k)k - p(k))X|_0^L.
\end{aligned}$$

Thus we obtain the first variational formula by using Lemma 1.

$$\begin{aligned}
&\frac{d}{dw}\int_0^{L(w)} p(k)ds|_{w=0} \\
&= -\int_0^L(p^{(3)}(k)(k')^2 + p''(k)k'' + 4p'(k)k - 2p(k))|W, X'|ds \\
&\quad +[|W, p'(k)X'' + (p'(k)X')'| + |p'(k)X', \nabla_{X'}W|]_0^L \\
&\quad +|W, (3p'(k)k - p(k))X|_0^L.
\end{aligned} \tag{2.17}$$

Therefore the Euler-Lagrange equation is

$$p^{(3)}(k)(k')^2 + p''(k)k'' + 4p'(k)k - 2p(k) = 0. \tag{2.18}$$

Definition 6 *A starlike curve in the affine plane* $\mathbf{R}^2$ *with the centroaffine arclength parameter* s *is called a generalized centroaffine elastica if it satisfies the equation(2.18).*

2.2 Integration of the generalized centroaffine elastica in $\mathbf{R}^2$

For k is a constant, we know that k satisfies $2p'(k)k - p(k) = 0$ from equation(2.18). We can give a formula directly, since the structure equation is a liner system with constant coefficients. The results are ellipse, hyperbola and straight line for k is positive, negative and zero respectively(See section 2.3.2). Now we assume k is not constant. Multiplying equation(2.18) by $2p''(k)k'$, we get

$$2p''(k)p^{(3)}(k)(k')^3 + 2p''(k)^2k''k' + 8p''(k)p'(k)k'k - 4p''(k)p(k)k' = 0. \quad (2.19)$$

Integrating it, we can obtain a first integral of the equation(2.18)

$$p''(k)^2(k')^2 + 4p'(k)(p'(k)k - p(k)) = c. \quad (2.20)$$

Here c is the integral constant. From this first integral, we can express the centroaffine curvature $k(s)$ by quadratures

$$\int \sqrt{\frac{p''(k)^2}{c - 4p'(k)(p'(k)k - p(k))}}dk = \pm \int ds. \quad (2.21)$$

For the nonconstant centroaffine curvature starlike generalized centroaffine elastica, we will use the Killing field and the classification of the conjugate class of $sl(2, \mathbf{R})$ to solve the structure equation(2.6). We need the following definition:

Definition 7 *Let $X(t)$ be a starlike curve in the affine plane $\mathbf{R}^2$. We call a vcctor field W Killing along the curve X if it annihilates the velocity ds/dt and centroaffine curvature k. This means*

$$W(\frac{ds}{dt}) = 0, \qquad W(k) = 0.$$

We set the Killing field along the generalized centroaffine elastica X having the form $W = f_1(s)X(s) + f_2(s)X'(s)$, then from Lemma 1 we have

$$\begin{aligned}
W' &= (f_1' - kf_2)X + (f_1 + f_2')X'; \\
W'' &= (f_1'' - kf_1 - 2kf_2' - k'f_2)X + (2f_1' + f_2'' - kf_2)X'; \\
W(\tfrac{ds}{dt}) &= -g\tfrac{ds}{dt} = (|W, X'| + |X, W'|)\tfrac{ds}{dt} = (2f_1 + f_2')\tfrac{ds}{dt}; \\
W(k) &= |W', X''| + |X', W''| + 3gk \\
&= -f_1'' + 2kf_1 + 3kf_2' + k'f_2 + 3gk.
\end{aligned}$$

Therefore the functions f_1 and f_2 must satisfy the following equations.

$$\begin{aligned}
&2f_1 + f_2' = 0; \\
&f_1'' - 2kf_1 - 3kf_2' - k'f_2 = 0.
\end{aligned} \quad (2.22)$$

From these equations and equation(2.18), we found $W = -\frac{1}{2}p''(k)k'X + p'(k)X'$ is Killing along the generalized centroaffine elastica X.

Definition 8 *A vector field W on the affine plane $\mathbf{R}^2$ is Killing with respect to $SL(2,\mathbf{R})$ if its flow generates a one-parameter subgroup of $SL(2,\mathbf{R})$.*

We can express f_1'' as a linear combination of f_1', f_1, f_2 from the second equation. Therefore, these two equations of f_1, f_2 can be written as a system of first order linear equations of f_1', f_1, f_2 in the following form

$$(f_1', f_1, f_2)' = (f_1', f_1, f_2)A. \tag{2.23}$$

Here A is a 3×3 matrix. Therefore the dimension of the solution space of equation(2.22) is 3. This dimension agrees with the dimension of the group $SL(2,\mathbf{R})$. Thus a vector field W which is Killing along the generalized centroaffine elastica X can extend to a Killing field on $\mathbf{R}^2$, the field which is invariant under the infinitesimal action of $SL(2,\mathbf{R})$. We denote it by $\tilde{W}$. We consider the integral curve $\tilde{X}$ of $\tilde{W}$ near the vertex P_0 of X, the point of which $k(s)$ has an extremum. We have the theorem

Theorem 3 *Let X be a generalized centroaffine elastica in the affine plane $\mathbf{R}^2$ and $P_0 = X(s_0)$ a vertex of X. Then W is tangent to X at P_0, the integral curve $\tilde{X}$ in $\mathbf{R}^2$ of $\tilde{W}$ through P_0 has centroaffine curvature $\tilde{k} = (p(k)-p'(k)k)/p'(k)$ and $\tilde{k}$ keeps the same sign at the vertex of the generalized centroaffine elastica.*

Proof We denote $\tilde{k}$ the centroaffine curvature of $\tilde{X}$, $\sigma(\tilde{X}) = \lfloor \tilde{X}, \tilde{W} \rfloor$ and $\tilde{X}'$ be the derivative of $\tilde{X}$ with respect to the centroaffine arclength of $\tilde{X}$. Then, from the definition of centroaffine arclength and centroaffine curvature of $\tilde{X}$, we know $\tilde{X}' = \tilde{W}/\sigma(\tilde{X}), \nabla_{\tilde{X}'}\tilde{X}' + \tilde{k}\tilde{X} = 0$. At the vertex P_0, $\nabla_{\tilde{X}'}\tilde{X}'$ can be considered as the covariant derivative of $\tilde{X}'$ along $X(s)$. Using $\sigma(X(s)) = p'(k), \tilde{X}'|_{X(s)} = (-\frac{1}{2}p''(k)k'X + p'(k)X')/p'(k)$, $k'(s_0) = 0$, we know at P_0

$$\nabla_{\tilde{X}'}\tilde{X}' = (p'(k)k - p(k))/p'(k)X. \tag{2.24}$$

This means $\tilde{k}(P_0) = (p(k) - p'(k)k)/p'(k)$. From the first integral(2.20), at the vertex of the generalized centroaffine elastica, we know the centroaffine curvature of the integral curve of the Killing field $\tilde{W}$ is $\tilde{k} = -c/(2p'(k))^2$. This completes the proof.

We will use the Killing field $\tilde{W}$ and the classification of the conjugate class of $sl(2,\mathbf{R})$ to solve the structure equation. Checking the eigenvalues, we have four conjugate classes in $sl(2,\mathbf{R})$.

$$1.\begin{pmatrix} 0 & -b \\ b & 0 \end{pmatrix}, \quad 2.\begin{pmatrix} b & 0 \\ 0 & -b \end{pmatrix}, \quad 3.\begin{pmatrix} 0 & 0 \\ 1 & 0 \end{pmatrix}, \quad 4.\begin{pmatrix} 0 & 0 \\ 0 & 0 \end{pmatrix}. \tag{2.25}$$

Here $b \neq 0$ is a positive constant. Using starlike property and the structure equation, we divide our problem into three cases and exclude the last matrix. We will study the case 1 and give the results for the other cases directly.

Case 1. $c < 0$, we pick $b = \sqrt{-c/4}$ and choose the first matrix in (2.25) as an element $M \in sl(2,\mathbf{R})$.

$$M = \begin{pmatrix} 0 & -b \\ b & 0 \end{pmatrix}. \tag{2.26}$$

We can choose a coordinate system (x, y) such that the Killing field generated by the 1-parameter Lie subgroup corresponding to M is

$$W = -by\frac{\partial}{\partial x} + bx\frac{\partial}{\partial y}. \tag{2.27}$$

Along thc generalized centroaffine elastica X, we have a linear system of first order differential equations.

$$\begin{cases} p'(k)x' - \frac{1}{2}p''(k)k'x = -by; \\ p'(k)y' - \frac{1}{2}p''(k)k'y = bx. \end{cases} \tag{2.28}$$

Introducing two new variables $U = x/p'(k)^{1/2}, V = y/p'(k)^{1/2}$, it becomes

$$p'(k)U' = -bV, \qquad p'(k)V' = bU. \tag{2.29}$$

Combining these two equations, we obtain the following second order differential equations.

$$p'(k)(p'(k)V')' + b^2V = 0; \tag{2.30}$$

$$p'(k)(p'(k)U')' + b^2U = 0. \tag{2.31}$$

We set a parameter transformation by $t = \int_{s0}^{s} 1/p'(k)ds$ and write $f(t) = V(s(t))$, then we have the equation

$$f''(t) + b^2f(t) = 0. \tag{2.32}$$

This equation has a general solution

$$f(t) = d_1 \cos(bt) + d_2 \sin(bt). \tag{2.33}$$

This leads to the generalized centroaffine elastica equations

$$\begin{cases} x(s) = p'(k(s))^{1/2}(d_2 \cos(\int \frac{b}{p'(k)} ds) - d_1 \sin(\int \frac{b}{p'(k)} ds)); \\ y(s) = p'(k(s))^{1/2}(d_1 \cos(\int \frac{b}{p'(k)} ds) + d_2 \sin(\int \frac{b}{p'(k)} ds)). \end{cases} \tag{2.34}$$

Here d_1, d_2 are constants and, by checking the starlike condition, satisfy $b(d_1^2 + d_2^2) = 1$.

Case 2. $c > 0$, we pick $b = \sqrt{c/4}$ and an element $M \in sl(2, \mathbf{R})$ as

$$M = \begin{pmatrix} b & 0 \\ 0 & -b \end{pmatrix}. \tag{2.35}$$

We can choose a coordinate system (x, y) such that the Killing field generated by the 1-parameter Lie subgroup corresponding to M is

$$W = bx\frac{\partial}{\partial x} - by\frac{\partial}{\partial y}. \tag{2.36}$$

Along the generalized centroaffine elastica X, we have a linear system.

$$\begin{cases} p'(k)x' - \frac{1}{2}p''(k)k'x = bx; \\ p'(k)y' - \frac{1}{2}p''(k)k'y = -by. \end{cases} \tag{2.37}$$

The solution is the centroaffine p-elastica equation

$$\begin{cases} x(s) = d_1 p'(k(s))^{1/2} \exp \int \frac{b}{p'(k)} ds; \\ y(s) = d_2 p'(k(s))^{1/2} \exp \int \frac{-b}{p'(k)} ds. \end{cases} \tag{2.38}$$

Here d_1, d_2 are constants and satisfy $2bd_1d_2 = -1$.

Case 3. $c = 0$, we pick an element $M \in sl(2, \mathbf{R})$ as

$$M = \begin{pmatrix} 0 & 0 \\ 1 & 0 \end{pmatrix}. \tag{2.39}$$

We can choose a coordinate system (x, y) such that the Killing field generated by the 1-parameter Lie subgroup corresponding to M is

$$W = x\frac{\partial}{\partial y}. \tag{2.40}$$

Along the generalized centroaffine elastica X, we have a linear system.

$$\begin{cases} p'(k)x' - \frac{1}{2}p''(k)k'x = 0; \\ p'(k)y' - \frac{1}{2}p''(k)k'y = x. \end{cases} \tag{2.41}$$

The solution is the generalized centroaffine elastica equation

$$\begin{cases} x(s) = p'(k(s))^{1/2}; \\ y(s) = p'(k(s))^{1/2}[d_1 + \int \frac{1}{p'(k)}ds]. \end{cases} \tag{2.42}$$

Here d_1 is a constant.

From these results, we have the theorem

Theorem 4 *The generalized centroaffine elastica of the affine plane* $\mathbf{R}^2$ *is integrable by quadratures, that means it can be expressed in terms of integrals as in (2.34), (2.38) and (2.42).*

2.3 Centro-affine elastica for $p(k) = k^2 + \lambda$ in the affine plane $\mathbf{R}^2$

Now we consider a special case for $p(k) = k^2 + \lambda$, λ is the Lagrange multiplier. The curvature energy functional(2.15) becomes $\int_0^L (k^2 + \lambda)ds$. The critical point of this energy functional is called a centroaffine elastica.

2.3.1 The solutions of motion equation for centroaffine elastica

At this time, the Euler-Lagrange equation is

$$k'' + 3k^2 - \lambda = 0. \tag{2.43}$$

For k is constant, we know $k = \pm\sqrt{\lambda/3}$ and λ must be nonnegative. When k is not a constant, multiplying equation(2.43) by k' and integrating it, we obtain the first integral of the Euler-Lagrange equation

$$\frac{1}{2}(k')^2 + k^3 - \lambda k = c. \tag{2.44}$$

The integral constant c is different from that in (2.20). Set $s_1 = \sqrt{2}s$, then

$$\frac{dk}{ds_1} = \frac{dk}{ds}\frac{ds}{ds_1} = \frac{1}{\sqrt{2}}\frac{dk}{ds}.$$

Thus

$$(\frac{dk}{ds_1})^2 = \frac{1}{2}(\frac{dk}{ds})^2 = -k^3 + \lambda k + c. \quad (2.45)$$

For convenience, we still write $\frac{dk}{ds_1} = k'$. So we have

$$(k')^2 = -k^3 + \lambda k + c. \quad (2.46)$$

We write

$$P(k) = -k^3 + \lambda k + c. \quad (2.47)$$

Since $\lim_{k\to\pm\infty} P(k) = \mp\infty$, the cubic polynomial $P(k)$ has at least one real root. Figure 2.1 shows the possible graphs of this cubic with k on the horizontal axis and $(k')^2 = P(k)$ on the vertical axis.

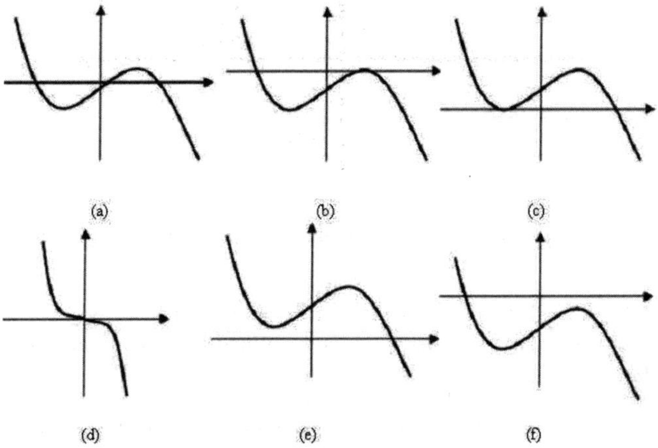

Figure 2.1: Possible configurations of $(k')^2 = P(k)$

Let $\alpha_1, \alpha_2, \alpha_3$ be the three roots of $P(k)$, we know

$$(\alpha_1-\alpha_2)^2(\alpha_2-\alpha_3)^2(\alpha_3-\alpha_1)^2 = 4\lambda^3 - 27c^2.$$

We assume two roots of $P(k)$ are imaginary, say

$$\alpha_2 = a+bi, \qquad \alpha_3 = a-bi.$$

Then

$$(\alpha_1-\alpha_2)^2(\alpha_2-\alpha_3)^2(\alpha_3-\alpha_1)^2 = -4b^2[(a-\alpha_1)^2+b^2]^2.$$

Hence we have three cases(see Figure 2.2):

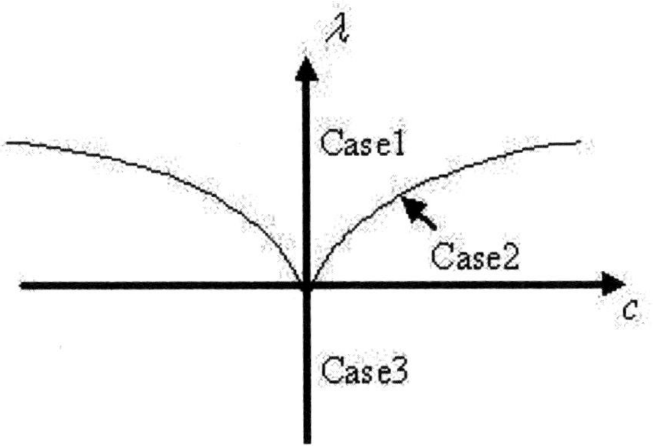

Figure 2.2: Root property on (c, λ) plane

Case 1. $4\lambda^3 - 27c^2 > 0$, the roots are real and unequal.
Case 2. $4\lambda^3 - 27c^2 = 0$, the roots are real and at least two of them are equal.
Case 3. $4\lambda^3 - 27c^2 < 0$, two roots are imaginary.

We can find the roots of $P(k)$ by Cartan formula. These roots are

$$\frac{2^{1/3}\lambda}{(27c+\sqrt{-108c^3+729\lambda^2})^{1/3}} + \frac{(27c+\sqrt{-108c^3+729\lambda^2})^{1/3}}{3\times 2^{1/3}},$$

$$\frac{-(1+i\sqrt{3})\lambda}{2^{2/3}(27c+\sqrt{-108c^3+729\lambda^2})^{1/3}}-\frac{(1-i\sqrt{3})(27c+\sqrt{-108c^3+729\lambda^2})^{1/3}}{6\times 2^{1/3}},$$

$$\frac{-(1-i\sqrt{3})\lambda}{2^{2/3}(27c+\sqrt{-108c^3+729\lambda^2})^{1/3}}-\frac{(1+i\sqrt{3})(27c+\sqrt{-108c^3+729\lambda^2})^{1/3}}{6\times 2^{1/3}}.$$

At least we have the follow relations.

$$\begin{aligned}\alpha_1+\alpha_2+\alpha_3 &= 0,\\ \alpha_1\alpha_2+\alpha_2\alpha_3+\alpha_3\alpha_1 &= -\lambda,\\ \alpha_1\alpha_2\alpha_3 &= c.\end{aligned}$$

For those real roots, we can find curves with these constant centroaffine curvature.

Now we are going to solve the the nonconstant centroaffine curvature by similar treatment for the solutions to the Kortteweg de Vries([9]).

Case 1. $4\lambda^3-27c^2>0$, the roots $\alpha_1,\alpha_2,\alpha_3$ are real and unequal. This is the case (a) of Figure 2.1. We may assume $\alpha_1<\alpha_2<\alpha_3$. Since $\alpha_1+\alpha_2+\alpha_3=0$, we have $\alpha_1<0,\alpha_3>0$.

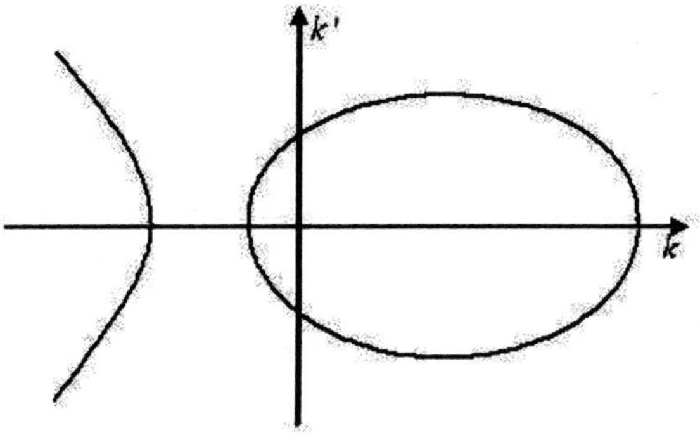

Figure 2.3: $(k')^2=-k^3+\lambda k+c$ on (k,k') plane

At this time, the differential equation

$$(k')^2=-k^3+\lambda k+c.$$

has the property as in shown Figure 2.3 and its solution can be written by the Jacobi elliptic sine function

$$k(s_1) = \alpha_3(1 - q^2\mathrm{sn}^2(ds_1, p)), \tag{2.48}$$

where $p^2 = (\alpha_3 - \alpha_2)/(\alpha_3 - \alpha_1), q^2 = (\alpha_3 - \alpha_2)/\alpha_3$ and $d = \sqrt{\alpha_3 q^2/4p^2} = \sqrt{\alpha_3 - \alpha_1}/2$. This well known solution also can be written as

$$k(s_1) = \alpha_3 - (\alpha_3 - \alpha_2)\mathrm{sn}^2(ds_1, p). \tag{2.49}$$

Therefore it is in the stable interval $k \in [\alpha_2, \alpha_3]$. It is a periodic solution with period $2K(p)/d$. We will discuss this solution later for the existence of closed centroaffine elastica.

For the unstable interval $(-\infty, \alpha_1]$, the function $\pm(P(k))^{-1/2} = \pm[(\alpha_1 - x)(x - \alpha_2)(x - \alpha_3)]^{-1/2}$ is integrable on the interval $(-\infty, \alpha_1]$. By using the integral formula 3.131(2) of Ref [15]

$$\int_k^{\alpha_1} \frac{dx}{\sqrt{(\alpha_1 - x)(\alpha_2 - x)(\alpha_3 - x)}} = \frac{2}{\sqrt{\alpha_3 - \alpha_1}} F(\arcsin\sqrt{\frac{\alpha_1 - k}{\alpha_2 - k}}, \sqrt{\frac{\alpha_3 - \alpha_2}{\alpha_3 - \alpha_1}}),$$

and the definition of the Jacobi elliptic sine function, we have

$$\sqrt{\frac{\alpha_1 - k}{\alpha_2 - k}} = \mathrm{sn}(\sqrt{d}s_1, p),$$

for $p = \sqrt{\frac{\alpha_3 - \alpha_2}{\alpha_3 - \alpha_1}}$ and $d = \frac{\sqrt{\alpha_3 - \alpha_1}}{2}$. At this time, we have assumed that $k(0) = \alpha_1$. We can obtain a solution

$$k(s_1) = \frac{\alpha_1 - \alpha_2\mathrm{sn}^2(ds_1, p)}{\mathrm{cn}^2(ds_1, p)}. \tag{2.50}$$

Since $k(0) = \alpha_1$, from Figure 2.3, we have $k(s_1) \le \alpha_1$ and $s_1 \in (-K(p)/d, K(p)/d)$. This implies $L < 2K(p)/d$. As $s_1 \to \pm K(p)/d$, $k(s_1) \to -\infty$.

Case 2. $4\lambda^3 - 27c^2 = 0$, the roots are real and at least two of them are equal.

(1). $\alpha_1 < \alpha_2 = \alpha_3$

At this time, $\lambda > 0$, $\alpha_1 = -2\sqrt{\lambda/3}$, $\alpha_2 = \alpha_3 = \sqrt{\lambda/3} = \alpha$, $c < 0$ and $k(s) \le -2\alpha$. This is the case (b) of Figure 2.1. The nonconstant centroaffine curvature curve has the same property as that of the unstable situation of case 1. We assume $k(0) = -2\alpha$, we have the following integral up to an orientation reversing.

$$\int_{-2\alpha}^{k(s_1)} \frac{dx}{(x-\alpha)\sqrt{-x-2\alpha}} = s_1.$$

By setting the integral transformation $t = \sqrt{-x-2\alpha}$, the left side becomes

$$\int_0^{\sqrt{\alpha_1 - k(s_1)}} \frac{2dt}{\alpha_2 - \alpha_1 + t^2} = \frac{2}{\sqrt{\alpha_2 - \alpha_1}} \arctan\sqrt{\frac{\alpha_1 - k(s_1)}{\alpha_2 - \alpha_1}}.$$

We have the following formula

$$k(s_1) = \alpha_2 + (\alpha_1 - \alpha_2)\sec^2(\frac{\sqrt{\alpha_2 - \alpha_1}}{2} s_1). \tag{2.51}$$

Or

$$k(s_1) = \sqrt{\frac{\lambda}{3}} - 3\sqrt{\frac{\lambda}{3}} \sec^2(ds_1). \tag{2.52}$$

Here $d = (3\lambda)^{1/4}/2$, $-\pi/2d < s_1 < \pi/2d$ and we get $L < \pi/(\sqrt{2}d)$.
(2). $\alpha_1 = \alpha_2 < \alpha_3$.

At this time, $\lambda > 0$, $\alpha_1 = \alpha_2 = -\sqrt{\lambda/3} = -\alpha$, $\alpha_3 = 2\sqrt{\lambda/3}$, $c > 0$. This is the case (c) of Figure 2.1. The configuration curve of the differential equation

$$(k')^2 = -k^3 + \lambda k + c.$$

has the property as shown in Figure 2.4.

We may assume $k(0) = \alpha_3$, we have the following integral up to an orientation reversing.

$$\int_{\alpha_3}^{k(s_1)} \frac{dx}{(x-\alpha_2)\sqrt{\alpha_3 - x}} = s_1.$$

By setting the integral transformation $t = \sqrt{\alpha_3 - x}$, the left side becomes

$$\int_0^{\sqrt{\alpha_3 - k(s_1)}} \frac{-2dt}{\alpha_3 - \alpha_2 - t^2} = \frac{1}{\sqrt{\alpha_3 - \alpha_2}} \ln \frac{\sqrt{\alpha_3 - \alpha_2} - \sqrt{\alpha_3 - k(s_1)}}{\sqrt{\alpha_3 - \alpha_2} + \sqrt{\alpha_3 - k(s_1)}}.$$

We have the following formula

$$k(s_1) = \alpha_2 + (\alpha_3 - \alpha_2)\text{sech}^2(\frac{\sqrt{\alpha_3 - \alpha_2}}{2} s_1). \tag{2.53}$$

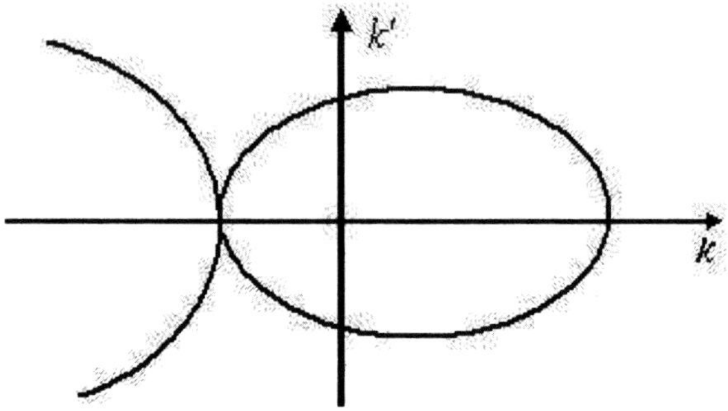

Figure 2.4: $(k')^2 = -(k-\alpha_1)^2(k-\alpha_3)$ on (k, k') plane

Or

$$k(s_1) = -\sqrt{\frac{\lambda}{3}} + 3\sqrt{\frac{\lambda}{3}}\operatorname{sech}^2(ds_1). \tag{2.54}$$

Here $d = (3\lambda)^{1/4}/2$ and $-\infty < s_1 < \infty$. This is the solution in the interval $k \in (\alpha_2, \alpha_3]$. The solution will have a maximum and be infinitely long with each tail asymptotic to a part of a hyperbola.

The solution in the interval $(-\infty, \alpha_2)$ will have the same property as the unstable situation in case 1 at $-\infty$. We assume $k(0) = 2\alpha_1$. Integrating as above, we have

$$\int_{\sqrt{\alpha_3-2\alpha_2}}^{\sqrt{\alpha_3-k(s_1)}} \frac{-2dt}{\alpha_3-\alpha_2-t^2} = s_1$$

$$= \frac{1}{\sqrt{\alpha_3-\alpha_2}}\left(\ln\frac{\sqrt{\alpha_3-k(s_1)}-\sqrt{\alpha_3-\alpha_2}}{\sqrt{\alpha_3-k(s_1)}+\sqrt{\alpha_3-\alpha_2}} - \ln\frac{\sqrt{\alpha_3-2\alpha_2}-\sqrt{\alpha_3-\alpha_2}}{\sqrt{\alpha_3-2\alpha_2}+\sqrt{\alpha_3-\alpha_2}}\right).$$

An elementary computation gives

$$k(s_1) = \alpha_2 - (\alpha_3-\alpha_2)\operatorname{csch}^2\left(\frac{\sqrt{\alpha_3-\alpha_2}}{2}s_1 + \frac{1}{2}\ln\frac{\sqrt{\alpha_3-2\alpha_2}-\sqrt{\alpha_3-\alpha_2}}{\sqrt{\alpha_3-2\alpha_2}+\sqrt{\alpha_3-\alpha_2}}\right). \tag{2.55}$$

Thus we obtain the formula

$$k(s_1) = -\sqrt{\frac{\lambda}{3}} - 3\sqrt{\frac{\lambda}{3}}\operatorname{csch}^2(ds_1 + \ln(2-\sqrt{3})). \tag{2.56}$$

For $-\infty < s_1 < \ln(2+\sqrt{3})/d$. As $s_1 \to \ln(2+\sqrt{3})/d$, $k(s_1) \to -\infty$ and as $s_1 \to -\infty$, $k(s_1) \to \alpha_1$.

(3). $\alpha_1 = \alpha_2 = \alpha_3$.

At this time, $\lambda = 0, c = 0$, then $\alpha_1 = \alpha_2 = \alpha_3 = 0$.This is the case (d) of Figure 2.1. We have

$$(k')^2 = -k^3.$$

Thus $k \leq 0$. We may assume $k(0) = -1$. A direct integration gives

$$k(s_1) = -\frac{4}{(s_1-2)^2}, \qquad -\infty < s_1 < 2. \tag{2.57}$$

Case 3. $4\lambda^3 - 27c^2 < 0$, two roots are imaginary. This is the case (e) and (f) of Figure 2.1. This case is similar to the unstable situation of case 1. By using the integral formula 3.138(8) of Ref [15]

$$\int_k^{\alpha_1} \frac{dx}{\sqrt{(\alpha_1 - x)[(x-a)^2+b^2]}} = \frac{1}{\sqrt{d}}F(2\operatorname{arccot}\sqrt{\frac{\alpha_1-k}{d}}, \sqrt{\frac{d-a+\alpha_1}{2d}}),$$

where $d = \sqrt{(\alpha_1-a)^2+b^2}$ and the definition of the Jacobi elliptic sine function, we have

$$\sin(2\operatorname{arccot}\sqrt{\frac{\alpha_1-k}{d}}) = \operatorname{sn}(\sqrt{d}s_1, p),$$

for $p = \sqrt{\frac{d-a+\alpha_1}{2d}}$. At this time, we have assumed that $k(0) = \alpha_1$. The formula $\sin 2\theta = 2\sin\theta\cos\theta$ gives

$$\sin(2\operatorname{arccot}\sqrt{\frac{\alpha_1-k}{d}}) = 2\sqrt{d}\frac{\sqrt{\alpha_1-k}}{\alpha_1-k+d}.$$

An elementary computation gives the formula of the centroaffine curvature

$$k(s_1) = \alpha_1 - d\frac{(1-\operatorname{cn}(\sqrt{d}s_1,p))^2}{\operatorname{sn}^2(\sqrt{d}s_1,p)}. \tag{2.58}$$

Therefore we can express the centroaffine curvature function $k(s)$ of the centroaffine elastica by the Jacobi Elliptic functions completely. But it may be unstable, that means $k(s)$ may be unbounded. The solutions (2.48), (2.52), (2.54) and (2.57) can be found in Ref [1].

2.3.2 The solutions of the structure equation for centroaffine elastica in the affine plane $\mathbb{R}^2$

Now we turn to the structure equation(2.6).

$$X''(s) = -k(s)X(s). \tag{2.59}$$

This is a linear system. We know it has global solutions. For the constant centroaffine curvature starlike elastica, from the equation of centroaffine elastica, we know $k = \pm\sqrt{\lambda/3}$ and $\lambda \geq 0$. We can give a formula for each of the following case.
(a). $k = 0$. Integrating $X''(s) = 0$ twice gives

$$X(s) = (c_1, c_2)^T s + (c_3, c_4)^T. \tag{2.60}$$

Here c_1, c_2, c_3, c_4 are constant and satisfy the condition

$$\det \begin{pmatrix} c_3 & c_1 \\ c_4 & c_2 \end{pmatrix} = 1.$$

This is a straight line.
(b). $k = r^2 = \sqrt{\lambda/3} > 0$. Then

$$X(s) = \cos(rs)(c_1, c_2)^T + \sin(rs)(c_3, c_4)^T. \tag{2.61}$$

Here c_1, c_2, c_3, c_4 are constant and satisfy the condition

$$\det \begin{pmatrix} c_1 & c_3 \\ c_2 & c_4 \end{pmatrix} = \frac{1}{r}.$$

Taking the coordinate system $\{x, y\}$ such that $(c_1, c_2) = (1, 0)$ and $(c_3, c_4) = (0, 1/r)$, we see that the curve is an ellipse $x^2 + r^2y^2 = 1$.
(c). $k = -r^2 = -\sqrt{\lambda/3} < 0$. We obtain the curve in the form

$$X(s) = \cosh(rs)(c_1, c_2)^T + \sinh(rs)(c_3, c_4)^T. \tag{2.62}$$

Here c_1, c_2, c_3, c_4 are constant and satisfy the condition

$$\det \begin{pmatrix} c_1 & c_3 \\ c_2 & c_4 \end{pmatrix} = \frac{1}{r}.$$

Choosing the coordinate system $\{x, y\}$ such that $(c_1, c_2) = (1, 0)$ and $(c_3, c_4) = (0, 1/r)$, the curve is a hyperbola $x^2 - r^2y^2 = 1$.

For the nonconstant centroaffine curvature starlike elastica, we found the vector field

$$W = -\frac{1}{2}k'X + kX'$$

is Killing along the centroaffine elastica X. As that done in section 2, we obtain the following results.

(1). $c > 0$, the centroaffine elastica equation is

$$\begin{cases} x(s) = d_1 k(s)^{1/2} \exp(\int \frac{\sqrt{c}}{\sqrt{2k}} ds), \\ y(s) = d_2 k(s)^{1/2} \exp \int \frac{\sqrt{c}}{\sqrt{2k}} ds. \end{cases}$$

Here d_1, d_2 are constants and satisfy $\sqrt{2c} d_1 d_2 = -1$.

(2). $c = 0$, the solution is

$$\begin{cases} x(s) = k(s)^{1/2}, \\ y(s) = k(s)^{1/2}[d + \int \frac{1}{k} ds]. \end{cases}$$

Here d are constant.

(3). $c < 0$, we get the solution

$$\begin{cases} x(s) = k(s)^{1/2}(d_2 \cos(\int \frac{\sqrt{-c}}{\sqrt{2k}} ds) - d_1(\int \frac{\sqrt{-c}}{\sqrt{2k}} ds)), \\ y(s) = k(s)^{1/2}(d_1 \cos(\int \frac{\sqrt{-c}}{\sqrt{2k}} ds) + d_2(\int \frac{\sqrt{-c}}{\sqrt{2k}} ds)). \end{cases}$$

Here d_1, d_2 are constants and satisfy $\sqrt{-c}(d_1^2 + d_2^2) = \sqrt{2}$.

We give a second method to solve the Frenet equation. We construct a coordinate system $\{x, y\}$ such that $\frac{\partial}{\partial x} = \tilde{W}$. We have

$$y' = |\tilde{W}, X'| = -\frac{1}{2}k',$$

$$y = -\frac{1}{2}(k + c_1).$$

From $|X, X'| = 1$, we know

$$\frac{1}{2}(k + c_1)x' - \frac{1}{2}k'x = 1,$$

$$x(s) = -\frac{1}{2}(k + c_1)[c_2 - \int_0^s \frac{4}{(k + c_1)^2} ds].$$

Here c_2, c_2 are constants. Therefore in this coordinate system, the centroaffine elastica has the parameter equation

$$\begin{aligned} x &= -\tfrac{1}{2}(k+c_1)[c_2 - \textstyle\int_0^s \frac{4}{(k+c_1)^2} ds], \\ y &= -\tfrac{1}{2}(k+c_1). \end{aligned}$$

2.3.3 Closed centroaffine elastica

In this section, we will look for the condition such that the centroaffine elastica is closed. This means that the equation of the curve is periodic. It implies that the centroaffine curvature $k(s)$ is periodic. For $k(s)$ is constant, we know that the ellipse is periodic from (2.61).

Now we consider the case the centroaffine curvature $k(s)$ is nonconstant. Checking all these cases, we found that the periodic centroaffine curvature is the stable solution(2.48) and we only need to find the condition for the integral $\int b/kds$ is a rational ratio of π in one period of $k(s)$. Only the case 3 provides the possibility. We check the equation

$$k(kV')' + b^2 V = 0.$$

We set a parameter transformation by

$$t = \int_a^s \frac{1}{k(u)} du.$$

And write $f(t) = V(s(t))$, then the above equation becomes

$$\frac{d^2 f}{dt^2} + b^2 f = 0.$$

This equation has the general solution

$$f(t) = d_1 \cos(bt) + d_2 \sin(bt).$$

We assume T the period of $V(s)$, then it has to satisfy the condition

$$b \int_0^T \frac{1}{k(u)} du = 2\pi n,$$

for some positive integer n. From the equation

$$\frac{1}{2}(k')^2 = -k^3 + \lambda k + c_1.$$

We know its solution can be written by the Jacobi elliptic sine function

$$k(s) = \alpha_3(1 - q^2\mathrm{sn}^2(\sqrt{2}ds, p)).$$

Here $\alpha_1 < \alpha_2 < \alpha_3$ are the roots of the cubic equation $-k^3 + \lambda k + c_1 = 0$ and $p^2 = (\alpha_3 - \alpha_2)/(\alpha_3 - \alpha_1), q^2 = (\alpha_3 - \alpha_2)/\alpha_3$ and $d = \sqrt{\alpha_3 q^2/4p^2} = \sqrt{\alpha_3 - \alpha_1}/2$. Using p and α_3 as basic arguments, we can obtain the formulae.

$$\alpha_1 = \frac{-2 + p^2}{1 + p^2}\alpha_3, \qquad \alpha_2 = \frac{1 - 2p^2}{1 + p^2}\alpha_3,$$

$$q^2 = \frac{3p^2}{1 + p^2}, \qquad 2d = \sqrt{\frac{3}{1 + p^2}\alpha_3},$$

$$-2b^2 = c_1 = \alpha_1\alpha_2\alpha_3 = \frac{(1 - 2p^2)(p^2 - 2)}{(1 + p^2)^2}\alpha_3^3.$$

Therefore we have the formula

$$b\int_0^{\frac{\sqrt{2}K(p)}{d}} \frac{1}{k}ds = 2[\frac{(1 - 2p^2)(2 - p^2)}{3(1 + p^2)}]^{\frac{1}{2}} \int_0^{K(p)} (1 - \frac{3p^2}{1 + p^2}\mathrm{sn}^2(t, p))^{-1}dt.$$

Figure 2.5 is the distribution of this integral for $p^2 \in (0, 0.5)$.

This integral is $\sqrt{2/3}\pi$ at $p = 0$ and about π at $p = \sqrt{1/2}$. By the intermediate value theorem, we can find a positive integer l and $p^2 \in (0, 1/2)$ such that the above integral has the value $2\pi n/l$ for a given positive integer n. We have the following theorem

Theorem 5 . *For every pair of integers l, n such that $1/6 < n^2/l^2 < 1/4$, there exists a closed nonconstant centroaffine curvature elastica having the rotating index n in the affine plane $\mathbf{R}^2$.*

The weakness of this result is that the upper bound is gotten by the numerical calculation. If we only need the existence of a closed nonconstant centroaffine curvature elastica, it is enough to prove the above integral is nonconstant. This is obvious. The above results extend that in Refs.([20],[21]).

Now we are going to study the periodic solution by using the Lamé's equation and get a special result which we will use in chapter 4. We go back to the stable solution of case 1 of the Euler-Lagrange equation

$$k(s) = \alpha_3(1 - q^2\mathrm{sn}^2(d\sqrt{2}s, p)).$$

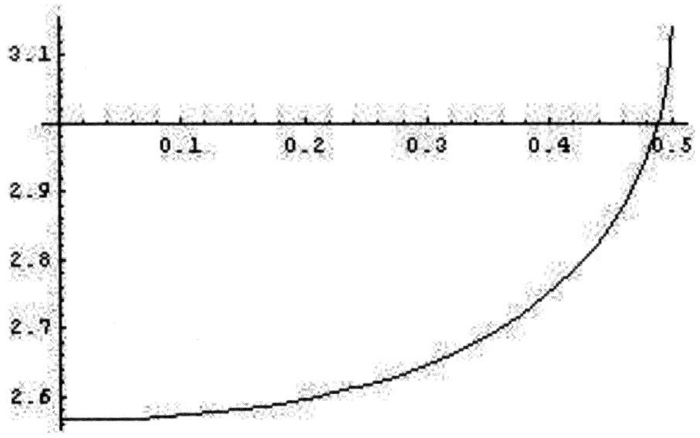

Figure 2.5: Distribution of the integral for $p^2 \in (0, 0.5)$

This function has a period $\sqrt{2}K(p)/d$. Here

$$K(p) = \int_0^{\frac{\pi}{2}} \frac{d\theta}{\sqrt{1 - p^2 \sin^2 \theta}}$$

is the complete elliptic integral of first kind. Now we have the equation

$$X'' + \alpha_3(1 - q^2 \mathrm{sn}(d\sqrt{2}s, p))X = 0. \tag{2.63}$$

Changing the parameter $s_2 = d\sqrt{2}s$, we obtain a standard Lamé's equation

$$X''(s_2) + [\mu - m(m+1)p^2 \mathrm{sn}^2(s_2, p)]X(s_2) = 0. \tag{2.64}$$

Here $\mu = \alpha_3/(2d^2) = 2(1+p^2)/3$, and $m(m+1) = \alpha_3 q^2/(2d^2p^2) = 2$. We hope this equation has periodic solution $X(s_2)$ with period $2lK$ for some p, here l is a positive integer. This question can be transferred to the following problem: Does the following Lamé's equation have two linearly independent solutions with period $2lK$?

$$y''(x) + [\mu - 2p^2 \mathrm{sn}^2(x, p)]y(x) = 0. \tag{2.65}$$

This linear equation has two continuously differentiable solutions $y_1(x)$ and $y_2(x)$ which are uniquely determined by the initial conditions:

$$y_1(0) = 1, y_1'(0) = 0, y_1(0) = 0, y_2(0) = 1.$$

These solutions are called normalized solutions of Lamé's equation. Since $\mu - 2p^2sn^2(x,p)$ is an even function, if $y(x)$ is a solution of this equation then $y(-x)$ is also a solution. Since the initial conditions for $y_1(-x)$ and $y_1(x)$ coincide and, similarly, those for $y_2(x)$ and $-y_2(-x)$ are identical, it follows that $y_1(x)$ is even and $y_2(x)$ is odd.

Theorem 6 : *Let $y_1(x)$ and $y_2(x)$ be the normalized solutions of Lamé's equation, then there exists a nontrivial periodic solution which is*

$$\begin{array}{lll} \text{even and of period } 2lK & \text{if and only if} & y_1'(lK) = 0, \\ \text{odd and of period } 2lK & \text{if and only if} & y_2(lK) = 0. \end{array}$$

Proof: If $y(x)$ is a periodic solution of Lamé's equation, the functions

$$u(x) = y(x) + y(-x), \qquad v(x) = y(x) - y(-x).$$

are solutions too, and $u(x)$ is even, $v(x)$ is odd. $u(x)$ and $v(x)$ cannot both be trivial unless $y(x)$ is trivial. Therefore if $y(x)$ is nontrivial, there is either one even solution or one odd solution. Since $y_1(x), y_2(x)$ are two linearly independent solutions of Lamé's equation. We have uniquely

$$y(x) = c_1y_1(x) + c_2y_2(x).$$

Here c_1, c_2 are constants. Since a both even and odd function is identical zero, we obtain the fact that an even solution of Lamé's equation is a multiple of $y_1(x)$ and that an odd solution must be a multiple of $y_2(x)$.

Assume $y(x)$ be a nontrivial, even, periodic solution of Lamé's equation with period $2lK$. Then $y_1(x)$ is also periodic with period $2lK$ and the same is true about $y_1'(x)$. Thus $y_1'(lK) = y_1'(-lK)$. Since $y_1'(x)$ is an odd function, it follows that $y_1'(lK) = -y_1'(-lK)$. Therefore $y_1'(lK) = 0$. Conversely, if $y_1'(lK) = 0$, then $y_1'(-lK) = 0$, and $y_1(-lK) = y_1(lK)$. This means $y_1(x)$ satisfies the same condition at $x = -lK$ and $x = lK$. Therefore $y_1(x)$ is periodic with period $2lK$.

The proof of the second statement is similar to that of the first one. ∎

Now we notice

$$\frac{2}{3}(1 - 2p^2) \leq \mu - m(m+1)p^2\text{sn}^2(s_2, p) \leq \frac{2}{3}(1 + p^2).$$

The number of zeros of y_1, y_1', y_2, y_2' in the interval $0 \le x \le lK$ is then maximized or minimized, respectively, by the number of zeros of the corresponding solutions of

$$z''(x) + \frac{2}{3}(1+p^2)z(x) = 0, \qquad z''(x) + \frac{2}{3}(1-2p^2)z(x) = 0.$$

Now we consider the zeros of these two equations

$$z_1 = \cos\sqrt{\frac{2}{3}(1+p^2)}x, \qquad z_2 = \frac{1}{\sqrt{\frac{2}{3}(1+p^2)}}\sin\sqrt{\frac{2}{3}(1+p^2)}x,$$

$$z_3 = \cos\sqrt{\frac{2}{3}(1-2p^2)}x, \qquad z_4 = \frac{1}{\sqrt{\frac{2}{3}(1-2p^2)}}\sin\sqrt{\frac{2}{3}(1-2p^2)}x.$$

Here in z_3, z_4, p is in $[0, \sqrt{2}/2]$. By calculating(From Figure 2.6), we know $\sqrt{2(1+p^2)/3}K(p) < \pi$,if $p < p_1 = 0.935$ and $\sqrt{2(1+p^2)/3}2K(p) < \pi$, if $p < p_2 = 0.4$, we have the following corollary

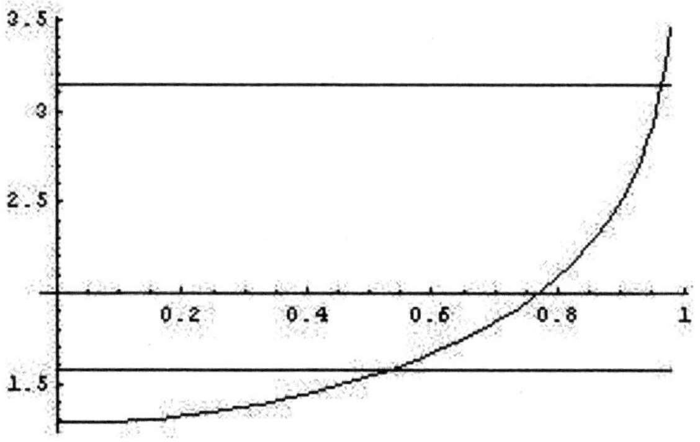

Figure 2.6: $\sqrt{2(1+p^2)/3}K(p)$, π and $\pi/2$

Corollary 1 . *The Lamé's equation (2.65) has no periodic solution with period $2K(p)$ if $p < p_1$ and has no periodic solution with period $4K(p)$ if $p < p_2$.*

We need further results about the periodic solution. We define the characteristic equation associated with Lamé's equation by

$$\rho^2 - [y_1(2K) + y_2'(2K)]\rho + 1 = 0.$$

This equation has two roots ρ_1 and ρ_2 written as

$$\rho_1 = \exp(i2\alpha K), \qquad \rho_2 = \exp(-i2\alpha K).$$

Here the number α is called the characteristic exponent.

We cite a form of Floquet's theorem about this Lamé's equation from Ref. [3].

Floquet's Theorem.

(1). If the roots ρ_1 and ρ_2 of the characteristic equation are different from each other, then Lamé's equation has two linearly independent solutions

$$z_1(x) = e^{i\alpha x} p_1(x), \qquad z_2(x) = e^{-i\alpha x} p_2(x).$$

Here $p_1(x)$ and $p_2(x)$ are periodic with period $2K$.

(2). If $\rho_1 = \rho_2$, then Lamé's equation has a nontrivial solution which is periodic with period $2K$ (when $\rho_1 = \rho_2 = 1$) or $4K$ (when $\rho_1 = \rho_2 = -1$). Let $p(x)$ denote such a periodic solution and let $y(x)$ be another solution linearly independent of $p(x)$. Then

$$y(x + 2K) = \rho_1 y(x) + \theta p(x)$$

Here θ is a constant and $\theta = 0$ is equivalent to

$$y_1(2K) + y_2'(2K) = \pm 2, \qquad y_2(2K) = 0, \; y_1'(2K) = 0.$$

Theorem 7 . *If Lamé's equation (2.65) has a nontrivial periodic solution $y(x)$ with a smallest period $2lK, l > 2$, then all solutions are periodic with period $2lK$.*

Proof: From Floquet's theorem, if $\rho_1 = \rho_2$, since $y(x)$ and $p(x)$ are linearly independent, we have

$$y(x) = y(x + 2lK) = \rho_1^l y(x) + l\theta p(x).$$

Hence $\theta = 0$, and $y(x)$ has period $2K$ or $4K$. This is a contradiction. We have $\rho_1 \neq \rho_2$ and every solution of Lamé's equation can be uniquely written as a linear combination of $z_1(x)$ and $z_2(x)$

$$y(x) = c_1 z_1(x) + c_2 z_2(x).$$

Since $y(x)$ has period $2lK$, we have

$$y(x) = y(x + 2lK) = c_1 e^{i2\alpha lK} z_1(x) + c_2 e^{-i2\alpha lK} z_2(x).$$

That is

$$c_1(e^{i2\alpha lK} - 1)z_1(x) + c_2(e^{-i2\alpha lK} - 1)z_2(x) = 0.$$

Since $z_1(x)$ and $z_2(x)$ are linearly independent and $y(x)$ is nontrivial, we know $\exp(i2\alpha lK) = 1$, l is the smallest positive integer such that $\alpha lK = \pi$ and both $z_1(x)$ and $z_2(x)$ are periodic with period $2lK$. ∎

Corollary 2 . *If Lamé's equation (2.65) has two nontrivial periodic solutions and one has the period $2K$ or $4K$, then the other has the same period.*

Checking the elementary Lame polynomial, we know $\text{sn}(x,p)$, $\text{dn}(x,p)$ can not be the solutions of Lame equation. For $\text{cn}(x,p)$, we have

$$\text{cn}(x)(p^2\text{sn}^2(x) - \text{dn}^2(x)) + (\mu - 2p^2\text{sn}^2(x))\text{cn}(x) = 0.$$

That is $\mu = \frac{2}{3}(p^2+1) = 1$, therefore $p^2 = 1/2$. From the recurrence relations, we know this is the only case of finite order solutions. We know the second normalized solution satisfies the equation

$$y_2(x + 2K) = -y_2(x) + 2\sqrt{2}y_2'(K)\text{cn}(x,p).$$

We have proved the Wronskian of the equation

$$y_1(x)y_2'(x) - y_1'(x)y_2(x) = 1.$$

Here $y_1(x)$, $y_2(x)$ are the normalized solutions of Lame's equation. In this case, $y_1(x) = \text{cn}(x)$. We have a first order nonlinear equation

$$\text{cn}(x)y_2'(x) - (\text{cn}(x))'y_2(x) = 1.$$

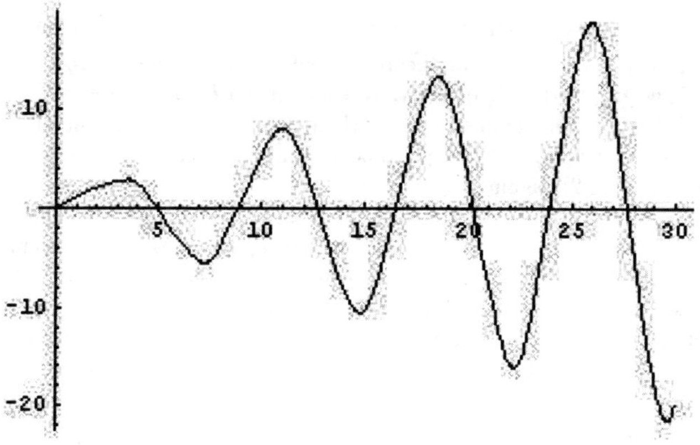

Figure 2.7: $y_2(x) = x\text{cn}(x) - 2E(x)\text{cn}(x) + 2\text{dn}(x)\text{sn}(x)$

We have the solution

$$\begin{aligned} y_2(x) &= e^{\int_0^x \frac{\text{cn}'(t)}{\text{cn}(t)}dt} \int_0^x e^{-\int_0^s \frac{\text{cn}'(t)}{\text{cn}(t)}dt} \frac{1}{\text{cn}(s)} ds \\ &= \text{cn}(x) \int_0^x \frac{1}{\text{cn}^2(x)} dx \\ &= x\text{cn}(x) - 2E(x)\text{cn}(x) + 2\text{dn}(x)\text{sn}(x). \end{aligned}$$

Here $E(x) = \int_0^x \text{dn}^2(x)dx$ are the normal elliptic integral of the second kind.

In this case, $y_2'(K) = [2E(K) - K]/\sqrt{2} \approx 0.59907$. Hence $y_2(x)$ is not a periodic solution with period $4K$(see Figure 2.7), and we have

$$y_2(x + 2K) = -y_2(x) + 2[E(2K) - K]\text{cn}(x).$$

We denote $x_i(p)$ the i-th positive zero of $y_1'(x,p)$ counting from zero. Then $x_i(p)$ are continuous functions of p, so is $s_i(p) = x_i(p)/K(p)$. We want to find some p such that $s_i(p)$ is an integer, then, by theorem 1, this $y_1(x,p)$ is a periodic solution with period $2s_1(p)K(p)$. Since $y_1'(x,0) = -\sqrt{2/3}\sin(\sqrt{2/3}x)$ and $y_1'(x, \sqrt{1/2}) = -\text{sn}(x, \sqrt{1/2})\text{dn}(x, \sqrt{1/2})$, we know $s_i(\sqrt{1/2}) = 2i$, $s_i(0) = i\sqrt{6}$. $i(\sqrt{6} - 2) > 1$ when $i > 2$, by the continuity of $s_i(p)$, there exists some $p \in (0, \sqrt{1/2})$ such that $s_i(p)$ could be any integer in $(2i, \sqrt{6}i)$. By corollary 1, there is no solution of period $2K$ for $p \in (0, \sqrt{1/2})$. For these odd $l = s_i(p) \in (2i, \sqrt{6}i), i > 2$, by theorem 2 and corollary 2, $y_1(x,p), y_2(x,p)$ are periodic with period $2lK(p)$. For these even integer $l = s_i(p) = 4n+2, n > 1$, if $2lK(p)$ is not the smallest period, then 2(2n+1)K(p) is the smallest period, then $y_1(x,p), y_2(x,p)$ are periodic with period $lK(p)$. At this point, we proved theorem 2.3 again.

Since we did not prove our Lamé's equation has no periodic solution of period $4K(p)$ when $p^2 < 1/2$ (we guess it is true) we use the above argument for $l = 4n+2$. This argument can continue for the further case. Using numerical analysis, we can obtain some approximate results. $s_3(p) = 7$ has a root near $p^2 = 0.3525$(see Figure 2.8), $s_4(p) = 9$ has a root near $p^2 = 0.4238$(see Figure 2.9), $s_5(p) = 11$ has a root near $p^2 = 0.4523$, $s_5(p) = 12$ has a root near $p^2 = 0.2515$, $\cdots$.

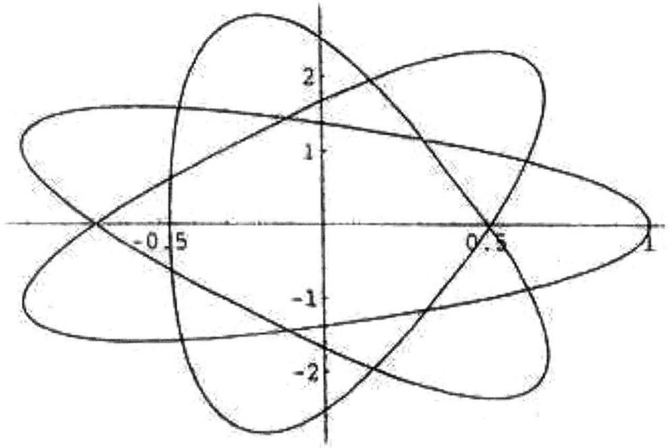

Figure 2.8: Centro-affine elastica for $p^2 = 0.3525$

For the solutions with period $4lK(p)$, we have the following result which should have an application to the existence of closed nonconstant affine curvature elastica.

Theorem 8 . *If all solutions of Lamé's equation are periodic with period*

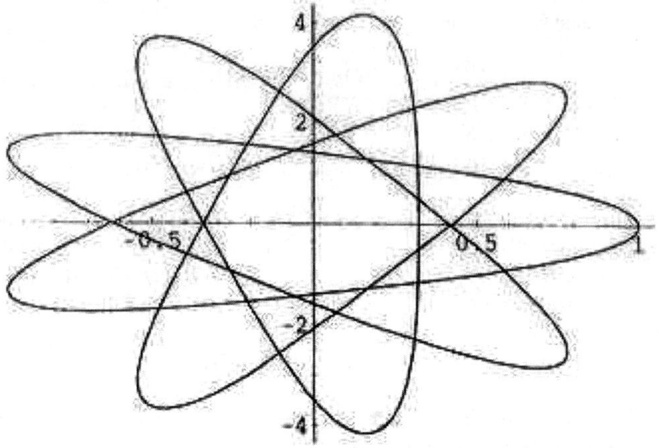

Figure 2.9: Centro-affine elastica for $p^2 = 0.4238$

$4lK$, l is a positive integer, but no solution is periodic with period $2lK$, then these solutions have the property

$$y(x + 2lK) = -y(x).$$

Proof: Since $2K$ is the basic period of the coefficients of Lamé's equation, we compare the initial condition at $x = 0$ and know

$$\begin{aligned} y_1(x + 2lK) &= y_1(2lK)y_1(x) + y_1'(2lK)y_2(x), \\ y_2(x + 2lK) &= y_2(2lK)y_1(x) + y_2'(2lK)y_2(x). \end{aligned}$$

We consider the values of $y_1(x), y_1'(x), y_2(x)$ and $y_2'(x)$ at $x = -lK$, we obtain a system of linear equations of $y_1(2lK)$, $y_1'(2lK)$, $y_2(2lK)$ and $y_2'(2lK)$

$$\begin{aligned} y_1(lK)y_1(2lK) - y_2(lK)y_1'(2lK) &= y_1(lK), \\ -y_1'(lK)y_1(2lK) + y_2'(lK)y_1'(2lK) &= y_1'(lK), \\ y_1(lK)y_2(2lK) - y_2(lK)y_2'(2lK) &= y_2(lK), \\ -y_1'(lK)y_2(2lK) + y_2'(lK)y_2'(2lK) &= y_2'(lK). \end{aligned}$$

Solving this system of linear equations, we obtain

$$\begin{aligned} y_1(2lK) = &\ y_1(lK)y_2'(lK) + y_1'(lK)y_2(lK), \\ y_1'(2lK) = &\ 2y_1(lK)y_1'(lK), \\ y_2(2lK) = &\ 2y_2(lK)y_2'(lK), \\ y_2'(2lK) = &\ y_1(lK)y_2'(lK) + y_1'(lK)y_2(lK). \end{aligned}$$

Since $y_1(x)$ has period $4lK$, we have $y_1'(2lK) = 0$ and $y_1'(lK) \neq 0$, otherwise $y_1(x)$ has period $2lK$. Then $y_1(lK) = 0$. Therefore $y_1(2lK) = -1$. Similarly, we have $y_2(2lK) = 0, y_2(lK) \neq 0, y_2'(lK) = 0$ and $y_2'(2lK) = -1$. Hence we have the relation

$$y_1(x + 2lK) = -y_1(x), \qquad y_2(x + 2lK) = -y_2(x).$$

All solutions are the linear combination of $y_1(x)$ and $y_2(x)$. We know all solutions have the above property.∎

2.3.4 Second variation formula

We consider the second variation with variation vector field $W = fX$. We do the derivative.

$$\begin{aligned} \nabla_{X'(s)}W = &\ f_s X(s) + fX'(s), \\ \nabla^2_{X'(s)}W = &\ f_{ss}X(s) + 2f_s X'(s) + fX''(s). \end{aligned}$$

From the computation done before, we know at $w = 0$

$$\begin{aligned} g = &\ -|X(t), X'(t)|^{-1}W(|X(t), X'(t)|) \\ = &\ -|X(t), X'(t)|^{-1}(|W(t), X'(t)| + |X(t), \tfrac{\partial}{\partial t}W(t)|) \\ = &\ -2f, \end{aligned}$$

$$\begin{aligned} W(k) = &\ |\nabla_{X'(s)}W, X''(s)| + |X'(s), \nabla^2_{X'(s)}W| + 3gk \\ = &\ -f_{ss} - 4fk, \end{aligned}$$

$$\begin{aligned} W(k_{ss}) = &\ WX'X'(k) \\ = &\ [W, X']X'(k) + X'[W, X'](k) + X'X'W(k) \\ = &\ -4fk_{ss} - 2f_s k_s + [W(k)]_{ss} \\ = &\ -f^{(4)} - 4f_{ss}k - 10f_s k_s - 8fk_{ss}. \end{aligned}$$

Therefore the second variational formula is

$$\begin{aligned}
\frac{d^2}{dw^2} & \int_0^{L(w)}(k^2+\lambda)ds|_{w=0} \\
= & -2\int_0^L |W, W(k_{ss}+3k^2-\lambda)X'|ds \\
= & \ 2\int_0^L f[f^{(4)}+10f_{ss}k+10f_sk_s+8fk_{ss}+24fk^2]ds \\
= & \ 2\int_0^L f[f^{(4)}+10f_{ss}k+10f_sk_s+8f\lambda]ds \\
= & \ 2\int_0^L[(f_{ss})^2-10k(f_s)^2+8\lambda f^2]ds.
\end{aligned}$$

In the last equality, we use $f(0)=f'(0)=f(L)=f'(L)=0$. For those constant centroaffine curvature elastica, the above formula can be written as

$$\frac{d^2}{dw^2}\int_0^{L(w)}(k^2+\lambda)ds|_{w=0}=2\int_0^L f[f^{(4)}+10f_{ss}k+24fk^2]ds.$$

Definition 9 . *Let $(x(s), y(s))$ be a constant centroaffine curvature elastica. s_0 is the first positive double zero of the nontrivial solution of the equation*

$$f^{(4)}+10f_{ss}k+24fk^2=0, \qquad f(0)=0, \qquad f'(0)=0.$$

s_0 is called the first conjugate point along the elastica with respect to 0.

For the case $k=0$, from the formula

$$\frac{d^2}{dw^2}\int_0^{L(w)}(k^2+\lambda)ds|_{w=0}=2\int_0^L(f_{ss})^2ds\geq 0.$$

Since there is no linear function with two zeroes, the above equality holds if and only if f is a zero function. Therefore we know this elastica is stable. When $k<0$, we may assume $k=-1$, for convenience, exactly we only need to multiply s by $\sqrt{|k|}$. Since $f(0)=f(L)=0$, we can extend it to be a periodic function with period L. We have Fourier series of the form

$$f(s)=\frac{a_0}{2}+\sum_{n=1}^{\infty}[a_n\cos(\frac{2n\pi}{L}s)+b_n\sin(\frac{2n\pi}{L}s)].$$

Computing f'', we obtain the formula

$$\frac{d^2}{dw^2}\int_0^{L(w)}(k^2+\lambda)ds|_{w=0}=L[12a_0^2+\sum_{n=1}^{\infty}((\frac{2n\pi}{L})^2+4)((\frac{2n\pi}{L})^2+6)(a_n^2+b_n^2)].$$

Therefore, we have the stability for $k < 0$. For $k > 0$, Similarly, we may assume $k = 1$, we consider the linear ordinary differential equation

$$f^{(4)} + 10f^{(2)} + (24 + c)f = 0.$$

With the boundary condition $f(0) = f(L) = f'(0) = f'(L) = 0$, here c is a constant with the range $0 \leq c < 1$. The general solution of this equation is

$$c_1 \cos \sqrt{5 + \sqrt{1 - c}}s + c_2 \sin \sqrt{5 + \sqrt{1 - c}}s$$

$$+c_3 \cos \sqrt{5 - \sqrt{1 - c}}s + c_4 \cos \sqrt{5 - \sqrt{1 - c}}s.$$

Then the boundary condition is a system of linear equation of c_i. The condition for the existence of nonzero solutions is the following equation

$$\sqrt{24 + c} - 5 \sin(\sqrt{5 - \sqrt{1 - c}}L) \sin(\sqrt{5 + \sqrt{1 - c}}L)$$

$$-\sqrt{24 + c} \cos(\sqrt{5 - \sqrt{1 - c}}L) \cos(\sqrt{5 + \sqrt{1 - c}}L) = 0.$$

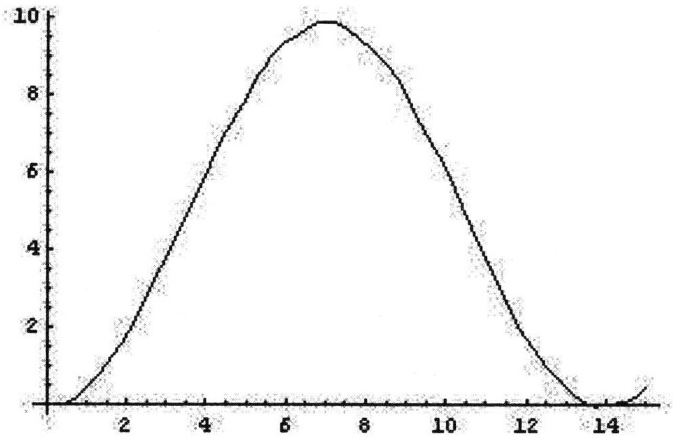

Figure 2.10: Conjugate point for centroaffine elastica

For $c = 0$, the first positive zero $L(0)$ of this equation is near $L_0 = 13.55$(see Figure 2.10). This is a single zero. Since these zero is a continuous function of c, for an arbitrary small $c > 0$, there is a zero $L(c)$ near $L(0)$. At this time, we have

$$\frac{d^2}{dw^2}\int_0^{L(w)}(k^2+\lambda)ds|_{w=0} = -2c\int_0^{L(c)} f^2 ds < 0.$$

For $L > L(0)$, by the continuity, we can find a $c > 0$ such that $L > L(c)$ and we define $f = 0$ on $(L(c), L]$, we obtain the unstability for $L > L(0)$, but this is in the function space $B_1 = \{f \in C^1[0, L_1] | f(0) = f'(0) = f(L_1) = f'(L_1) = 0\}$. We hope the stability for $L < L(0)$. We consider the case $L = 4\pi$, using Fourier series, we know

$$\frac{d^2}{dw^2}\int_0^{L(w)}(k^2+\lambda)ds|_{w=0} = L[12a_0^2 + \sum_{n=1}^{\infty}((\frac{2n\pi}{L})^2 - 4)((\frac{2n\pi}{L})^2 - 6)(a_n^2 + b_n^2)].$$

It would be zero only if $f(s) = a_4\cos(2s) + b_4\sin(2s)$. But $f(0) = f(4\pi) = f'(0) = f'(4\pi) = 0$ if and only if $a_4 = b_4 = 0$. Therefore we know the stability of the case $L = 4\pi$. This is a special case. Now we first prove a localized lemma.

Lemma 2 *For sufficiently small number $L > 0$, the integral*

$$\int_0^L [(f_{ss})^2 - 10k(f_s)^2 + 8\lambda f^2]ds$$

is positive, for fixed λ, k and any nontrivial function $f \in C^2[0, L]$ and $f(0) = 0, f'(0) = 0$.

Proof:Using Schwarz inequality, we have

$$f^2(s) = (\int_0^s f ds)^2 \le s\int_0^s f'^2 ds \le s\int_0^L f'^2 ds.$$

One more integral from 0 to L, we know

$$\int_0^L f^2 ds \le \frac{L^2}{2}\int_0^L f'^2 ds.$$

Using this inequality, we obtain

$$\int_0^L [(f_{ss})^2 - 10k(f_s)^2 + 8\lambda f^2]ds \ge (1 - 5|k|L^2 - 2|\lambda|L^4)\int_0^L f_{ss}^2 ds.$$

For sufficiently small L, we have $1 - 5|k|L^2 - 2|\lambda|L^4 > 0$. We obtain the result, same as in the case $k = 0$.■

We can find a largest positive number $L_1 \in (0, L(0)]$(if no,we set $L_1 = \infty$) such that when $0 < L < L_1$

$$\frac{d^2}{dw^2}\int_0^{L(w)}(k^2+\lambda)ds|_{w=0} > 0.$$

We want to prove $L_1 = L(0)$. We consider the index form

$$I(f,g) = \int_0^{L_1}[f_{ss}g_{ss} - 10f_sg_s + 24fg]ds.$$

We denote the set $B_2 = \{f \in C^2[0, L_1] | f(0) = f'(0) = f(L_1) = f'(L_1) = 0\}$ and $B_4 = \{f \in B,\ f \in C^4[0, L_1]\}$. When $f \in B_4$, we have

$$I(f,g) = \int_0^{L_1}[f^{(4)} + 10f_{ss} + 24f]gds.$$

Especially, when $f \in B_2$, we have

$$I(f,f) = \frac{d^2}{dw^2}\int_0^{L(w)}(k^2+\lambda)ds|_{w=0}.$$

Since the index form I is bilinear, we consider the function $f + \epsilon g$ for $\epsilon > 0$ and assume $f \in B_4$ and $I(f,f) = 0$, we have

$$I(f,f) + 2\epsilon I(f,g) + \epsilon^2 I(g,g) = I(f+\epsilon g, f+\epsilon g) \geq 0.$$

Divided by ϵ, we obtain

$$2I(f,g) + \epsilon I(g,g) \geq 0.$$

Set $\epsilon \to 0$, we know $I(f,g) \geq 0$. Similarly, we consider $f - \epsilon g$, we get $I(f,g) \leq 0$. Therefore we have $I(f,g) = 0$ for all $g \in B_2$. We construct a function $h \in B_2$ and $h(s) > 0$ when $s \in (0, L_1)$,for example $h(s) = s^2(s - L_1)^2$ and set $g = h[f^{(4)} + 10f_{ss} + 24f]$. We obtain

$$I(f,g) = \int_0^{L_1}[f^{(4)} + 10f_{ss} + 24f]^2hds.$$

Therefore we have

$$f^{(4)} + 10f_{ss} + 24f = 0.$$

Since $f \in B_2$, we have either $f = 0$ or $L_1 = L(0)$ and we get the stability for the case $L < L(0)$ Therefore, we have the theorem.

Theorem 9 *The constant centroaffine curvature elastica* $(x(s), y(s)), s \in [0, L]$ *is stable if* L *is less than the first conjugate point. If* $k \leq 0$*, the first conjugate point is* ∞*. If* $k > 0$*, the first conjugate point is beyond two period of the elastica, the ellipse.*

Now we consider the regularity of the function f. In the above proof, we assume $f \in B_4$, that means $f \in C^4[0, L_1]$. We will prove that if $f \in B_2$ and $I(f, f) = 0$, then $f \in B_4$. Since in the proof of $I(f, g) = 0$ for all $g \in B_2$, we only use $f \in B_2$. That is for $f \in B_2$ and all $g \in B_2$, we have

$$\int_0^{L_1} [f_{ss}g_{ss} - 10f_s g_s + 24fg]ds = 0.$$

Now we claim that $f_{ss} \in C^2[0, L_1]$,that is $f \in C^4[0, L_1]$. Set

$$A(s) = \int_0^s f(t)dt.$$

Then $A' = f$ and $A \in C^3[0, L_1]$. Integrating by parts, we find

$$\begin{aligned} \textstyle\int_0^{L_1} fgds = & \ A(s)g(s)|_0^{L_1} - \textstyle\int_0^{L_1} A(s)g'(s)ds \\ = & \ -\textstyle\int_0^{L_1} A(s)g'(s)ds. \end{aligned}$$

Set

$$B(s) = \int_0^s [10f_t(t) + 24A(t)]dt.$$

Then $B' = 10f_s + 24A$ and $B \in C^2[0, L_1]$. Integrating by parts, we get

$$\begin{aligned} \textstyle\int_0^{L_1} [10f_s(s) + 24A(s)]g_s ds = & \ B(s)g_s(s)|_0^{L_1} - \textstyle\int_0^{L_1} B(s)g_{ss}(s)ds \\ = & \ -\textstyle\int_0^{L_1} B(s)g_{ss}(s)ds. \end{aligned}$$

Therefore we have

$$\int_0^{L_1} [f_{ss}(s) + B(s)]g_{ss}(s)ds = 0.$$

for any $g \in B_2$. Let c_1 and c_2 be two numbers defined by the following equations

$$\begin{aligned} \textstyle\int_0^{L_1} [f_{ss}(s) + B(s) - c_1 - c_2 s]ds = & \ 0 \\ \textstyle\int_0^{L_1} ds \int_0^s [f_{tt}(t) + B(t) - c_1 - c_2 t]dt = & \ 0. \end{aligned}$$

Since this is a system of linear equations of c_1, c_2, the determinant of the coefficient matrix is $-L_1^4/12$, which is not zero, there is a unique solution c_1, c_2. We set

$$h(s) = \int_0^s dr \int_0^r [f_{tt}(t) + B(t) - c_1 - c_2 t]dt.$$

Then $h \in C^2[0, L_1]$ and satisfies the conditions $h(0) = h(L_1) = 0$, $h'(0) = h'(L_1) = 0$. Integrating by parts, we obtain

$$\begin{aligned}
&\int_0^{L_1} [f_{ss}(s) + B(s) - c_1 - c_2 s]^2 ds \\
&= \int_0^{L_1} [f_{ss}(s) + B(s) - c_1 - c_2 s] h''(s) ds \\
&= \int_0^{L_1} [f_{ss}(s) + B(s)] h''(s) ds - c_1 h'(s)|_0^{L_1} - c_2 \int_0^{L_1} s h''(s) ds \\
&= -c_2 s h'(s)|_0^{L_1} + c_2 h(s)|_0^{L_1} = 0.
\end{aligned}$$

It follows that $f_{ss}(s) + B(s) - c_1 - c_2 s = 0$. This means $f_{ss} \in C^2[0, L_1]$. Therefore $f \in C^4[0, L_1]$.

Chapter 3

Centroaffine elastica in affine space $\mathbf{R}^3$

This chapter is devoted to the study of the variational properties of space curves invariant under the group of volume-preserving homogeneous affine transformations or centroaffine transformations in the punctured space. The group is generated by the action of the unimodular linear group $SL(3, \mathbf{R})$. We may expect to characterize the centroaffine properties of critical space curves related to energy functionals by the centroaffine arclength parameter, centroaffine curvature and centroaffine torsion.

3.1 Definition and variation formula of the energy functional in $\mathbf{R}^3$

We consider curves in the punctured affine space $\mathbf{R}^3$([33], [38]). Let $\{x, y, z\}$ be a standard coordinate system of the affine space $\mathbf{R}^3$, and $X : I \to \mathbf{R}^3$ be a smooth curve. We write it as a column vector

$$X(t) = (x(t), y(t), z(t))^T, \tag{3.1}$$

here $x(t), y(t), z(t)$ are smooth functions of t defined on a certain interval I. We identify the tangent space of $\mathbf{R}^3$ with $\mathbf{R}^3$ and write the tangent vector as $X'(t) = dX/dt = (x'(t), y'(t), z'(t))^T$ or simply X', and $(X(t), X'(t), X''(t))$

be the 3×3 matrix

$$\begin{pmatrix} x(t) & x'(t) & x''(t) \\ y(t) & y'(t) & y''(t) \\ z(t) & z'(t) & z''(t) \end{pmatrix}. \tag{3.2}$$

Definition 10 *A curve $X : I \to \mathbf{R}^3$ is called starlike if*

$$|X(t), X'(t), X''(t)| = \det(X(t), X'(t), X''(t)) \neq 0 \tag{3.3}$$

for any $t \in I$. A parameter s is called a centroaffine arclength parameter of the starlike curve X if

$$(X(s), X'(s), X''(s)) \in SL(3, \mathbf{R}),$$

that is $|X(s), X'(s), X''(s)| = 1$, for any $s \in I$.

Definition 11 *Let s be a centroaffine arclength parameter of a curve $X : I = [0, L] \to \mathbf{R}^3$, here L is the centroaffine arclength of X defined as in (3.8). We define the centroaffine curvature and centroaffine torsion of the starlike curve X by*

$$k(s) = |X(s), X''(s), X^{(3)}(s)|; \tag{3.4}$$

$$\tau(s) = |X'(s), X''(s), X^{(3)}(s)|. \tag{3.5}$$

Differentiating $|X(s), X'(s), X''(s)| = 1$, we obtain $|X(s), X'(s), X^{(3)}(s)| = 0$. This means $X(s)$, $X'(s)$, $X^{(3)}(s)$ are linearly dependent and from (3.4) and (3.5), we have the structure equation.

$$X^{(3)}(s) = \tau(s)X(s) - k(s)X'(s). \tag{3.6}$$

If $X(t)$ is a starlike curve and t be an arbitrary parameter of X. Same as in chapter 2, we have the formula of the velocity

$$\frac{ds}{dt} = |X(t), X'(t), X''(t)|^{\frac{1}{3}}. \tag{3.7}$$

We may reparametrize X by the centroaffine arclength, using the formula

$$s(t) = \int_{t_0}^{t} |X(t), X'(t), X''(t)|^{\frac{1}{3}} dt + s(t_0). \tag{3.8}$$

If t is also a centroaffine arclength parameter, then $ds/dt = 1$. That means $s = t + c$, for a constant c. This shows that a starlike curve $X(t)$ can admit a centroaffine arclength parameter uniquely up to a constant, and the starlike property and the centroaffine curvature and centroaffine torsion are invariant under the action of $SL(3, \mathbf{R})$ on $\mathbf{R}^3$.

Similar to the case of $\mathbf{R}^2$, we denote ∇ the usual flat, torsion-free affine connection of $\mathbf{R}^3$. Then we have the same structural equation. We denote $X_w(t)$ a variation $X(w,t) : (-\varepsilon, \varepsilon) \times I \to \mathbf{R}^3$ with $X(0,t) = X(t)$ and we have the similar notation. Equation(3.7) yields

$$\begin{aligned} W(\tfrac{ds}{dt}) = &\ \tfrac{1}{3}|X(t), X'(t), X''(t)|^{-\frac{2}{3}} W(|X(t), X'(t), X''(t)|) \\ = &\ -g\tfrac{ds}{dt}. \end{aligned} \tag{3.9}$$

Here we write $g = -\frac{1}{3}|X(t), X'(t), X''(t)|^{-1} W(|X(t), X'(t), X''(t)|)$ and same as in $\mathbf{R}^2$, we have the formula

$$[W, X'(s)] = gX'(s) \tag{3.10}$$

and the derivatives of X along W

$$\begin{aligned} \nabla_W X'(s) = &\ \nabla_{X'(s)} W + gX'(s), \\ \nabla_W \nabla_{X'(s)} X'(s) = &\ \nabla^2_{X'(s)} W + g_s X'(s) + 2gX''(s), \\ \nabla_W \nabla_{X'(s)} X''(s) = &\ \nabla_{X'(s)} \nabla_W X''(s) + \nabla_{[W, X'(s)]} X''(s) \\ = &\ \nabla_{X'(s)} (\nabla^2_{X'(s)} W + g_s X'(s) + 2gX''(s)) + gX^{(3)}(s) \\ = &\ \nabla^3_{X'(s)} W + g_{ss} X'(s) + 3g_s X''(s) + 3gX^{(3)}(s). \end{aligned}$$

By a direct computation, We have the following lemma.

Lemma 3 *Using the above notation, we have the following formulas:*

1. $[X'(t), W(t)] = 0$.
2. $W(\frac{ds}{dt}) = -g\frac{ds}{dt}$, *where*
   $g = -\frac{1}{3}|X(t), X'(t), X''(t)|^{-1} W(|X(t), X'(t), X''(t)|)$.
3. $[W, X'(s)] = gX'(s)$.
4. $W(k) = |W, X''(s), X^{(3)}(s)| + |X(s), \nabla^2_{X'(s)} W, X^{(3)}(s)|$
   $+|X(s), X''(s), \nabla^3_{X'(s)} W| + 5gk - g_{ss}$.
5. $W(\tau) = |W', X''(s), X^{(3)}(s)| + |X'(s), W'', X^{(3)}(s)|$
   $+|X'(s), X''(s), W^{(3)}| + 6g\tau$.

We consider the curvature energy functional defined on a class of starlike curves in $\mathbf{R}^3$.

$$F(X) = \int_0^{L(w)} f(k,\tau)ds. \tag{3.11}$$

Here $f(k,\tau)$ is a smooth function of k and τ. When we restrict on the curve $X(t) = X(0,t)$, since $s = t$, we will drop t, s in $X'(t), X'(s)$ respectively.

$$\begin{aligned}
&\tfrac{d}{dw}\textstyle\int_0^{L(w)} f(k,\tau)ds|_{w=0}\\
&= \textstyle\int_0^L[(f_k W(k) + f_\tau W(\tau))\frac{ds}{dt} + f(k,\tau)W(\frac{ds}{dt})]dt\\
&= \textstyle\int_0^L[f_k(|W,X'',X^{(3)}| + |X,W'',X^{(3)}| + |X,X'',W^{(3)}|)\\
&\quad +f_\tau(|W',X'',X^{(3)}| + |X',W^{(2)},X^{(3)}| + |X',X'',W^{(3)}|)\\
&\quad +f_k(5gk - g_{ss}) + 6f_\tau\tau g - f(k,\tau)g]ds.
\end{aligned} \tag{3.12}$$

Here we have used f_k and f_τ to denote the partial derivative of $f(k,\tau)$ with respect to k and τ, respectively, the prime $(\ )'$ means the usual derivative with respect to s.

We will give $X(w,t)$ a boundary condition such that $W(0,0) = W(0,L) = 0$, $\nabla_{X'}W(0,0) = \nabla_{X'}W(0,L) = 0$, $\nabla^2_{X'}W(0,0) = \nabla^2_{X'}W(0,L) = 0$, $\nabla^3_{X'}W(0,0) = \nabla^3_{X'}W(0,L) = 0$. We calculate the integral part by part.

$$\begin{aligned}
&\textstyle\int_0^L f_k(|W,X'',X^{(3)}| + |X,W'',X^{(3)}| + |X,X'',W^{(3)}|)ds\\
&= \textstyle\int_0^L[f_k(|W,X'',X^{(3)}| - |W'',X,X^{(3)}| - |W'',X',X''|\\
&\quad -|W'',X,X^{(3)}|) - f_k'|W'',X,X''|]ds\\
&= \textstyle\int_0^L[f_k(|W,X'',X^{(3)}| + 2k|W'',X,X'| - |W'',X',X''|)\\
&\quad -f_k'|W'',X,X''|]ds\\
&= \textstyle\int_0^L[f_k(|W,X'',X^{(3)}| - 2k|W',X,X''| + |W',X',X^{(3)}|)\\
&\quad -2(kf_k)'|W',X,X'| + 2f_k'|W',X',X''| + f_k'|W',X,X^{(3)}|\\
&\quad +f_k''|W',X,X''|]ds\\
&= \textstyle\int_0^L[f_k|W,X'',X^{(3)}| + (f_k'' - 2kf_k)|W',X,X''|\\
&\quad +2f_k'|W',X',X''| - (2(kf_k)' + f_k'k + f_\tau\tau)|W',X,X'|]ds\\
&= \textstyle\int_0^L[f_k|W,X'',X^{(3)}| - (f_k'' - 2kf_k)'|W,X,X''|\\
&\quad -(f_k'' - 2kf_k)|W,X',X''| - (f_k'' - 2kf_k)|W,X,X^{(3)}|\\
&\quad -2f_k''|W,X',X''| + (2(kf_k)' + f_k'k + f_\tau\tau)'|W,X,X'|\\
&\quad -2f_k'|W,X',X^{(3)}| + (2(kf_k)' + f_k'k + f_\tau\tau)|W',X,X''|]ds\\
&= \textstyle\int_0^L[(-f_k^{(3)} + 4(f_kk)' + f_k')|W,X,X''| - 3(f_k'' - f_kk)|W,X',X''|\\
&\quad +(4f_k''k + f_k'(5k' + 3tau) + p_k(\tau' + 2k'' - 2k^2))|W,X,X'|]ds.
\end{aligned}$$

Similarly we have

$$\begin{aligned}&\int_0^L f_\tau(|W',X'',X^{(3)}|+|X',W'',X^{(3)}|+|X',X'',W^{(3)}|)ds\\&=\int_0^L[(5(f_\tau\tau)'+f_\tau'\tau)|W,X,X''|\\&\quad-(f_\tau^{(3)}-3f_\tau\tau+(f_\tau k)')|W,X',X''|\\&\quad+(4f_\tau''\tau+5f_\tau'\tau'+2f_\tau(\tau''-k\tau))|W,X,X'|]ds.\end{aligned}$$

We denote $h=5f_kk+6f_\tau\tau-f-f_k''$, then at $w=0$, we have

$$\begin{aligned}&\int_0^L(f_k(5gk-g_{ss})+6f_\tau\tau g-f(k,\tau)g)ds\\&=-\tfrac{1}{3}\int_0^L h(|\nabla_WX,X',X''|+|X,\nabla_WX'(t),X''|+|X,X',\nabla_WX''(t)|)ds\\&=-\tfrac{1}{3}\int_0^L[h|W,X',X''|-|\nabla_{X'}W,hX,X''|+|\nabla_{X'}^2W,hX,X'|)ds\\&=-\tfrac{1}{3}\int_0^L[h|W,X',X''|-2|\nabla_{X'}W,hX,X''|-|\nabla_{X'}W,h'X,X'|)ds\\&=-\tfrac{1}{3}\int_0^L[3h|W,X',X''|+3h'|W,X,X''|+(h''-2kh)|W,X,X'|]ds\\&=\int_0^L[(f_k''-5f_kk-6f_\tau\tau+f)|W,X',X''|\\&\quad+(f_k^{(3)}-5(f_kk)'-6(f_\tau\tau)'+f')|W,X,X''|\\&\quad+\tfrac{1}{3}(f_k^{(4)}-7f_k''k-9f_k'k'-f_k(4k''-10k^2)\\&\quad-6(f_\tau\tau)''+(f_\tau\tau')'+12f_\tau k\tau-2fk)|W,X,X'|]ds.\end{aligned}$$

In the first equality, we use $|X(t)\ X'(t)\ X''(t)|=1$ at $w=0$. In the second equality, we use $\nabla_WX'(t)=\nabla_{X'(t)}W$. Combining these three formulae, we obtain the first variational formula.

$$\begin{aligned}&\tfrac{d}{dw}\int_0^{L(w)}f(k,\tau)ds|_{w=0}\\&=\int_0^L[(-f_\tau^{(3)}-2f_k''-(f_\tau k)'-2f_kk-3f_\tau\tau+f)|W,X',X''|\\&\quad+\tfrac{1}{3}(6f_\tau''\tau+4f_\tau'\tau'+f_\tau(\tau''+6k\tau)-2fk+f_k^{(4)}+5f_k''k\\&\quad+3f_k'(2k'+3\tau)+f_k(2k''+4k^2+3\tau')|W,X,X'|]ds.\end{aligned}\tag{3.13}$$

Therefore the Euler-Lagrange equation is

$$\begin{aligned}&-f_\tau^{(3)}-2f_k''-(f_\tau k)'-2f_kk-3f_\tau\tau+f=0,\\&6f_\tau''\tau+4f_\tau'\tau'+f_\tau(\tau''+6k\tau)-2fk+f_k^{(4)}\\&+5f_k''k+3f_k'(2k'+3\tau)+f_k(2k''+4k^2+3\tau')=0.\end{aligned}\tag{3.14}$$

For arbitrary function $f(k,\tau)$, it is difficult to solve this equation. We will treat special case in the following sections.

3.2 Integration of the generalized centroaffine elastica in $\mathbf{R}^3$

In this section, we consider the Lagrange $f(k,\tau)$ depends only on the centroaffine curvature k. That means $f(k,\tau) = p(k)$. At this time, the Euler-Lagrange equation is

$$\begin{aligned}&p^{(5)}(k)(k')^4 + 6p^{(4)}(k)k''(k')^2 + p^{(3)}(k)(4k^{(3)}k' + 3(k'')^2\\&+5(k')^2k) + p''(k)(k^{(4)} + 5k''k + 6(k')^2 + 9k'\tau)\\&+p'(k)(2k'' + 4k^2 + 3\tau') - 2p(k)k = 0;\\&2p^{(3)}(k)(k')^2 + 2p''(k)k'' + 2p'(k)k - p(k) = 0.\end{aligned} \tag{3.15}$$

Definition 12 *A starlike curve in* $\mathbf{R}^3$ *with the centroaffine arclength parameter* s *is called a generalized centroaffine elastica if it satisfies the equation(3.15).*

For k is constant, we know k satisfies $2p'(k)k - p(k) = 0$ and the first equation of equation (3.15) becomes $p'(k)(4k^2+3\tau') - 2p(k)k = 3p'(k)\tau' = 0$. If $p'(k) \neq 0$, we know τ is a constant. At this time, the structure equation (3.6) is a liner system with constant coefficients. Its solution depends on its characteristic equation

$$r^3 + kr - \tau = 0. \tag{3.16}$$

We divide it into the following cases.
I. $\tau = 0$. At least one characteristic root is zero.
(1). $k = 0$. Integrating $X^{(3)}(s) = 0$ three times, we obtain

$$X(s) = \frac{1}{2}\mathbf{a}s^2 + \mathbf{b}s + \mathbf{c}, \tag{3.17}$$

where $\mathbf{a}$, $\mathbf{b}$ and $\mathbf{c}$ are constant vectors such that $|\mathbf{c}, \mathbf{b}, \mathbf{a}| = 1$. By taking

$$\mathbf{a} = (0,0,1)^T, \quad \mathbf{b} = (0,1,0)^T, \quad \mathbf{c} = (1,0,0)^T.$$

Then (3.17) becomes

$$X(s) = (1, s, \frac{1}{2}s^2)^T. \tag{3.18}$$

This is a parabola $z = y^2/2$ on the plane $x = 1$.
(2). $k = a^2 > 0$. We have the equation

$$X^{(3)}(s) + a^2X'(s) = 0. \tag{3.19}$$

Its characteristic equation is

$$r^3 + a^2 r = 0. \tag{3.20}$$

The characteristic roots are $ai, -ai, 0$. Therefore the general solution of the structure equation is

$$X(s) = \cos(as)\mathbf{a} + \sin(as)\mathbf{b} + \mathbf{c}, \tag{3.21}$$

where $\mathbf{a}$, $\mathbf{b}$ and $\mathbf{c}$ are constant vectors such that $|\mathbf{c}, \mathbf{b}, \mathbf{a}| = -1/a^3$. By choosing

$$\mathbf{a} = (0, 0, -1/a^2)^T, \quad \mathbf{b} = (0, 1/a, 0)^T, \quad \mathbf{c} = (1, 0, 0)^T,$$

we have

$$X(s) = (1, \frac{1}{a}\sin(as), -\frac{1}{a^2}\cos(as))^T. \tag{3.22}$$

This is an ellipse $y^2 + a^2z^2 = 1/a^2$ on the plane $x = 1$.

(3). $k = -a^2 < 0$. Similar to the above positive constant curvature case, we have the solution.

$$X(s) = (1, \frac{1}{a}\sinh(as), \frac{1}{a^2}\cosh(as))^T. \tag{3.23}$$

This is a hyperbola $x^2 - a^2y^2 = -1/a^2$ on the plane $x = 1$.

II. $\tau \neq 0$.

(4). If equation (3.16) has three distinct real roots α_1, α_2 and α_3, then $\alpha_1 + \alpha_2 + \alpha_3 = 0$ and $-4k^3 - 27\tau^2 > 0$. A particular solution of the structure equation (3.6) is

$$X(s) = (d_1 e^{\alpha_1 s}, d_2 e^{\alpha_2 s}, d_1 e^{\alpha_3 s})^T \tag{3.24}$$

where d_1, d_2 and d_3 are constants. The condition $|X(s), X'(s), X''(s)| = 1$ implies $d_1 d_2 d_3 \sqrt{-4k^3 - 27\tau^2} = 1$. This curve is on the cubic surface

$$xyz = d_1 d_2 d_3 = \frac{1}{\sqrt{-4k^3 - 27\tau^2}}.$$

(5). If the characteristic equation (3.16) has a double root $\alpha_1 = \alpha_2 = \alpha$, then $\alpha_3 = -2\alpha$ and $k = -3\alpha^2$, $\tau = -2\alpha^3$. The general solution of the structure equation (3.6) is

$$X(s) = \mathbf{a}e^{\alpha_1 s} + \mathbf{b}se^{\alpha_1 s} + \mathbf{c}e^{-2\alpha_1 s}, \tag{3.25}$$

where **a**, **b** and **c** are constant vectors such that $3k|\mathbf{a}, \mathbf{b}, \mathbf{c}| = -1$. By choosing

$$\mathbf{a} = (1,0,0)^T, \quad \mathbf{b} = (0,1,0)^T, \quad \mathbf{c} = (0,0,-\frac{1}{3k})^T,$$

we have

$$X(s) = (e^{\alpha s}, se^{\alpha s}, -\frac{1}{3k}e^{-2\alpha s}). \tag{3.26}$$

(6). If the characteristic equation (3.16) has only one real root α_1 and the other two roots are

$$\alpha_2 = a + bi, \qquad \alpha_3 = a - bi, \qquad b \neq 0.$$

Then $\alpha_1 = -2a$. A particular solution of the structure equation (3.6) is

$$X(s) = (d_1 e^{-2as}, d_2 e^{as}\cos(bs), d_3 e^{as}\sin(bs))^T, \tag{3.27}$$

where d_1, d_2 and d_3 are constants. The condition $|X(s), X'(s), X''(s)| = 1$ implies $d_1 d_2 d_3 b(9a^2 + b^2) = 1$.

Now we assume k is not constant. We can use the second equation of the Euler-lagrange equation to reduce the order of the first equation. Then the Euler-Lagrange equation becomes the following form

$$\begin{aligned} 3p''(k)k'(k' + 2\tau) + p'(k)(k'' + 2\tau') &= 0; \\ 2p^{(3)}(k)(k')^2 + 2p''(k)k'' + 2p'(k)k - p(k) &= 0. \end{aligned} \tag{3.28}$$

Multiplying the first equation by $p'(k)^2$ and the second by $p''(k)k'$, we obtain the first integrals of the equation(3.15)

$$\begin{aligned} &p'(k)^3(k' + 2\tau) = c_1, \\ &p''(k)^2(k')^2 + p'(k)(p'(k)k - p(k)) = c_2. \end{aligned} \tag{3.29}$$

Here c_1, c_2 are the integral constants. The second equation is similar to that of $\mathbf{R}^2$, we can express the centroaffine curvature k by quadratures

$$\int \sqrt{\frac{p''(k)^2}{c_2 - p'(k)(p'(k)k - p(k))}}dk = \pm \int ds, \tag{3.30}$$

and then we can get the τ by the formula

$$\tau = \frac{c_1}{2p'(k)^3} - \frac{k'}{2}. \tag{3.31}$$

We will use the Killing field and the classification of the conjugate class of $sl(3, \mathbf{R})$ to solve the structure equation(3.6). We need the following definition:

Definition 13 *Let $X(t)$ be a starlike curve in $\mathbf{R}^3$. We call a vector field W Killing along $X(t)$ if it annihilates ds/dt, k, τ.*

We set the Killing field along the generalized centroaffine elastica $X(s)$ having the form $W = f_1(s)X(s) + f_2(s)X'(s) + f_3(s)X''(s)$, then we have

$$\begin{aligned}
W' = &\ (f_1' + \tau f_3)X + (f_1 + f_2' - kf_3)X' + (f_2 + f_3')X'';\\
W'' = &\ (f_1'' + \tau f_2 + 2\tau f_3' + \tau' f_3)X\\
&+(2f_1' + f_2'' - kf_2 - 2kf_3' + (\tau - k')f_3)X'\\
&+(f_1 + 2f_2' + f_3'' - kf_3)X'';\\
W^{(3)} = &\ (f_1^{(3)} + \tau f_1 + 3\tau f_2' + \tau' f_2 + 3\tau f_3'' + 3\tau' f_3' + (\tau'' - k\tau)f_3)X\\
&+(3f_1'' - kf_1 + f_2^{(3)} - 3kf_2' + (\tau - k')f_2 - 3kf_3''\\
&+3(\tau - k')f_3' + (2\tau' - k'' + k^2)f_3)X'\\
&+(3f_1' + 3f_2'' - kf_2 + f_3^{(3)} - 3kf_3' + (\tau - 2k')f_3)X''.
\end{aligned}$$

From the definition of s, k and τ or lemma, we have

$$\begin{aligned}
W(\tfrac{ds}{dt}) = &\ -g\tfrac{ds}{dt} = \tfrac{1}{3}(|W, X', X''| + |X, W', X''| + |X, X', W''|)\tfrac{ds}{dt}\\
= &\ \tfrac{1}{3}(3f_1 + 3f_2' + f_3'' - 2kf_3)\tfrac{ds}{dt};\\
W(k) = &\ |W, X'', X^{(3)}| + |X, W'', X^{(3)}| + |X, X'', W^{(3)}| - g'' + 5gk\\
= &\ -3f_1'' + 3kf_1 - f_2^{(3)} + 5kf_2' + k'f_2 + 4kf_3'' - 3(\tau - k')f_3'\\
&+(k'' - 2\tau' - 2k^2)f_3 - g'' + 5gk;
\end{aligned}$$

$$\begin{aligned}
W(\tau) = &\ |W', X'', X^{(3)}| + |X', W'', X^{(3)}| + |X', X'', W^{(3)}| + 6g\tau\\
= &\ f_1^{(3)} + kf_1' + 3\tau f_1 + 6\tau f_2' + \tau' f_2\\
&+4\tau f_3'' + 3\tau' f_3' + (\tau'' - 2k\tau)f_3 + 6g\tau.
\end{aligned}$$

Therefore f_1, f_2, f_3 must satisfy the following equations.

$$\begin{aligned}
&3f_1 + 3f_2' + f_3'' - 2kf_3 = 0;\\
&3f_1'' - 3kf_1 + f_2^{(3)} - 5kf_2' - k'f_2 - 4kf_3''\\
&+3(\tau - k')f_3' - (k'' - 2\tau' - 2k^2)f_3 = 0;\\
&f_1^{(3)} + kf_1' + 3\tau f_1 + 6\tau f_2' + \tau' f_2 + 4\tau f_3'' + 3\tau' f_3' + (\tau'' - 2k\tau)f_3 = 0.
\end{aligned} \tag{3.32}$$

From these equations and equation(3.15), we found $W = -p''(k)k'X + p'(k)X'$ is Killing along the generalized centroaffine elastica X.

Definition 14 *A vector field W on $\mathbf{R}^3$ is Killing with respect to $SL(3, \mathbf{R})$ if its flow generates a one-parameter subgroup of $SL(3, \mathbf{R})$.*

We can express f_3'' as a linear combination of f_3, f_2', f_1 from the first equation, $f_1^{(3)}$ as a linear combination of $f_3', f_3, f_2', f_2, f_1', f_1$ from the second equation, $f_2^{(3)}$ as a linear combination of $f_3', f_3, f_2', f_2, f_1'', f_1', f_1$. Therefore, these three equations can be written as a system of first order linear equations of $f_3', f_3, f_2'', f_2', f_2, f_1'', f_1', f_1$ in the following form

$$(f_3', f_3, f_2'', f_2', f_2, f_1'', f_1', f_1)' = (f_3', f_3, f_2'', f_2', f_2, f_1'', f_1', f_1)A. \tag{3.33}$$

Here A is a 8×8 matrix. Therefore the dimension of the solution space of equation(3.32) is 8. This dimension agrees with the dimension of group $SL(3, \mathbf{R})$. Thus a vector field W which is Killing along X can extend to a Killing field on $\mathbf{R}^3$, the field which is invariant under the infinitesimal action of $SL(3, \mathbf{R})$.

We denote the Lie algebra of $SL(3, \mathbf{R})$ by $sl(3, \mathbf{R})$. For any $M \in sl(3, \mathbf{R})$, let $\lambda_1, \lambda_2, \lambda_3$ be its eigenvalues, At least one of them is real and $\lambda_1 + \lambda_2 + \lambda_3 = 0$. We classify it into the following cases:

1. $\lambda_1 = -2a, \lambda_2 = a + bi, \lambda_3 = a - bi, b \neq 0$.
2. $\lambda_1 = a, \lambda_2 = b, \lambda_3 = -a - b$, each invariant subspace has dimension 1.
3. $\lambda_1 = \lambda_2 = a, \lambda_3 = -2a$, the dimension of the invariant subspacc corresponding to a is 2.
4. All roots are zero and the dimension of the invariant subspace is 3.

The centroaffine elastic curves are determined by the two parameters c_1 and c_2. We divide the (c_1, c_2) plane into four pieces by using the classification of the conjugate class of $\tilde{W}$ in the Lie algebra $sl(3, \mathbf{R})$

The Killing field $\tilde{W}$ is one of the Killing fields generated by the 1-parameter subgroup corresponding to one of the four cases above. The normal forms of these cases are:

$$\begin{pmatrix} -2a & 0 & 0 \\ 0 & 0 & a^2+b^2 \\ 0 & -1 & 2a \end{pmatrix}, \quad \begin{pmatrix} a & 0 & 0 \\ 1 & a & 0 \\ 0 & 0 & -2a \end{pmatrix},$$
$$\begin{pmatrix} a & 0 & 0 \\ 0 & b & 0 \\ 0 & 0 & -a-b \end{pmatrix}, \quad \begin{pmatrix} 0 & 0 & 0 \\ 1 & 0 & 0 \\ 0 & 1 & 0 \end{pmatrix}. \tag{3.34}$$

Here a and $b \neq 0$ are real constants. Using starlike property and the structure equation, we will get the parametric expression of the generalized centroaffine elastica. We will study it case by case.

Case 1. We can choose a coordinate system (x, y, z) such that the Killing field generated by the 1-parameter Lie subgroup corresponding to the first matrix is

$$W = -2ax\frac{\partial}{\partial x} + (a^2 + b^2)z\frac{\partial}{\partial y} + (-y + 2az)\frac{\partial}{\partial z} \tag{3.35}$$

Along the generalized centroaffine elastica X, we have a linear system.

$$\begin{cases} p'(k)x' - p''(k)k'x = -2ax; \\ p'(k)y' - p''(k)k'y = (a^2 + b^2)z; \\ p'(k)z' - p''(k)k'z = -y + 2az. \end{cases} \tag{3.36}$$

For the first equation, we have the solution

$$x(s) = d_3 p'(k(s))e^{\int \frac{-2a}{p'(k)}ds}. \tag{3.37}$$

The other two equations become the following form by dividing by $p'(k)^2$

$$\begin{aligned} (\tfrac{y}{p'(k)})' &= (a^2 + b^2)\tfrac{z}{p'(k)^2}, \\ (\tfrac{z}{p'(k)})' &= -\tfrac{y}{p'(k)^2} + 2a\tfrac{z}{p'(k)^2}. \end{aligned} \tag{3.38}$$

Introducing two new variables $U = y/p'(k), V = z/p'(k)$, the last two equations become

$$\begin{aligned} p'(k)U' &= (a^2 + b^2)V, \\ p'(k)V' &= -U + 2aV. \end{aligned} \tag{3.39}$$

Combining these two equations, we obtain the following second order equation.

$$p'(k)(p'(k)V')' - 2ap'(k)V' + (a^2 + b^2)V = 0. \tag{3.40}$$

We set a parameter transformation by $t = \int_{s_0}^{s} 1/p'(k)ds$ And write $f(t) = V(s(t))$, then we have the equation

$$f''(t) - 2af'(t) + (a^2 + b^2)f(t) = 0. \tag{3.41}$$

This equation has the general solution

$$f(t) = e^{at}(d_1 \cos(bt) + d_2 \sin(bt)). \tag{3.42}$$

This leads to the generalized centroaffine elastica equation

$$\begin{cases} x(s) = d_3 p'(k(s))e^{\int \frac{-2a}{p'(k)}ds}; \\ y(s) = p'(k(s))e^{\int \frac{a}{p'(k)}ds}((ad_1 - bd_2)\cos(\int \frac{b}{p'(k)}ds) \\ \qquad +(ad_2 + bd_1)\sin(\int \frac{b}{p'(k)}ds)); \\ z(s) = p'(k(s))e^{\int \frac{a}{p'(k)}ds}(d_1 \cos(\int \frac{b}{p'(k)}ds) + d_2 \sin(\int \frac{b}{p'(k)}ds)). \end{cases} \tag{3.43}$$

Here d_1, d_2 and d_3 are constants. We can choose d_1, d_2, d_3 such that s is the centroaffine arclength parameter according to the equality

$$|X, X', X''| = -b^2(9a^2 + b^2)(d_1^2 + d_2^2)d_3 = 1. \tag{3.44}$$

Calculating the centroaffine curvature and torsion and using the Euler-Lagrange equation and its first integrals, we obtain

$$\begin{aligned} k(s) &= \tfrac{-3a^2+b^2+c_2}{p'(k)^2} + k(s); \\ \tau(s) &= -\tfrac{4a(a^2+b^2)+c_1}{2p'(k)^3} + \tau(s). \end{aligned} \tag{3.45}$$

Therefore we have the condition.

$$\begin{cases} 3a^2 - b^2 = c_2; \\ -4a(a^2 + b^2) = c_1. \end{cases} \tag{3.46}$$

This implies $16c_2^3 - 27c_1^2 = -16(9a^2 + b^2)b^2 < 0$.
Case 2. We can choose a coordinate system (x, y, z) such that the Killing field generated by the 1-parameter Lie subgroup corresponding to the third matrix is

$$W = ax\frac{\partial}{\partial x} + by\frac{\partial}{\partial y} - (a + b)z\frac{\partial}{\partial z}. \tag{3.47}$$

Along the generalized centroaffine elastica X, we get a linear system.

$$\begin{cases} p'(k)x' - p''(k)k'x = ax; \\ p'(k)y' - p''(k)k'y = by; \\ p'(k)z' - p''(k)k'z = -(a + b)z. \end{cases} \tag{3.48}$$

The generalized centroaffine elastica equation is

$$\begin{cases} x(s) = d_1 p'(k(s))e^{\int \frac{a}{p'(k)}ds}; \\ y(s) = d_2 p'(k(s))e^{\int \frac{b}{p'(k)}ds}; \\ z(s) = d_3 p'(k(s))e^{\int \frac{-a-b}{p'(k)}ds}. \end{cases} \tag{3.49}$$

Here d_1, d_2 and d_3 are constants. We can choose d_1, d_2, d_3 such that s is the centroaffine arclength parameter according to the equality

$$|X, X', X''| = (b - a)(2a + b)(a + 2b)d_1 d_2 d_3 = 1. \tag{3.50}$$

This excludes the case that any two of $a, b, -(a+b)$ are equal. Calculation reveals that

$$\begin{aligned} k(s) &= -\tfrac{a^2+ab+b^2-c_2}{p'(k)^2} + k(s); \\ \tau(s) &= -\tfrac{2ab(a+b)+c_1}{2p'(k)^3} + \tau(s). \end{aligned} \tag{3.51}$$

Therefore we have the condition.

$$\begin{cases} c_2 = a^2 + ab + b^2; \\ c_1 = -2ab(a+b). \end{cases} \tag{3.52}$$

This implies $16c_2^3 - 27c_1^2 = 4(a-b)^2(2a+n)^2(a+2b)^2 > 0$.

Case 3. We can choose a coordinate system (x, y, z) such that the Killing field generated by the 1-parameter Lie subgroup corresponding to the second matrix is

$$W = ax\frac{\partial}{\partial x} + (x+ay)\frac{\partial}{\partial y} - 2az\frac{\partial}{\partial z}. \tag{3.53}$$

Along the generalized centroaffine elastica X, we get a linear system.

$$\begin{cases} p'(k)x' - p''(k)k'x = ax; \\ p'(k)y' - p''(k)k'y = x + ay; \\ p'(k)z' - p''(k)k'z = -2az. \end{cases} \tag{3.54}$$

The generalized centroaffine elastica is

$$\begin{cases} x(s) = d_1 p'(k(s))e^{\int \frac{a}{p'(k)}ds}; \\ y(s) = p'(k(s))e^{\int \frac{a}{p'(k)}ds}(d_2 + d_1 \int \frac{1}{p'(k)}ds); \\ z(s) = d_3 p'(k(s))e^{\int \frac{-2a}{p'(k)}ds}. \end{cases} \tag{3.55}$$

Here d_1, d_2 and d_3 are constants. We can choose d_1, d_2, d_3 such that s is the centroaffine arclength parameter according to the equality

$$|X, X', X''| = 9a^2 d_1^2 d_3 = 1. \tag{3.56}$$

This excludes the case $a = 0$. Calculation reveals that

$$\begin{aligned} k(s) &= \tfrac{-3a^2+c_2}{p'(k)^2} + k(s); \\ \tau(s) &= -\tfrac{4a^3+c_1}{2p'(k)^3} + \tau(s). \end{aligned} \tag{3.57}$$

Therefore we have the condition.

$$\begin{cases} c_2 = 3a^2; \\ c_1 = -4a^3. \end{cases} \tag{3.58}$$

This implies $16c_2^3 - 27c_1^2 = 0$ and $(c_1, c_2) \neq (0,0)$.
Case 4. We can choose a coordinate system (x, y, z) such that the Killing field generated by the 1-parameter Lie subgroup corresponding to the last matrix is

$$W = x\frac{\partial}{\partial y} + y\frac{\partial}{\partial z}. \tag{3.59}$$

Along the generalized centroaffine elastica X, we get a linear system.

$$\begin{cases} p'(k)x' - p''(k)k'x = 0; \\ p'(k)y' - p''(k)k'y = x; \\ p'(k)z' - p''(k)k'z = y. \end{cases} \tag{3.60}$$

The generalized centroaffine elastica is

$$\begin{cases} x(s) = p'(k(s)); \\ y(s) = p'(k(s))(d_1 + \int \frac{1}{p'(k)}ds); \\ z(s) = p'(k(s))(\int(d_1 + \int \frac{1}{p'(k)}ds)/p'(k)ds + d_2). \end{cases} \tag{3.61}$$

Here d_1, d_2 are constants. We can choose $d_1 = 1$ such that s is the centroaffine arclength parameter according to the equality

$$|X, X', X''| = d_1^3 = 1. \tag{3.62}$$

Calculation reveals that

$$\begin{aligned} k(s) = &\ \tfrac{c_2}{p'(k)^2} + k(s); \\ \tau(s) = &\ -\tfrac{c_1}{2p'(k)^3} + \tau(s). \end{aligned} \tag{3.63}$$

This implies that $c_1 = 0, c_2 = 0$.

When we look at the regions on the plane (c_1, c_2) corresponding to these cases, we see that the origin $(0,0)$ represents case 4, the curve $16c_2^3 - 27c_1^2 = 0$ minus the origin represents case 3, the region to the below of the curve $16c_2^3 - 27c_1^2 = 0$ represents case 1 and the remaining region represents case 2. This gives the bifurcation diagram in the (c_1, c_2) plane(Figure 3.1) and means that the point (c_1, c_2) determines the type of the Killing field.

From these results, we have

Theorem 10 *The generalized centroaffine elastica of* $\mathbf{R}^3$ *is integrable by quadratures. that means it can be expressed in terms of integral as in (3.43), (3.49), (3.55) and (3.61).*

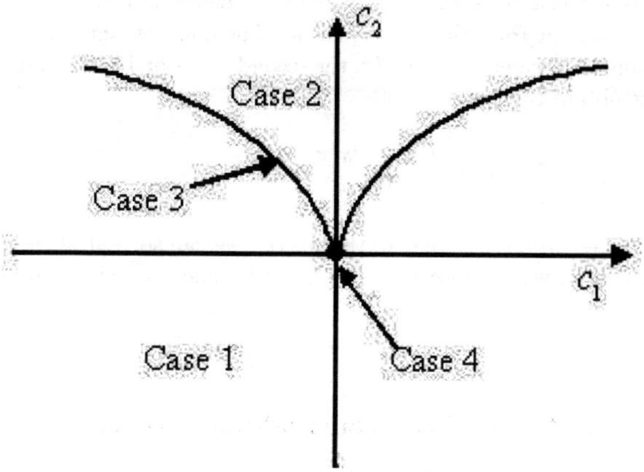

Figure 3.1: Bifurcation diagram in the (c_1, c_2) plane

3.3 Centroaffine elastica for $p(k) = k^2 + \lambda$ in the affine space $\mathbf{R}^3$

We consider a special case for $p(k) = k^2 + \lambda$. The critical point of the energy functional $\int_0^L (k^2 + \lambda) ds$ is called a centroaffine elastica.

3.3.1 Motion equation of the centroaffine elastica

The Euler-Lagrange equation is

$$\begin{aligned} &k^{(4)} + 7kk'' + 6(k')^2 + 9\tau k' + 3\tau' k + 3k^3 - \lambda k = 0, \\ &4k'' + 3k^2 - \lambda = 0. \end{aligned} \tag{3.64}$$

We can use the second equation to reduce the order of the first equation. Then the Euler-Lagrange equation becomes the following form

$$\begin{aligned} &kk'' + 3(k')^2 + 6\tau k' + 2\tau' k = 0, \\ &4k'' + 3k^2 - \lambda = 0. \end{aligned} \tag{3.65}$$

If k is a constant, then $k = \pm\sqrt{\lambda/3}$. When k is a nonzero constant, τ is also constant from the first equation. For nonconstant curvature $k(s)$, multiplying the first equation by k^2, the second equation by k' and integrating them, we obtain the first integrals

$$\begin{aligned} &k^3k' + 2k^3\tau' = 2c_1, \\ &2(k')^2 + k^3 - \lambda k = c_2. \end{aligned} \tag{3.66}$$

The second equation is similar to that of $\mathbf{R}^2$, we can solve it along that line. After solving k, we can insert it into the first equation and we obtain

$$\tau(s) = -\frac{1}{2}k' + \frac{c_1}{k^3}. \tag{3.67}$$

3.3.2 Solutions of the centroaffine elastica

We consider the structure equation

$$X^{(3)}(s) = \tau(s)X(s) - k(s)X'(s). \tag{3.68}$$

This is a linear equation of $X(s)$, we could solve it completely. For the nonconstant centroaffine curvature starlike elastica, we found the vector field $W = -k'X + kX'$ is Killing along the centroaffine elastica X. As that done in section 2, we obtain the following results.
(1). $c_2^3 - 54c_1^2 < 0$. Choosing a, b such that

$$c_2 = 6a^2 - 2b^2, \qquad c_1 = -2a(a^2 + b^2).$$

The centroaffine elastica equation is

$$\begin{cases} x(s) = d_3k(s)e^{\int \frac{-2a}{k}ds}, \\ y(s) = k(s)e^{\int \frac{a}{k}ds}((ad_1 - bd_2)\cos(\int \frac{b}{k}ds) \\ \qquad\qquad +(ad_2 + bd_1)\sin(\int \frac{b}{k}ds)), \\ z(s) = k(s)e^{\int \frac{a}{k}ds}(d_1\cos(\int \frac{b}{k}ds) + d_2\sin(\int \frac{b}{k}ds)). \end{cases} \tag{3.69}$$

Here d_1, d_2, d_3 are constants and satisfy $-b^2(9a^2 + b^2)(d_1^2 + d_2^2)d_3 = 1$.
(2). $c_2^3 - 54c_1^2 > 0$. Choosing a, b such that

$$c_2 = 2(a^2 + ab + b^2), \qquad c_1 = -ab(a + b).$$

The centroaffine elastica equation is

$$\begin{cases} x(s) = d_1 k(s) e^{\int \frac{a}{k} ds}, \\ y(s) = d_2 k(s) e^{\int \frac{b}{k} ds}, \\ x(s) = d_3 k(s) e^{-\int \frac{(a+b)}{k} ds}. \end{cases} \tag{3.70}$$

Here d_1, d_2, d_3 are constants and satisfy $(b-a)(2a+b)(a+2b)d_1 d_2 d_3 = 1$.
(3). $c_2^3 - 54c_1^2 = 0$ and $(c_1, c_2) \neq (0,0)$. Choosing a, b such that $a = -\sqrt[3]{c_1/2}$
The centroaffine elastica equation is

$$\begin{cases} x(s) = d_1 k(s) e^{-\int \frac{2a}{k} ds}, \\ y(s) = d_2 k(s) e^{\int \frac{a}{k} ds}, \\ x(s) = k(s) e^{\int \frac{a}{k} ds} (d_3 + d_2 \int \frac{1}{k} ds). \end{cases} \tag{3.71}$$

Here d_1, d_2, d_3 are constants and satisfy $9a^2 d_1 d_2^2 = 1$.
(4). $(c_1, c_2) = (0,0)$. The centroaffine elastica equation is

$$\begin{cases} x(s) = k(s), \\ y(s) = k(s)(\int \frac{1}{k} ds + d_2), \\ x(s) = k(s)(\int (\int \frac{1}{k} ds + d_2)/k ds + d_3). \end{cases} \tag{3.72}$$

Here d_2, d_3 are constants

From these results, we have

Theorem 11 *The centroaffine elastica of* $\mathbf{R}^3$ *is integrable.*

When we look at the distribution of the region corresponding to these cases on the plane (c_1, c_2), we found the origin $(0,0)$ represents case (4), the curve $c_2^3 - 54c_1^2 = 0$ except the origin represents case (3), the region below the curve $c_2^3 - 54c_1^2 = 0$ represents case (1), and the remaining region represents case (2). This means that the point (c_1, c_2) determines the type of the centroaffine elastica.

3.3.3 Closed centroaffine elastica

We look for the condition such that the centroaffine elastica is closed. This means that the equation of the curve is periodic. It implies that the centroaffine curvature $k(s)$ and centroaffine torsion $\tau(s)$ are periodic. We may assume $\lambda \neq 0$. This forces $k \not\equiv 0$. Then $k(s)$ can be classified into two cases: constant or nonconstant. Checking all these cases, we only need to find the condition when the integral $\int 1/kds$ is periodic and only the case 1 provides the possibility. From the expression of $k(s)$ (using the elliptic function), we know that $a = 0$.

For the case k is constant, we have the solution

$$\begin{cases} x(s) = d_1, \\ y(s) = d_2 \cos(\frac{b}{k}s) + d_3 \sin(\frac{b}{k}s), \\ z(s) = d_4 \cos(\frac{b}{k}s) + d_5 \sin(\frac{b}{k}s). \end{cases} \tag{3.73}$$

Here d_1, d_2, d_3, d_4, d_5 are constants and satisfy the condition

$$\frac{d_1(d_2d_5 - d_3d_4)b^3}{k^3} = 1.$$

This solution has the period $|k/(2\pi b)|$. From the relations $\lambda = 3k^2 > 0, c_1 = -2k^3 = -2b^2$, we know $k > 0$ and the period is $1/(2\pi\sqrt{k})$. The centroaffine elastica is an ellipse on the plane $x = d_1$ up to the action of $SL(2, \mathbf{R}^3$.

Now we consider the case k is nonconstant. We check the equation

$$k(kV')' + b^2V = 0.$$

Same as in chapter 2, we assume T the period of $V(s)$, then it has to satisfy the condition

$$b\int_0^T \frac{1}{k}ds = 2\pi n.$$

for some positive integer n. From the equation

$$2(k')^2 = -k^3 + \lambda k + c_1.$$

We know its solution can be written by the Jacobi elliptic sine function

$$k(s) = \alpha_3(1 - q^2\mathrm{sn}^2(\frac{d}{\sqrt{2}}s, p)).$$

Then we have the same formula

$$b\int_0^{2\sqrt{2}K(p)/d} \frac{1}{k}ds = 2[\frac{(1-2p^2)(2-p^2)}{3(1+p^2)}]^{1/2}\int_0^{K(p)}(1 - \frac{3p^2}{1+p^2}\mathrm{sn}^2(t,p))^{-1}dt.$$

Using the same arguments as in the affine plane $\mathbf{R}^2$, we have the following theorem

Theorem 12 *There exists lots of closed nonconstant centroaffine curvature elastica in the affine space* $\mathbf{R}^3$.

3.4 The variation of centroaffine torsion for Starlike curves

We consider the energy functional

$$\int_0^{L(w)}(\tau+\lambda)ds$$
$$=\int_I(|X'(w,s),X''(w,s),X^{(3)}(w,s)|+\lambda)|X(w,t),X'(w,t),X''(w,t)|^{\frac{1}{3}}dt.$$

We use the formula

$$W(\tau)=|W',X'',X^{(3)}|+|X',W'',X^{(3)}|+|X',X'',W^{(3)}|+6g\tau,$$

and give a suitable boundary condition, we get

$$\frac{d}{dw}\int_0^{L(w)}(\tau+\lambda)ds|_{w=0}$$
$$=\frac{1}{3}\int_0^L[(\tau''+4k\tau-2k\lambda)|W,X,X'|-3(k'+2\tau-\lambda)|W,X',X''|]ds.$$

The Euler-Lagrange equation is

$$\begin{aligned}\tau''+4k\tau-2k\lambda=&\ 0,\\ k'+2\tau-\lambda=&\ 0.\end{aligned}\tag{3.74}$$

From the second equation, we get

$$\tau=\frac{\lambda-k'}{2}.\tag{3.75}$$

Do the derivative twice and use the first equation, we get

$$k^{(3)}+2kk'=0.\tag{3.76}$$

If one of the k,τ is constant, both of them are constants. From the Euler-Lagrange equation, we know $\tau=\lambda/2$ and k could be any constant. If k is not constant, integrating the above equation, we get

$$(k')^2=-\frac{4}{3}k^3+c_1k+c_2.\tag{3.77}$$

Here c_1,c_2 are constants. This could be solved by the Jacobi elliptic functions.

3.5 Hamiltonian flows on the space of starlike curves in the affine space $\mathbf{R}^3$

Ulrich Pinkall([35]) studies the Hamiltonian flows on the space of starlike plane curves by using the natural symplectic structure on this space. In this section, we try to find a similar structure on the the space of closed curves in the centroaffine space $\mathbf{R}^3$ and study the behavior of some flows about this structure.

We consider the infinite dimensional manifold

$$M = \hat{M}/\{\text{parameter shifts } [t \to \gamma(t)] \to [t \to \gamma(t-\alpha]\}, \tag{3.78}$$

$$\hat{M} = \{2\pi - \text{periodic } \gamma : \mathbf{R} \to \mathbf{R}^3 | \det(\gamma, \gamma', \gamma'') = 1\}. \tag{3.79}$$

We may consider $\hat{M}$ as a subspace of the vector space $V = \mathrm{Map}(S^1, \mathbf{R}^3)$. Let us see the tangent vector of $\hat{M}$. A vector

$$X = a_1\gamma + a_2\gamma' + a_3\gamma'' \in T_\gamma V \tag{3.80}$$

is tangent to $\hat{M}$ if and only if X is the variation vector field $\dot{\gamma} = \frac{d}{dt}|_{t=0}\gamma_t$, here $a_1, a_2, a_3 : S^1 \to \mathbf{R}$ are smooth functions.

$$\begin{aligned}
\dot{\gamma} = &\ a_1\gamma + a_2\gamma' + a_3\gamma'';\\
\dot{\gamma}' = &\ (a_1' + a_3\tau)\gamma + (a_1 + a_2' - a_3k)\gamma' + (a_2 + a_3')\gamma'';\\
\dot{\gamma}'' = &\ (a_1'' + a_2\tau + 2a_3'\tau + a_3\tau')\gamma + [2a_1' + a_2'' - a_2k - 2a_3'k + a_3(\tau - k')]\gamma'\\
&+(a_1 + 2a_2' + a_3'' - a_3k)\gamma''.
\end{aligned}$$

We have

$$\begin{aligned}
0 = &\ \tfrac{d}{dt}\det(\gamma, \gamma', \gamma'')|_{t=0}\\
= &\ \det(\dot{\gamma}, \gamma', \gamma'') + \det(\gamma, \dot{\gamma}', \gamma'') + \det(\gamma, \gamma', \dot{\gamma}'')\\
= &\ 3a_1 + 3a_2' + a_3'' - 2a_3k.
\end{aligned} \tag{3.81}$$

Therefore the tangent vectors $X \in T_\gamma\hat{M}$ are of the form

$$X = -(3a_2' + a_3'' - 2a_3k)\gamma + 3a_2\gamma' + 3a_3\gamma'' \tag{3.82}$$

for some C^∞-function $a_2, a_3 : S^1 \to \mathbf{R}$. This gives a canonical trivialization of $T\hat{M}$. Let

$$Y = -(3b_2' + b_3'' - 2b_3k)\gamma + 3b_2\gamma' + 3b_3\gamma'' \tag{3.83}$$

be another tangent vector to $\hat{M}$. We define an anti-symmetric 2-form ω on $T\hat{M}$ by

$$\omega(X,Y) = \oint \det(\gamma', X, Y) = 3\oint [(3a_2' + a_3'')b_3 - a_3(3b_2' + b_3'')]. \tag{3.84}$$

Thus $\omega(X,Y)$ vanishes for all $X \in T_\gamma \hat{M}$ if and only if $Y = c_2\gamma' + c_3(2k\gamma + 3\gamma'')$, for constants c_2, c_3. On $\hat{M}$, from $\det(\gamma, \gamma', \gamma'') = 1$, we have $[X, \gamma'] = 0$, then $\nabla_X \gamma' = \nabla_{\gamma'} X = X'$. Let X_1, X_2, X_3 be three tangent vectors of $\hat{M}$, using the formula

$$\begin{aligned} d\omega(X_1, X_2, X_3) = \; & X_1(\omega(X_2, X_3)) - X_2(\omega(X_1, X_3)) + X_3(\omega(X_1, X_2)) \\ & -\omega([X_1, X_2], X_3) + \omega([X_1, X_3], X_2) - \omega([X_2, X_3], X_1) \end{aligned} \tag{3.85}$$

and

$$\nabla_{X_i} X_j - \nabla_{X_j} X_i = [X_i, X_j], \tag{3.86}$$

we have

$$d\omega(X_1, X_2, X_3) = \oint [|X_1', X_2, X_3| - |X_2', X_1, X_3| + |X_3', X_1, X_2| = 0. \tag{3.87}$$

Therefore we have the theorem

Theorem 13 *There exists a closed 2-form ω on the space M.*

Under the deformation

$$\begin{aligned} X = \dot{\gamma} = \; & -(3a_2' + a_3'' - 2a_3 k)\gamma + 3a_2\gamma' + 3a_3\gamma''; \\ \dot{\gamma}' = \; & [-3a_2'' - a_3^{(3)} + 2a_3' k + a_3(2k' + 3\tau)]\gamma \\ & -(a_3'' + a_3 k)\gamma' + 3(a_2 + a_3')\gamma''; \\ \dot{\gamma}'' = \; & [-3a_2^{(3)} + 3a_2\tau - a_3^{(4)} + 2a_3'' k + a_3'(4k' + 6\tau) + a_3(2k'' + 3\tau'))\gamma \\ & -(3a_2'' + 3a_2 k + 2a_3^{(3)} + 2a_3' k - a_3(k' + 3\tau)]\gamma' \\ & +(3a_2' + 2a_3'' - a_3 k)\gamma''; \\ \dot{\gamma}^{(3)} = \; & [-3a_2^{(4)} + 6a_2'\tau + 3a_2\tau' - a_3^{(5)} + 2a_3^{(3)} k + a_3''(6k' + 8\tau) \\ & +a_3'(6k'' + 9\tau') + a_3(2k^{(3)} - k\tau + 3\tau'')]\gamma \\ & +[-6a_2^{(3)} - 6a_2' k + 3a_2(\tau - k') - 3a_3^{(4)} - 2a_3'' k + 3a_3'(3\tau + k') \\ & +a_3(k^2 + 3k'' + 6\tau')]\gamma' + (-3a_2 k - 3a_3' k + 3a_3\tau)\gamma''. \end{aligned}$$

$\dot{k}$ and $\dot{\tau}$ are given by

$$\begin{aligned}\dot{k} = & \det(\dot{\gamma},\gamma'',\gamma^{(3)}) + \det(\gamma,\dot{\gamma}'',\gamma^{(3)}) + \det(\gamma,\gamma'',\dot{\gamma}^{(3)}) \\ = & 6a_2^{(3)} + 6a_2'k + 3a_2k' + 3a_3^{(4)} + 3a_3''k - 3a_3'(3\tau + k') - 3a_3(k'' + 2\tau'); \\ \dot{\tau} = & -3a_2^{(4)} - 3a_2''k + 9a_2'\tau + 3a_2\tau' - a_3^{(5)} + a_3^{(3)}k + 3a_3''(2k' + 3\tau) \\ & + a_3'(6k'' + 2k^2 + 9\tau') + a_3(2k^{(3)} + 2kk' + 3\tau'').\end{aligned} \tag{3.88}$$

We choose a Hamiltonian function $H : M \to \mathbf{R}$, $\gamma \mapsto \oint k$. We can generate a Hamiltonian flow.

$$X_\gamma H = \dot{H} = \oint \dot{k} = \oint -3a_2k' + 3a_3(\tau' + k'') = \omega(X, Y_H). \tag{3.89}$$

The Hamiltonian vector field Y_H is given by

$$Y_H = \frac{1}{3}(2k'' + 2k^2 + 3\tau')\gamma - (k' + \tau)\gamma' + k\gamma''. \tag{3.90}$$

The flow of X_H is given by the evolution equation

$$\dot{\gamma} = \frac{1}{3}(2k'' + 2k^2 + 3\tau')\gamma - (k' + \tau)\gamma' + k\gamma''. \tag{3.91}$$

Under this flow, k and τ satisfy the evolution equation

$$\begin{aligned}\dot{k} &= -k^{(4)} - 2kk'' - 2(k')^2 - 2\tau^{(3)} - 4k\tau' - 4k'\tau; \\ \dot{\tau} &= \tfrac{2}{3}k^{(5)} + 2kk^{(3)} + 4k'k'' + \tfrac{4}{3}k^2k' + \tau^{(4)} + 2k\tau'' + 2k'\tau' - 4\tau\tau'.\end{aligned} \tag{3.92}$$

We are interested in these vector fields which keep the hamiltonian function H as a conservation law. The Hamiltonian vector field is a special one. The vector field $-k'\gamma + k\gamma'$ is interesting to the completely integrable systems.

3.6 Flow Generated by the vector field $-k'X + kX'$

The vortex filament equation $\partial\gamma/\partial t = kB$ on the 3-dimensional Euclidean space describes the motion of a very thin isolated vortex filament $\gamma = \gamma(s,t)$ of radius ϵ in an incompressible unbounded fluid by its own induction, where s is the arclength measured along the filament, t the time, k the geodesic curvature, and B the unit binormal vector. Hasimoto discovered soliton solutions of this equation([17], [25]). The vector field kB is the curvature binormal vector of γ. We are interested in the fact that this vector field is a Killing field along the elastica([27]). We will study the flow given by the Killing field which we obtain in the 3-dimensional affine space.

We will consider the localized induction system generated by the field $-k'X + kX'$ in the case of 3-dimensional affine space. Here k is the centroaffine curvature. Let $X : I \times J \longrightarrow \mathbf{R}^3$, $(s,t) \longmapsto X(s,t)$ be a smooth surface. Here I, J are given intervals, we may assume $0 \in J$. For a fixed t, s is the centroaffine arclength parameter of the curve $s \longmapsto X(s,t)$. We will try to consider the evolution system

$$\frac{\partial X}{\partial t} = W$$

Here W is the vector field $-\partial k/\partial s\ X(s,t) + k\partial X/\partial s$.

We consider W as a variational vector field on the starlike curve $X_t(s) = X(s,t)$. We have the variation formulae for the centroaffine curvature and centroaffine torsion as before

$$\begin{aligned} W(k) &= 2k^{(3)} + 3kk' \\ W(\tau) &= k\tau' + 3k'\tau - k^{(4)} - kk'' \end{aligned}$$

Here we use $k^{(n)}$ to represent the n-th partial derivative of k with respect to the centroaffine arclength parameter s. The first equation is the KdV equation([13]). This gives the KdV equation a new geometric interpretation. Using the first equation, we could find a sequence of functions $f_n(s)$ such that the integral of f_n under variation with the variational vector field W is invariant. Which means it is a complete integral. The first several functions are

$$\begin{aligned} f_0 = &\ 1 \\ f_1 = &\ k \\ f_2 = &\ k^2 \\ f_3 = &\ k^3 - 2(k')^2 \\ f_4 = &\ k^4 - 8k(k')^2 + \tfrac{16}{5}(k'')^2 \\ f_5 = &\ k^5 - 20k^2(k')^2 + 16k(k'')^2 - \tfrac{32}{7}(k^{(3)})^2 \\ f_6 = &\ k^6 - 40k^3(k')^2 - \tfrac{40}{3}(k')^4 + 48k^2(k'')^2 + \tfrac{640}{21}(k'')^3 - \tfrac{192}{7}k(k^{(3)})^2 \\ &+\tfrac{128}{21}(k^{(4)})^2 \\ f_7 = &\ k^7 - 70k^4(k')^2 - \tfrac{280}{3}k(k')^4 + 112k^3(k'')^2 224(k')^2(k'')^2 + \tfrac{640}{3}k(k'')^3 \\ &-96k^2(k^{(3)})^2 - \tfrac{640}{3}k''(k^{(3)})^2 + 128k(k^{(4)})^2 - \tfrac{256}{33}(k^{(5)})^2 \\ &\cdots \end{aligned}$$

The corresponding phenomenon for LIE is described in ([25]), where the sequence of conserved integrals is derived by a recursion operator. We found the integrals above by an ad hoc method. It would be interesting to determine a recursion operator in the present situation. See ([26]) for further discussion of geometric invariants of integrable flows.

The integrals $I_i = \int f_i ds$ are geometrical functionals. They form a family of conservation laws of the flow. This implies we could try to solve this evolution problem. We write the variation formulae of k and τ as two standard partial differential equations

$$\begin{array}{ll} \frac{\partial k}{\partial t} & -3k\frac{\partial k}{\partial s} - 2\frac{\partial^3 k}{\partial s^3} = 0 \\ \frac{\partial \tau}{\partial t} & -k\frac{\partial \tau}{\partial s} = 3\frac{\partial k}{\partial s}\tau - \frac{\partial^4 k}{\partial s^4} - k\frac{\partial^2 k}{\partial s^2} \end{array}$$

The first equation is the Korteweg-de Vries(or KdV) equation. For an initial condition, we can solve it by using inverse scattering method. We can transform this equation under

$$\frac{i}{2}s \longrightarrow s, \qquad \frac{i}{4}t \longrightarrow t$$

to yield a special form of the KdV equation

$$\frac{\partial k}{\partial t} - 6k\frac{\partial k}{\partial s} + \frac{\partial^3 k}{\partial s^3} = 0$$

For this equation, we may assume the initial condition has the form $k(s,0) = g(s)$ and $g(s)$ satisfies the conditions

$$\int_{-\infty}^{\infty} |g(s)|ds < \infty, \qquad \int_{-\infty}^{\infty} (1+|s|)|g(s)|ds < \infty$$

Let $g(s)$ be the potential function for the Sturm-Liouville equation

$$\psi_{ss} + (\lambda - g(s))\psi = 0, \qquad -\infty < s < \infty$$

where the λ is the eigenvalue. The scattering data can be described by the behaviors of the eigenfunction ψ in the form

$$\psi(s,l) \sim \begin{cases} e^{-ils} + b(l)e^{ils}, & \text{as } s \to +\infty \\ a(l)e^{-ils} & \text{as } s \to -\infty \end{cases}$$

for $\lambda > 0$, with $l = \lambda^{1/2}$ for the continuous spectrum, and

$$\psi(s) \sim c_n \exp(-l_n s)$$

as $s \to +\infty$, for $\lambda < 0$, with $l_n = (-\lambda)^{1/2}$ for each discrete eigenvalue $(n = 1, 2, \cdots, N)$. The time evolution of these scattering data is then given by

$$l_n = c, c_n(t) = c_n \exp(4l_n^3 t), b(l,t) = b(l)\exp(8il^3 t)$$

We define a function

$$F(x,t) = \sum_{n=1}^{N} c_n^2 \exp(8l_n^3 t - l_n x) + \frac{1}{2\pi}\int_{-\infty}^{\infty} b(l)\exp(8il^3 t + ilx)dl$$

Let $K(x,z,t)$ be the solution of the Marchenko equation

$$K(x,z,t) + F(x+z,t) + \int_{x}^{\infty} K(x,y,t)F(y+z,t)dy = 0$$

Then the solution of the KdV equation can be expressed as

$$k(s,t) = -2\frac{d}{ds}K(s,s,t)$$

For the second equation, we use the solution of the first equation. At this time, k is a function of t and s. We cite a standard proposition for the solution of linear partial differential equations of the first order([6]).

Theorem 14 *For the linear partial differential equation of the form*

$$P\frac{\partial z}{\partial x} + Q\frac{\partial z}{\partial y} = R$$

here P, Q and R are functions of x, y and z, we can consider the following symmetric system of ordinary differential equations

$$\frac{dx}{P} = \frac{dy}{Q} = \frac{dz}{R}$$

If u, v are two first integrals of this system, then $u = f(v)$ is the general solution of the partial differential equation. Here f is an arbitrary function.

Now we consider the system of ordinary differential equations

$$\frac{dt}{1} = \frac{ds}{-k} = \frac{d\tau}{3k'\tau - k^{(4)} - kk''}$$

We can obtain two first integrals

$$\begin{array}{ll} t & +\int \frac{ds}{k} = C_1 \\ \tau & = \frac{1}{k^3}[C_2 + \int (k^{(4)} + kk'')k^2 ds] \end{array}$$

Therefore we have the solution

$$\tau = \frac{1}{k^3}[f(t + \int \frac{ds}{k}) + \int (k^{(4)} + kk'')k^2 ds]$$

Here f can be determined by the initial condition. We have proved:

Theorem 15 *The evolution of k and τ under the flow generated by $-k'X + kX'$ is completely integrable. In particular, the evolution of k is governed by the Korteweg - de Vries equation, a completely integrable equation.*

For this topic, we could do more work. At least, from the problem settlement. we know that the centroaffine elastica is a solution to this problem.

Chapter 4

Affine elastica in the affine plane $\mathbf{R}^2$

In this chapter, we will study the variational properties of plane curves invariant under the group of area-preserving affine transformations or equiaffine transformations. The group is generated by the action of the unimodular linear group $SL(2, \mathbf{R})$ and the translation group $\mathbf{R}^2$. We may expect to characterize the equiaffine properties of critical plane curves with respect to certain curvature energy functionals by the affine arclength parameter and affine curvature.

4.1 Definition and variation formula of affine elastica in the affine plane $\mathbf{R}^2$

We denote (x, y) a coordinate system of the affine plane $\mathbf{R}^2$, and $X : I \to \mathbf{R}^2$ be a smooth curve. We write it as a column vector

$$X(t) = (x(t), y(t))^T$$

where $x(t), y(t)$ are smooth functions of t defined on a certain interval I. We write the tangent vector as $X'(t) = dX/dt = (x'(t), y'(t))^T$ or simply X', similarly for $X'', \cdots, X^{(n)}$ and $(X'(t), X''(t))$ be the 2×2 matrix

$$\begin{pmatrix} x'(t) & x''(t) \\ y'(t) & y''(t) \end{pmatrix}$$

Definition 15 *A regular curve $X : I \to \mathbf{R}^2$ is called nondegenerate if*

$$|X'(t), X''(t)| = \det(X'(t), X''(t)) \neq 0,$$

for any $t \in I$. A parameter s is called an affine arclength parameter of a nondegenerate curve X if

$$A(s) = (X'(s), X''(s)) \in SL(2, \mathbf{R})$$

that is $|X'(s), X''(s)| = 1$, for any $s \in I$.

One may easily verify that the condition for a nondegenerate curve is independent of parametrization. Let t be an arbitrary parameter of X. We are going to look for an affine arclength parameter.

$$\begin{aligned} X'(t) &= X'(s)\tfrac{ds}{dt}, \\ X''(t) &= X''(s)(\tfrac{ds}{dt})^2 + X'(s)\tfrac{d^2s}{dt^2}. \end{aligned}$$

We have the relation

$$|X'(t), X''(t)| = |X'(s), X''(s)|(\frac{ds}{dt})^3 = (\frac{ds}{dt})^3.$$

Thus we have the formula of velocity

$$\frac{ds}{dt} = |X'(t), X''(t)|^{\frac{1}{3}}$$

and hence we obtain the affine arclength formula of a segment of the curve X between $X(t_0)$ and $X(t)$ for any parameter t

$$s(t) = \int_{t_0}^{t} |X'(t), X''(t)|^{\frac{1}{3}} dt. \tag{4.1}$$

When we use this $s(t)$ as a parameter of the curve X, it is an affine arclength parameter. If t is also an affine arclength parameter of this curve, then $ds/dt = |X'(t), X''(t)|^{1/3} = 1$. Therefore $s = t + c$, for a constant c. We have proven that a nondegenerate curve $X(t)$ can admit an affine arclength parameter uniquely up to a constant.

To see the geometric meaning of the condition for a nondegenerate curve, we may assume $|X'(t), X''(t)| > 0$ for all $t \in I$. Then for small Δt, by Taylor expansion, we have

$$X(t + \Delta t) - X(t) = X'(t)\Delta t + \frac{1}{2}X''(t)(\Delta t)^2 + \alpha(t, \Delta t),$$

where $\alpha(t, \Delta t)$ is an infinitesimal vector of degree 2 with respect to Δt. Then, $\beta(t, \Delta t) = |X'(t), \alpha(t, \Delta t)| \to 0$ as $\Delta \to 0$. Hence, for sufficient small Δt, we get

$$|X'(t), X(t+\Delta t) - X(t)| = \frac{1}{2}(\Delta t)^2(|X'(t), X''(t)| + \beta(t, \Delta t)) > 0.$$

This means that the secant from $X(t)$ to $X(t+\Delta t)$ lies on one side of the tangent line at $X(t)$. Thus our condition means that a nondegenerate affine curve has no inflection points.

Definition 16 *Let s be an affine arclength parameter of a curve $X : I = [0, L] \to \mathbf{R}^2$, here L is the affine arclength of X. We define the affine curvature of the curve X by*

$$k(s) = |X''(s), X^{(3)}(s)|. \tag{4.2}$$

Differentiating $|X'(s), X''(s)| = 1$, we obtain $|X'(s), X^{(3)}(s)| = 0$. This means $X'(s)$, $X^{(3)}(s)$ are linearly dependent. Hence we have

$$X^{(3)}(s) = -k(s)X'(s). \tag{4.3}$$

From equations (4.1) and (4.2), we know that the affine arclength parameter and affine curvature of a nondegenerate curve are invariant under any equiaffine transformation of the plane $\mathbf{R}^2$. This means if X(s) is a nondegenerate curve with affine arclength parameter s and f is any equiaffine transformation of the plane $\mathbf{R}^2$, then the curve $Y(s) = f(X(s))$ is a nondegenerate curve with affine arclength parameter s and the affine curvature of $Y(s)$ is equal to the affine curvature of $X(s)$ for each value of s.

We denote ∇ the usual affine connection of $\mathbf{R}^2$. This is a torsion-free and flat connection. We have the structural equation

$$\begin{aligned} \nabla_Y Z - \nabla_Z Y - [Y, Z] &= 0, \\ \nabla_Y \nabla_Z V - \nabla_Z \nabla_Y V - \nabla_{[Y,Z]} V &= 0. \end{aligned} \tag{4.4}$$

For vector fields Y, Z, V on $\mathbf{R}^2$. We will write $X''(t) = \nabla_{\frac{\partial}{\partial t}} X'(t) = \nabla_{X'} X'(t)$. First we consider the arclength variation.

The letter X will also denote a variation $X = X_w(t) = X(w, t) : (-\varepsilon, \varepsilon) \times I \to \mathbf{R}^2$ with $X(0, t) = X(t)$. Associated with such a variation is the variation vector field $W = W(t) = (\partial X / \partial w)(0, t) = dX(\frac{\partial}{\partial w})(0, t)$ along the curve $X(t)$. We will also write $X = X(w,t), W = W(w,t) = dX(\frac{\partial}{\partial w})(w,t)$, $X'(t) = X'(w,t) = dX(\frac{\partial}{\partial t})(w,t)$. We know

$$[X'(w,t), W(w,t)] = [dX(\frac{\partial}{\partial t}), dX(\frac{\partial}{\partial w})] = dX([\frac{\partial}{\partial t}, \frac{\partial}{\partial w}]) = 0. \tag{4.5}$$

Therefore, from the torsion tensor equation(first equation of (4.4)), we have

$$\nabla_{X'(t)} W = \nabla_W X'(t).$$

We denote s the affine arclength parameter and write $X(s), k(w,s)$ for the corresponding reparametrizations and $s \in [0, L]$, where L is the affine arclength of $X(s)$. We may assume $t = s$ be the affine arclength parameter of $X(t)$ and then $I = [0, L]$.

$$W(\frac{ds}{dt}) = \frac{1}{3}|X'(t), X''(t)|^{-\frac{2}{3}} W(|X'(t), X''(t)|) = -g\frac{ds}{dt}. \tag{4.6}$$

Here we write $g = -\frac{1}{3}|X'(t), X''(t)|^{-1} W(|X'(t), X''(t)|)$.

$$\begin{aligned} W(|X'(t), X''(t)|) = & \ |\nabla_W X'(t), X''(t)| + |X'(t), \nabla_W X''(t)| \\ = & \ |\nabla_{X'(t)} W, X''(t)| + |X'(t), \nabla_{X'(t)} \nabla_W X'(t)| \\ = & \ |\nabla_{X'(t)} W, X''(t)| + |X'(t), \nabla^2_{X'(t)} W| \\ = & \ \nabla_{X'(t)} [2|W, X''(t)| + |X'(t), \nabla_{X'(t)} W|] + 2|X^{(3)}(t), W|. \end{aligned}$$

$$\begin{aligned} & \tfrac{d}{dw} \textstyle\int_I |X', X''|^{\frac{1}{3}} dt|_{w=0} \\ & = \tfrac{2}{3} \textstyle\int_0^L |X^{(3)}(s), W| ds + \tfrac{1}{3}[2|W, X''| + |X', \nabla_{X'} W|]_0^L. \end{aligned}$$

We give $X(w,t)$ a good boundary condition such that $W(0,0) = W(0,L) = 0$, $\nabla_{X'} W(0,0) = \nabla_{X'} W(0,L) = 0$. Then we get

$$\frac{d}{dw} \int_I |X', X''|^{\frac{1}{3}} dt|_{w=0} = \frac{2}{3} \int_0^L |X^{(3)}(s), W| ds. \tag{4.7}$$

Therefore the Euler-Lagrange equation is

$$X^{(3)}(s) = 0. \tag{4.8}$$

This means the affine curvature $k(s) = 0$. Integrating it, we know

$$X(s) = \frac{1}{2} s^2 \mathbf{a} + s\mathbf{b} + \mathbf{c}, \tag{4.9}$$

where $\mathbf{a}, \mathbf{b}$ and $\mathbf{c}$ are constant vectors such that $|\mathbf{b}, \mathbf{a}| = 1$. By changing the coordinate system (x, y) such that

$$\mathbf{b} = (1, 0)^T, \quad \mathbf{a} = (0, 1)^T, \quad \mathbf{c} = (0, 0)^T.$$

We see that the curve is a parabola

$$y = \frac{1}{2}x^2. \tag{4.10}$$

We may assume $W = h(s)X'(s) + f(s)X''(s)$. Formula (4.7) becomes

$$\frac{d}{dw}\int_I |X', X''|^{\frac{1}{3}} dt|_{w=0} = -\frac{2}{3}\int_0^L k(s)f(s)ds. \tag{4.11}$$

This means the part $h(s)X'(s)$ of W has no effect on the variation of the total affine arclength. It only represents an infinitesimal reparametrisation of the curve. At this step, we will work out the second variational formula of the affine arclength with the variational field $W = f(s)X''(s)$. From the initial condition, we know $f(0) = f(L) = f'(0) = f'(L) = 0$.

$$\begin{aligned} W(|X^{(3)}(t), W|) = & \ |\nabla_W X^{(3)}(t), W| + |X^{(3)}(t), \nabla_W W| \\ = & \ |\nabla_{X'(t)} \nabla_W X''(t), W| + |X^{(3)}(t), \nabla_W W| \\ = & \ \nabla_{X'(t)} |\nabla_W X''(t), W| - |\nabla^2_{X'(t)} W, \nabla_{X'(t)} W| \\ & +|X^{(3)}(t), \nabla_W W|. \end{aligned}$$

$$\begin{aligned} & \nabla^2_W(|X'(t), X''(t)|^{\frac{1}{3}}) = \tfrac{1}{3}\nabla_W(|X'(t), X''(t)|^{\frac{-2}{3}}\nabla_W(|X'(t), X''(t)|) \\ & = \tfrac{1}{3}|X'(t), X''(t)|^{\frac{-2}{3}}\nabla_W(\nabla_{X'(t)}(2|W, X''(t)| + |X'(t), \nabla_{X'(t)}W|) \\ & \quad +2|X^{(3)}(t), W|) - \tfrac{2}{9}|X'(t), X''(t)|^{\frac{-5}{3}}(\nabla_W(|X'(t), X''(t)|))^2 \\ & = \tfrac{1}{3}|X'(t), X''(t)|^{\frac{-2}{3}}(\nabla_{X'(t)}\nabla_W(2|W, X''(t)| + |X'(t), \nabla_{X'(t)}W|) \\ & \quad +2(\nabla_{X'(t)}|\nabla_W X''(t), W| - |\nabla^2_{X'(t)}W, \nabla_{X'(t)}W| + |X^{(3)}(t), \nabla_W W|) \\ & \quad -\tfrac{2}{9}|X'(t), X''(t)|^{\frac{-5}{3}}(\nabla_{X'(t)}[2|W, X''(t)| + |X'(t), \nabla_{X'(t)}W|] \\ & \quad +2|X^{(3)}(t), W|)^2. \end{aligned}$$

Then we have the second variational formula of the affine arclength with the variational field $W = f(s)X''(s)$

$$\frac{d^2}{dw^2}L(w)|_{w=0} = -\frac{2}{9}\int_0^L (f'')^2 ds \leq 0. \tag{4.12}$$

Since there is no nonzero linear function with two zeroes, the above equality holds if and only if f is zero. We know the solution curve is stable. But the solution curve obtains the maximum affine arclength, not the minimum as usual.

For simplicity, we write $X^{(n)}(s) = X^{(n)}(w,s)$, $X^{(n)}(t) = X^{(n)}(w,t)$. Equation (4.5) yields

$$0 = [W, X'(t)] = [W, X'(s)\frac{ds}{dt}] = \frac{ds}{dt}[W, X'(s)] + W(\frac{ds}{dt})X'(s).$$

We have

$$[W, X'(s)] = -\frac{dt}{ds}W(\frac{ds}{dt})X'(s) = gX'(s). \tag{4.13}$$

$$\begin{aligned}
\nabla_W X''(s) = &\ \nabla_W \nabla_{X'(s)} X'(s) \\
= &\ \nabla_{X'(s)} \nabla_W X'(s) + \nabla_{[W,X'(s)]} X'(s) \\
= &\ \nabla_{X'(s)} (\nabla_{X'(s)} W + [W, X'(s)]) + gX''(s) \\
= &\ \nabla^2_{X'(s)} W + g_s X'(s) + 2gX''(s),
\end{aligned}$$

$$\begin{aligned}
\nabla_W X^{(3)}(s) = &\ \nabla_W \nabla_{X'(s)} X''(s) \\
= &\ \nabla_{X'(s)} \nabla_W X''(s) + \nabla_{[W,X'(s)]} X''(s) \\
= &\ \nabla_{X'(s)} (\nabla^2_{X'(s)} W + g_s X'(s) + 2gX''(s)) + gX^{(3)}(s) \\
= &\ \nabla^3_{X'(s)} W + g_{ss} X'(s) + 3g_s X''(s) + 3gX^{(3)}(s).
\end{aligned}$$

Thus we have

$$\begin{aligned}
W(k(s)) = &\ |\nabla_W X''(s), X^{(3)}(s)| + |X''(s), \nabla_W X^{(3)}(s)| \\
= &\ |\nabla^2_{X'(s)} W, X^{(3)}(s)| + |X''(s), \nabla^3_{X'(s)} W| - g_{ss} + 5gk.
\end{aligned} \tag{4.14}$$

Now we consider the energy functional

$$\int_\gamma p(k)ds = \int_I p(k)\frac{ds}{dt}dt. \tag{4.15}$$

Here $p(k)$ is a smooth function of k, s is the affine arclength parameter of the curve $X_w(t) = X(w,t)$. When we refine on the curve $X(t) = X(0,t)$, since $s = t$, we will drop t, s in $X'(t), X'(s)$ respectively.

$$\begin{aligned}
&\tfrac{d}{dw}\textstyle\int_0^L p(k)ds|_{w=0} \\
&= \textstyle\int_0^L [p'(k)(|\nabla^2_{X'} W, X^{(3)}| + |X'', \nabla^3_{X'} W| - g_{ss} + 5gk) - p(k)g]ds.
\end{aligned}$$

Now we give $X(w,t)$ a boundary condition such that $W(0,0) = W(0,L) = 0$, $\nabla_{X'} W(0,0) = \nabla_{X'} W(0,L) = 0$, $\nabla^2_{X'} W(0,0) = \nabla^2_{X'} W(0,L) = 0$, $\nabla^3_{X'} W(0,0)$

$= \nabla^3_{X'}W(0,L) = 0$. We calculate the integral part by part.

$$\begin{aligned}
&\int_0^L p'(k)(|\nabla^2_{X'}W, X^{(3)}| + |X'', \nabla^3_{X'}W|)ds\\
&= \int_0^L(|\nabla^2_{X'}W, p'(k)X^{(3)}| + |p'(k)X'', \nabla^3_{X'}W|)ds\\
&= \int_0^L(|\nabla^2_{X'}W, p'(k)'X'' - 2kp'(k)X'|)ds + |p'(k)X'', \nabla^2_{X'}W|_0^L\\
&= -\int_0^L(|\nabla_{X'}W, (p'(k)'' - 2kp'(k))X'' - (p'(k)'k + 2(kp'(k))')X'|)ds\\
&\quad +[|p'(k)X'', \nabla^2_{X'}W| + |\nabla_{X'}W, p'(k)'X'' - 2kp'(k)X'|]_0^L\\
&= \int_0^L(|W, (p'(k)^{(3)} - 4(kp'(k))' - p'(k)'k)X''\\
&\quad -(p'(k)''k - 2k^2p'(k) + (p'(k)'k)' + 2(kp'(k))'')X'|)ds\\
&\quad +[|p'(k)X'', \nabla^{(3)}_{X'}W| + |\nabla_{X'}W, p'(k)'X'' - 2kp'(k)X'|\\
&\quad -|W, (p'(k)'' - 2kp'(k))X'' - (p'(k)'k + 2(kp'(k))')X'|]_0^L,
\end{aligned}$$

$$\begin{aligned}
\int_0^L p'(k)g_{ss}ds &= p'(k)g_s|_0^L - \int_0^L p'(k)'g_s ds\\
&= [p'(k)g_s - p'(k)'g]_0^L + \int_0^L p'(k)''g ds,
\end{aligned}$$

$$\begin{aligned}
&\int_0^L(5p'(k)k - p'(k)'' - p(k))gds\\
&= -\tfrac{1}{3}\int_0^L(5p'(k)k - p'(k)'' - p(k))(|\nabla_{X'}W, X''| + |X', \nabla^2_{X'}W|)ds\\
&= -\tfrac{1}{3}\int_0^L |\nabla_{X'}W, (5(p'(k)k)' - p'(k)^{(3)} - p'(k)k')X'\\
&\quad +2(5p'(k)k - p'(k)'' - p(k))X''|ds\\
&\quad -\tfrac{1}{3}|(5p'(k)k - p'(k)'' - p(k))X', \nabla_{X'}W|_0^L\\
&= \tfrac{1}{3}\int_0^L |W, (5(p'(k)k)'' - p'(k)^{(4)} - (p'(k)k')')X' + 3(5(p'(k)k)'\\
&\quad -p'(k)^{(3)} - p'(k)k')X'' - 2k(5p'(k)k - p'(k)'' - p(k))X'|ds\\
&\quad -\tfrac{1}{3}[|(5p'(k)k - p'(k)'' - p(k))X', \nabla_{X'}W| + |W, (5(p'(k)k)'\\
&\quad -p'(k)^{(3)} - p'(k)k')X' + 2(5p'(k)k - p'(k)'' - p(k))X''|]_0^L.
\end{aligned}$$

Then we obtain

$$\begin{aligned}
&\tfrac{d}{dw}\int_0^L p(k)ds|_{w=0} = \tfrac{1}{3}\int_0^L |[(p'(k)'' + 4p'(k)k - 2p(k))''\\
&+k(p'(k)'' + 4p'(k)k - 2p(k))]X', W|ds + B(\gamma, W)|_0^L.
\end{aligned}$$

The boundary term is

$$\begin{aligned}
&B(\gamma, W) = |p'(k)X'', \nabla^{(3)}_{X'}W| + |\nabla_{X'}W, p'(k)'X'' - 2kp'(k)X'|\\
&-|W, (p'(k)'' - 2kp'(k))X'' - (p'(k)'k + 2(kp'(k))')X'|\\
&-\tfrac{1}{3}[|(5p'(k)k - p'(k)'' - p(k))X', \nabla_{X'}W| + |W, (5(p'(k)k)' - p'(k)^{(3)}\\
&-p'(k)k')X' + 2(5p'(k)k - p'(k)'' - p(k))X''] + p'(k)g_s - p'(k)'g.
\end{aligned}$$

Therefore the Euler-Lagrange equation is

$$E = [(p'(k)'' + 4p'(k)k - 2p(k))'' + k(p'(k)'' + 4p'(k)k - 2p(k))]X' = 0. \quad (4.16)$$

Definition 17 *A nondegenerate affine curve in the affine plane* $\mathbf{R}^2$ *with the affine arclength parameter* s *is called a generalized affine elastica if it satisfies the above Euler-Lagrange equation (4.16).*

For any constant vector field W, along a generalized affine elastica, the variation formula becomes

$$\frac{d}{dw}\int_0^L p(k)ds|_{w=0} = -\frac{1}{3}(|W\ J(L)| - |W\ J(0)|) = 0.$$

Here

$$J(s) = -(p'(k)^{(3)} + 4(p'(k)k)' - 2p(k)')X' + (p'(k)'' + 4p'(k)k - 2p(k))X''.$$

This variation formula continues to hold when L is replaced with any intermediate value L', $0 < L' < L$. It follows that $|W, J|$ is constant on $[0, L]$. But W is an arbitrary constant vector field, so we obtain that J is a constant vector field along a generalized affine elastica.

The Euler-Lagrange equation implies $p'(k)'' + 4p'(k)k - 2p(k)$ satisfies the second order linear differential equation $z'' + kz = 0$. Equation (4.3) shows that $X'(s) = (x'(s), y'(s))^T$ is the solution of this equation. The nondegenerate condition implies that $x'(s)$, $y'(s)$ are linearly independent. Therefore the general solution of the differential equation $z'' + kz = 0$ is $c_1x'(s) + c_2y'(s)$.

Theorem 16 *A nondegenerate curve* $X = (x, y)^T$ *in the affine plane* $\mathbf{R}^2$ *is a generalized affine elastica if and only if*

$$p'(k)'' + 4p'(k)k - 2p(k) = c_1x'(s) + c_2y'(s), \tag{4.17}$$

for constants c_1 *and* c_2.

The Euler-Lagrange equation (4.16) is not easy to solve. If $p'(k)'' + 4p'(k)k - 2p(k) = 0$, Multiplying by $2p''(k)k'$, we can get a first integral of this equation as in chapter 2

$$p''(k)^2(k')^2 + 4p'(k)(p'(k)k - p(k)) = c. \tag{4.18}$$

Then we can solve $k(s)$ by quadratures.

When $p'(k)'' + 4p'(k)k - 2p(k) \neq 0$, we do know the vector field J is a constant vector field along the generalized affine elastica(if it exists). If we assume we solved the Euler-Lagrange equation under the condition $p'(k)'' +$

$4p'(k)k - 2p(k) \neq 0$, using this Killing field J, we can solve the structure equation. We denote $A(s) = p'(k)'' + 4p'(k)k - 2p(k)$. We choose a coordinate system such that the Killing field has the form $J = a\partial/\partial x + b\partial/\partial y$, here a, b are constants. Then the structure equations have the form

$$\begin{aligned} Ax'' - A'x' &= a, \\ Ay'' - A'y' &= b. \end{aligned}$$

We have the solution

$$\begin{aligned} x &= \int_0^s A(c_1 \int_0^s \frac{1}{A^2} ds + d_1) ds + e_1, \\ y &= \int_0^s A(c_2 \int_0^s \frac{1}{A^2} ds + d_2) ds + e_2. \end{aligned} \tag{4.19}$$

Here d_1, d_2, e_1, e_2 are constants and satisfy the condition

$$ad_2 - bd_1 = 1.$$

4.2 Affine elastica for $p(k) = k^2 + \lambda$ in the affine plane $\mathbf{R}^2$

Now we consider a special case for $p(k) = k^2 + \lambda$, λ is the Lagrange multiplier. The curvature energy functional(4.15) becomes $\int_0^L (k^2 + \lambda) ds$. The critical point of this energy functional is called an affine elastica.

4.2.1 The solutions of motion equation for affine elastica

At this time, the Euler-Lagrange equation (4.16) becomes

$$E = 2[(k'' + 3k^2)'' + k(k'' + 3k^2 - \lambda)]X'.$$

Therefore,we have

$$(k'' + 3k^2)'' + k(k'' + 3k^2 - \lambda) = 0. \tag{4.20}$$

We consider the following equation whose solutions are also solutions of the above equation

$$k'' + 3k^2 - \lambda = 0. \tag{4.21}$$

We can solve this equation and then solve $X'(s)$ from the linear equation

$$X^{(3)} + k(s)X'(s) = 0. \tag{4.22}$$

That is the work we have done in chapter 2. Using one more integration, we partially solve the affine elastica. We now discuss some special cases. First we consider the constant affine curvature elastica.
(1). $k = 0$. $X^{(3)}(s) = 0$, integrating three times, we obtain

$$X(s) = \frac{1}{2}\mathbf{a}s^2 + \mathbf{b}s + \mathbf{c}, \tag{4.23}$$

where $\mathbf{a}$, $\mathbf{b}$ and $\mathbf{c}$ are constant vectors such that $|\mathbf{b}, \mathbf{a}| = 1$. By taking

$$\mathbf{a} = (0,1)^T, \quad \mathbf{b} = (1,0)^T, \quad \mathbf{c} = (0,0)^T.$$

Then (4.23) becomes

$$X(s) = (s, \frac{1}{2}s^2)^T. \tag{4.24}$$

This is a parabola $y = x^2/2$.
(2). $k = r^2 = \sqrt{\lambda/3} > 0$. We have the equation

$$X^{(3)}(s) + r^2 X'(s) = 0.$$

Its characteristic equation is

$$q^3 + r^2 q = 0.$$

The characteristic roots are $ri, -ri, 0$. Therefore the general solution of the structure equation is

$$X(s) = \cos(rs)\mathbf{a} + \sin(rs)\mathbf{b} + \mathbf{c}, \tag{4.25}$$

where $\mathbf{a}$, $\mathbf{b}$ and $\mathbf{c}$ are constant vectors such that $|\mathbf{a}, \mathbf{b}| = 1/r^3$. Choosing

$$\mathbf{a} = (0, -1/r^2)^T, \quad \mathbf{b} = (1/r, 0)^T, \quad \mathbf{c} = (0,0)^T,$$

we have

$$X(s) = (\frac{1}{r}\sin(rs), -\frac{1}{r^2}\cos(rs))^T. \tag{4.26}$$

This is an ellipse $x^2 + r^2y^2 = 1/r^2$.

(3). $k = -r^2 = -\sqrt{\lambda/3} < 0$. Similar to the above positive constant curvature case, we have the solution.

$$X(s) = (\frac{1}{r}\sinh(rs), \frac{1}{r^2}\cosh(rs))^T. \tag{4.27}$$

This is a hyperbola $x^2 - r^2y^2 = -1/r^2$.

For those nonconstant affine curvature elastica, we are interested in those closed elastica. At least for these periodic solution $X'(s_2)$ with period $4lK(p)$. We know from theorem 2.6.

$$X'(s_2 + 2lK(p)) = -X'(s_2).$$

Thus we have

$$X(4lK(p) + s_2) - X(s_2) = \int_{s_2}^{4lK(p)+4_2} X'(s)ds = 0.$$

Therefore we have

Theorem 17 . *There exists closed nonconstant affine curvature elastica in the affine plane* $\mathbf{R}^2$.

The position vector of the affine elastica can be written as $X(s) = \phi(s)X'(s) - \psi(s)X''(s)$, where $\psi(s)$ is the affine support function of X. Differentiating this relation, we get

$$X'(s) = \phi'(s)X'(s) + (\phi(s) - \psi'(s))X''(s) - \psi(s)X^{(3)}(s).$$

Thus we have

$$\phi - \psi' = 0, \qquad \phi' + k\psi - 1 = 0. \tag{4.28}$$

This gives the equation

$$\psi'' + k\psi = 1. \tag{4.29}$$

For closed affine elastica, we have

$$\begin{aligned}&\oint(k'' + 3k^2 - \lambda)ds = \oint(k'' + 3k^2 - \lambda)(\psi'' + k\psi)ds\\&= \oint(k'' + 3k^2)'' + k(k'' + 3k^2 - \lambda)\psi ds = 0.\end{aligned}$$

Therefore the affine arclength of a closed affine elastica satisfy the condition $L = 3\oint k^2ds/\lambda$.

4.2.2 Second variation of affine elastica in $\mathbf{R}^2$

We consider the second variation with variation vector field $W = fX''(s)$. At $w = 0$, we do the computation and know

$$g = -\frac{1}{3}(f'' - 2fk),$$

$$\begin{aligned}
W(k) = \ & |\nabla^2_{X'(s)}W, X^{(3)}(s)| + |X''(s), \nabla^3_{X'(s)}W| - g_{ss} + 5gk \\
= \ & k(f'' - kf) + (3f''k + 3f'k' + fk'' - fk^2) \\
& +\tfrac{1}{3}(f^{(4)} - 2(fk)'') - \tfrac{5}{3}k(f'' - 2fk) \\
= \ & \tfrac{1}{3}[f^{(4)} + 5kf'' + 5k'f' + (k'' + 4k^2)f],
\end{aligned}$$

$$\begin{aligned}
W(k') = \ & WX'(k) = [W, X'](k) + X'W(k) \\
= \ & -\tfrac{1}{3}(f'' - 2fk)k' + \tfrac{1}{3}[f^{(5)} + 5kf^{(3)} + 10k'f'' \\
& +(6k'' + 4k^2)f' + (k^{(3)} + 4(k^2)')f] \\
= \ & \tfrac{1}{3}[f^{(5)} + 5kf^{(3)} + 9k'f'' + (6k'' + 4k^2)f' + (k^{(3)} + 5(k^2)')f],
\end{aligned}$$

$$\begin{aligned}
W(k'') = \ & [W, X'](k') + X'W(k') \\
= \ & -\tfrac{1}{3}(f'' - 2fk)k'' + \tfrac{1}{3}[f^{(6)} + 5kf^{(4)} + 14k'f^{(3)} \\
& +(15k'' + 4k^2)f'' + (7k^{(3)} + 9(k^2)')f' + (k^{(4)} + 5(k^2)'')f] \\
= \ & \tfrac{1}{3}[f^{(6)} + 5kf^{(4)} + 14k'f^{(3)} + (14k'' + 4k^2)f'' \\
& +(7k^{(3)} + 9(k^2)')f' + (k^{(4)} + 5(k^2)'' + 2kk'')f],
\end{aligned}$$

$$\begin{aligned}
W(k^{(4)}) = \ & [W, X']X'(k'') + X'[W, X'](k'') + X'X'W(k'') \\
= \ & -\tfrac{1}{3}(f'' - 2fk)k^{(4)} - \tfrac{1}{3}(f'' - 2fk)'k^{(3)} - \tfrac{1}{3}(f'' - 2fk)k^{(4)} \\
& +\tfrac{1}{3}[f^{(6)} + 5kf^{(4)} + 14k'f^{(3)} + (14k'' + 4k^2)f'' \\
& +(7k^{(3)} + 9(k^2)')f' + (k^{(4)} + 5(k^2)'' + 2kk'')f]'' \\
= \ & \tfrac{1}{3}[f^{(8)} + 5kf^{(6)} + 24k'f^{(5)} + (47k'' + 4k^2)f^{(4)} \\
& +(48k^{(3)} + 17(k^2)')f^{(3)} + (27k^{(4)} + 27(k^2)'' + 2kk'')f'' \\
& +(9k^{(5)} + 19(k^2)^{(3)} + 4(kk'')' + 2kk^{(3)})f' \\
& +(k^{(6)} + 5(k^2)^{(4)} + 2(kk'')'' + 4kk^{(4)} + 2k'k^{(3)})f].
\end{aligned}$$

Therefore the second variational formula is

$$\begin{aligned}
& \tfrac{d^2}{dw^2} \int_0^{L(w)}(k^2 + \lambda)ds|_{w=0} \\
= \ & -\tfrac{2}{3}\int_0^L |W, W(k^{(4)} + 3(k^2)'' + kk'' + 3k^3 - \lambda k)X'|ds \\
= \ & \tfrac{2}{9}\int_0^L f[f^{(8)} + 12kf^{(6)} + 24k'f^{(5)} \\
& +(54k'' + 48k^2 - \lambda)f^{(4)} + (48k^{(3)} + 96(k^2)')f^{(3)} \\
& +(27k^{(4)} + 27(k^2)'' + 108(k')^2 + 135kk'' + 73k^3 - 5\lambda k)f'' \\
& +(9k^{(5)} + 9(k^2)^{(3)} + 55kk^{(3)} + 111k'k'' + 219k^2k' - 5\lambda k')f' \\
& +(k^{(6)} + 5(k^2)^{(4)} + 2(kk'')'' + 11kk^{(4)} + 14k'k^{(3)} + 60k'(k^2)' + 7(k'')^2 \\
& +51k''k^2 + 35k(k^2)'' + 36k^4 - \lambda k'' - 4\lambda k^2)f]ds.
\end{aligned}$$

For those constant affine curvature elastica, we know $\lambda = 3k^2$, the above formula can be written as

$$\begin{aligned}&\tfrac{d^2}{dw^2}\int_0^{L(w)}(k^2+\lambda)ds|_{w=0}\\&=\tfrac{2}{9}\int_0^L f[f^{(8)}+12kf^{(6)}+45k^2f^{(4)}+58k^3f''+24k^4f]ds.\end{aligned} \tag{4.30}$$

We transfer these initial condition to f and get $f(0) = f(L) = f'(0) = f'(L) = f''(0) = f''(L) = f^{(3)}(0) = f^{(3)}(L) = 0$.

Definition 18 *Let $(x(s), y(s))$ be a constant affine curvature elastica. s_0 is the first positive forth order zero of the nontrivial solution of the equation*

$$\begin{aligned}&f^{(8)}+12kf^{(6)}+45k^2f^{(4)}+58k^3f_{ss}+24fk^4=0,\\&f(0)=0,\ f'(0)=0,\ f''(0)=0,\ f^{(3)}(0)=0.\end{aligned}$$

s_0 is called the first conjugate point along the elastica with respect to 0.

For the case $k = 0$, from the formula

$$\frac{d^2}{dw^2}\int_0^{L(w)}(k^2+\lambda)ds|_{w=0}=\frac{2}{9}\int_0^L(f^{(4)})^2ds\geq 0.$$

Since there is no third degree polynomial with eight zeroes, the above equality holds if and only if f is zero. Therefore we know this elastica is stable. When $k < 0$, we may assume $k = -1$, for convenience, exactly we only need to multiply s by $\sqrt{|k|}$. Since $f(0) = f(L) = 0$, we can extend it to be a periodic function with period L. We have Fourier series of the form

$$f(s)=\frac{a_0}{2}+\sum_{n=1}^{\infty}[a_n\cos(\frac{2n\pi}{L}s)+b_n\sin(\frac{2n\pi}{L}s)].$$

Computing f'', we obtain the formula

$$\begin{aligned}&\tfrac{d^2}{dw^2}\int_0^{L(w)}(k^2+\lambda)ds|_{w=0}\\&=L[12a_0^2+\textstyle\sum_{n=1}^{\infty}((\frac{2n\pi}{L})^2+4)((\frac{2n\pi}{L})^2+6)((\frac{2n\pi}{L})^2+1)^2(a_n^2+b_n^2)]/9.\end{aligned}$$

Therefore, we have the stability for $k < 0$. For $k > 0$, Similarly, we assume $k = 1$, we consider the linear ordinary differential equation

$$f^{(8)}+12f^{(6)}+45f^{(4)}+58f^{(2)}+24f=0.$$

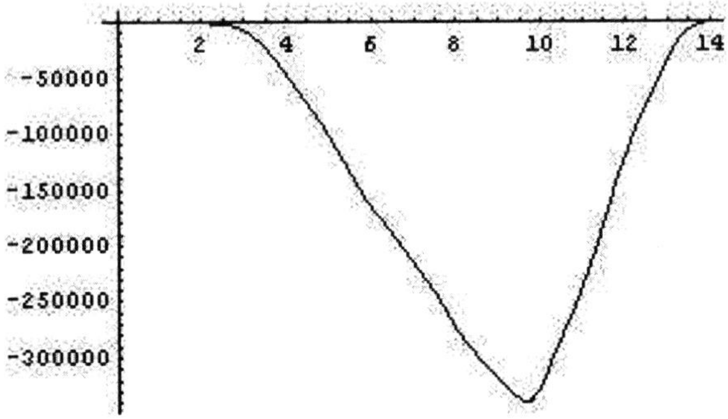

Figure 4.1: Conjugate point for affine elastica in $\mathbf{R}^2$

With the boundary condition $f(0) = f(L) = f'(0) = f'(L) = f''(0) = f''(L) = f^{(3)}(0) = f^{(3)}(L) = 0$. The general solution of this equation is

$$(c_1 + c_2 s)\cos s + (c_3 + c_4 s)\sin s + c_5 \cos 2s + c_6 \sin 2s + c_7 \cos \sqrt{6}s + c_8 \sin \sqrt{6}s.$$

Then the boundary condition is a system of linear equations of c_i. The condition for the existence of nonzero solutions is the determinant of the coefficient matrix is zero. The first positive zero $L(0)$ of this determinant is near $L_0 = 14.07$(see Figure 4.1). By the same method we used in the constant affine elastica, we have the theorem.

Theorem 18 *The constant affine curvature elastica* $(x(s), y(s))$, $s \in [0, L]$ *is stable if L is less than the first conjugate point. If* $k \leq 0$, *the first conjugate point is* ∞. *If* $k > 0$, *the first conjugate point is beyond two period of the elastica, the ellipse.*

4.3 Critical curves for $p(k) = k+\lambda$ in the affine plane $\mathbf{R}^2$

We consider the case for $p(k) = k + \lambda$, λ is the Lagrange multiplier. The curvature energy functional(4.15) becomes $\int_0^L (k + \lambda)ds$. At this time, the

Euler-Lagrange equation (4.16) becomes

$$k'' + k^2 - \lambda k = 0. \tag{4.31}$$

For k is constant, we know $k = \lambda$. When k is not a constant, multiplying equation(4.31) by k' and integrating it, we obtain the first integral of the Euler-Lagrange equation

$$\frac{3}{2}(k')^2 + k^3 - \frac{3\lambda}{2}k^2 = c. \tag{4.32}$$

c is the integral constant. Set $s_1 = \sqrt{2/3}s$, then

$$\frac{dk}{ds_1} = \frac{dk}{ds}\frac{ds}{ds_1} = \frac{\sqrt{3}}{\sqrt{2}}\frac{dk}{ds}.$$

Thus

$$(\frac{dk}{ds_1})^2 = \frac{3}{2}(\frac{dk}{ds})^2 = -k^3 + \frac{3\lambda}{2}k^2 + c. \tag{4.33}$$

For convenience, we still write $\frac{dk}{ds_1} = k'$. So we have

$$(k')^2 = -k^3 + \frac{3\lambda}{2}k^2 + c. \tag{4.34}$$

We write

$$P(k) = -k^3 + \frac{3\lambda}{2}k^2 + c. \tag{4.35}$$

Since $\lim_{k\to\pm\infty} P(k) = \mp\infty$, the cubic polynomial $P(k)$ has at least one real root.

Let $\alpha_1, \alpha_2, \alpha_3$ be the three roots of $P(k)$, a direct calculation gives

$$(\alpha_1 - \alpha_2)^2(\alpha_2 - \alpha_3)^2(\alpha_3 - \alpha_1)^2 = -\frac{27}{2}c(\lambda^3 + 2c).$$

We assume two roots of $P(k)$ are imaginary, say

$$\alpha_2 = a + bi, \qquad \alpha_3 = a - bi,$$

$b \neq 0$. Then

$$(\alpha_1 - \alpha_2)^2(\alpha_2 - \alpha_3)^2(\alpha_3 - \alpha_1)^2 = -4b^2[(a - \alpha_1)^2 + b^2]^2.$$

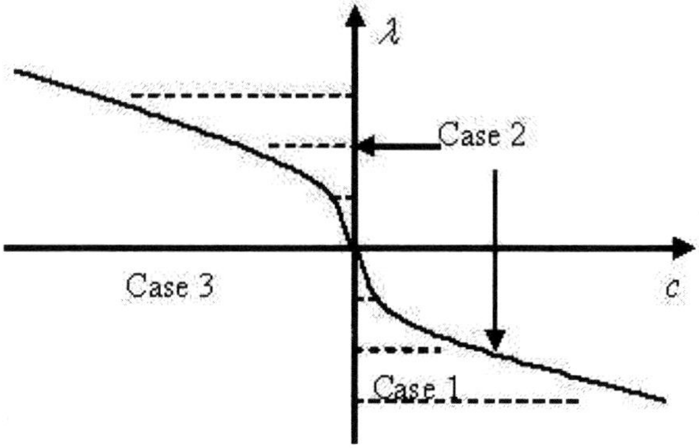

Figure 4.2: Root property on (c, λ) plane

Hence we have three cases (see Figure 4.2):
Case 1. $c(\lambda^3 + 2c) < 0$, the roots are real and distinct.
Case 2. $c(\lambda^3 + 2c) = 0$, the roots are real and at least two of them are equal.
Case 3. $c(\lambda^3 + 2c) > 0$, two roots are imaginary.

The roots of $P(k) = 0$ have the following relations.

$$\begin{aligned} \alpha_1 + \alpha_2 + \alpha_3 &= \tfrac{3}{2}\lambda, \\ \alpha_1\alpha_2 + \alpha_2\alpha_3 + \alpha_3\alpha_1 &= 0, \\ \alpha_1\alpha_2\alpha_3 &= c. \end{aligned} \tag{4.36}$$

For those real roots, we can find curves with these constant affine curvature. Now we are going to solve the the nonconstant affine curvature by similar treatment as in chapter 2.
Case 1. $c(\lambda^3 + 2c) < 0$, the roots $\alpha_1, \alpha_2, \alpha_3$ are real and distinct. We may assume $\alpha_1 < \alpha_2 < \alpha_3$. Since $\alpha_1\alpha_2 + \alpha_2\alpha_3 + \alpha_3\alpha_1 = 0$, we have $\alpha_1 < 0, \alpha_3 > 0$.

At this time, the differential equation

$$(k')^2 = -k^3 + \frac{3\lambda}{2}k^2 + c.$$

has two solutions satisfying $k \in [\alpha_2, \alpha_3]$ and $k \leq \alpha_1$, respectively.

For the stable solution $k \in [\alpha_2, \alpha_3]$, it can be written by the Jacobi elliptic sine function

$$k(s_1) = \alpha_3 - (\alpha_3 - \alpha_2)\mathrm{sn}^2(ds_1, p). \tag{4.37}$$

where $p^2 = (\alpha_3 - \alpha_2)/(\alpha_3 - \alpha_1)$ and $d = \sqrt{\alpha_3 - \alpha_1}/2$. It is a periodic solution with period $2K(p)/d$.

For the unstable interval $(-\infty, \alpha_1]$, we can obtain a solution

$$k(s_1) = \frac{\alpha_1 - \alpha_2\mathrm{sn}^2(ds_1, p)}{\mathrm{cn}^2(ds_1, p)}. \tag{4.38}$$

We have $k(s_1) \leq \alpha_1$ and $s_1 \in (-K(p)/d,\ K(p)/d)$. This implies $L < 2K(p)/d$. As $s_1 \to \pm K(p)/d$, $k(s_1) \to -\infty$.

Case 2. $c(\lambda^3 + 2c) = 0$, the roots are real and at least two of them are equal.

(1). $\alpha_1 < \alpha_2 = \alpha_3$.

At this time, equation (4.36) has the form

$$\alpha_1 + 2\alpha_2 = \frac{3}{2}\lambda, \quad \alpha_2(2\alpha_1 + \alpha_2) = 0, \quad \alpha_1\alpha_2^2 = c.$$

We divide it into two subcases.

(a). $c = 0$. This implies $\alpha_2 = 0$ and $\alpha_1 = \frac{3}{2}\lambda < 0$. The general solution is solution (2.51). In this case, we have the solution

$$k(s_1) = \frac{3}{2}\lambda \sec^2(\frac{\sqrt{-6\lambda}}{4}s_1). \tag{4.39}$$

(b). $c \neq 0$. This implies $2\alpha_1 + \alpha_2 = 0$ and $\alpha_1 = -\lambda/2$, $\alpha_2 = \alpha_3 = \lambda > 0$, $c = -\lambda^3/2 < 0$ and $k(s) \leq -\lambda/2$. Solution (2.51) gives the result

$$k(s_1) = \lambda - \frac{3}{2}\lambda \sec^2(\frac{\sqrt{6\lambda}}{4}s_1). \tag{4.40}$$

(2). $\alpha_1 = \alpha_2 < \alpha_3$.

At this time, equation (4.36) has the form

$$2\alpha_2 + \alpha_3 = \frac{3}{2}\lambda, \quad \alpha_2(\alpha_2 + 2\alpha_3) = 0, \quad \alpha_2^2\alpha_3 = c.$$

We divide it into two subcases.

(c). $c = 0$. This implies $\alpha_2 = 0$ and $\alpha_3 = \frac{3}{2}\lambda > 0$. The bounded solution is solution (2.53). In this case, we have the solution which has value in $[0, \frac{3}{2}\lambda]$

$$k(s_1) = \frac{3}{2}\lambda \mathrm{sech}^2(\frac{\sqrt{6\lambda}}{4}s_1). \tag{4.41}$$

We may assume $k(0) = -\alpha_3$. Similar to solution (2.55), we obtain another solution which has value in $(-\infty, 0]$

$$k(s_1) = -\frac{3}{2}\lambda \mathrm{csch}^2(\frac{\sqrt{6\lambda}}{4}s_1 + \ln(\sqrt{2} - 1)). \tag{4.42}$$

(d). $c \neq 0$. This implies $\alpha_2 + 2\alpha_3 = 0$ and $\alpha_3 = -\lambda/2$, $\alpha_1 = \alpha_2 = \lambda < 0$, $c = -\lambda^3/2 > 0$. Solution (2.53) gives the result which has value in $[\lambda, -\lambda/2]$

$$k(s_1) = \lambda - \frac{3}{2}\lambda \mathrm{sech}^2(\frac{\sqrt{-6\lambda}}{4}s_1). \tag{4.43}$$

We may assume $k(0) = 2\alpha_1$. Similar to solution (2.55), we obtain another solution which has value in $(-\infty, \lambda]$

$$k(s_1) = \lambda + \frac{3}{2}\lambda \mathrm{csch}^2(\frac{\sqrt{-6\lambda}}{4}s_1 + \ln\frac{\sqrt{10} - \sqrt{6}}{2}). \tag{4.44}$$

(3). $\alpha_1 = \alpha_2 = \alpha_3$.

At this time, $\lambda = 0, c = 0$, then $\alpha_1 = \alpha_2 = \alpha_3 = 0$. We may assume $k(0) = -1$. The solution is same as solution (2.57)

$$k(s_1) = -\frac{4}{(s_1 - 2)^2}, \qquad -\infty < s_1 < 2. \tag{4.45}$$

Case 3. $4\lambda^3 - 27c^2 < 0$, two roots are imaginary. The result is same as solution (2.58)

$$k(s_1) = \alpha_1 - d\frac{(1 - \mathrm{cn}(\sqrt{d}s_1, p))^2}{\mathrm{sn}^2(\sqrt{d}s_1, p)}, \tag{4.46}$$

for $d = \sqrt{(\alpha_1 - a)^2 + b^2}$ and $p = \sqrt{\frac{d - a + \alpha_1}{2d}}$.

Therefore we can express the centroaffine curvature function $k(s)$ of the centroaffine elastica by the Jacobi Elliptic functions completely.

For the closed critical curve X of the energy functional $\int_0^L (k + \lambda)ds$, Euler-Lagrange equation (4.31) gives

$$\oint_X k^2 ds = \oint_X (k'' + k^2)ds = \lambda \oint_X k ds.$$

Equality (4.29) and integrating by parts imply

$$\oint_X kds = \oint_X k(\psi'' + k\psi)ds = \oint_X (k'' + k^2)\psi ds$$
$$= \lambda \oint_X k\psi ds = \lambda \oint_X (1 - \psi'')ds = \lambda \oint_X ds.$$

These two equalities yield

$$\oint_X (k - \lambda)^2 ds = \oint_X k^2 ds - 2\oint_X \lambda k ds + \oint_X \lambda^2 ds = 0.$$

This implies $k = \lambda$. The property of closed curves implies $\lambda > 0$ and this curve is an ellipse. This means solution (4.37) can not produce closed critical curves. Therefore we have following theorem:

Theorem 19 *The only closed critical curve of the energy functional* $\int_0^L (k + \lambda)ds$ *is the ellipse.*

The newest information about the variations of the equiaffine curve can be founded in Ref.[40]. Furthermore the equiaffine curve theory has applications in human body movements[11]. Even if the Lagrange $p(k)$ is a nice function, it is difficult to solve the Euler-Lagrange equation (4.16) and the structure equation (4.3). For global properties of the closed critical curves, there has been much research by using the geometric heat equation([2], [36]). In particular, affine curvature was found to be highly significant in the study of curve evolution based on the use of invariant heat flow type diffusion equations.

Chapter 5

Affine elastica in the affine space $\mathbf{R}^3$

This chapter is devoted to the study of the variational properties of space curves invariant under the group of volume-preserving affine transformations or equiaffine transformations in $\mathbf{R}^3$. The group is generated by the action of the unimodular linear group $SL(3,\mathbf{R})$ and the translation group $\mathbf{R}^3$. We expect to characterize the equiaffine properties of critical space curves with respect to certain energy functionals by the affine arclength parameter, affine curvature and affine torsion.

5.1 Definition and variation formula of the affine elastica in the affine space $\mathbf{R}^3$

We consider curves in the affine space $\mathbf{R}^3$. We denote (x, y, z) a coordinate system of the affine space $\mathbf{R}^3$ and $X : I \to \mathbf{R}^3$ be a smooth curve. We write it as a column vector

$$X(t) = (x(t), y(t), z(t))^T, \qquad x \in I,$$

where $x(t), y(t)$ and $z(t)$ are smooth functions of t defined on a certain interval I. We write the tangent vector as $X'(t) = dX/dt = (x'(t), y'(t), z'(t))^T$ or simply X', similarly $(X'(t), X''(t), X^{(3)}(t))$ be the 3×3 matrix

$$\begin{pmatrix} x'(t) & x''(t) & x^{(3)}(t) \\ y'(t) & y''(t) & y^{(3)}(t) \\ z'(t) & z''(t) & z^{(3)}(t) \end{pmatrix}$$

Definition 19 *A regular curve* $X : I \to \mathbf{R}^3$ *is called nondegenerate if*

$$|X'(t), X''(t), X^{(3)}(t)| = \det(X'(t), X''(t), X^{(3)}(t)) \neq 0$$

for any $t \in I$. *A parameter* s *is called an affine arclength parameter of a nondegenerate curves* X *if*

$$A(s) = (X'(s), X''(s), X^{(3)}(s)) \in SL(3, \mathbf{R})$$

that is $|X'(s), X''(s), X^{(3)}(s)| = 1$, *for any* $s \in I$.

We can show that a nondegenerate curve $X(t)$ can admit an affine arclength parameter uniquely up to a constant.

Definition 20 *Let* s *be an affine arclength parameter of a nondegenerate curve* $X : I = [0, L] \to \mathbf{R}^3$, *here* L *is the affine arclength of* X *defined by (5.4) below. We define the affine curvature and affine torsion of the curve* X *by*

$$\begin{aligned} k(s) = & \ |X'(s), X^{(3)}(s), X^{(4)}(s)|, \\ \tau(s) = & \ |X''(s), X^{(3)}(s), X^{(4)}(s)|. \end{aligned} \tag{5.1}$$

Differentiating the equality $|X'(s), X''(s), X^{(3)}(s)| = 1$, we obtain $|X'(s), X''(s), X^{(4)}(s)| = 0$. This means $X'(s), X''(s), X^{(4)}(s)$ are linearly dependent. Since $X'(s), X''(s)$ are linearly independent, and from the definition of k, τ, we have the expression

$$X^{(4)}(s) = \tau(s)X'(s) - k(s)X''(s). \tag{5.2}$$

Let t be an arbitrary parameter of X. Same as done in chapter 2, we have the formula of the velocity

$$\frac{ds}{dt} = |X'(t), X''(t), X^{(3)}(t)|^{\frac{1}{6}}. \tag{5.3}$$

Hence we obtain the affine arclength formula of a segment of the curve X between $X(t_0)$ and $X(t)$ for any parameter t.

$$s(t) = \int_{t_0}^{t} |X'(t), X''(t), X^{(3)}(t)|^{\frac{1}{6}} dt. \tag{5.4}$$

This can be used as an affine arclength parameter. If t is also an affine arclength parameter of the curve X, then $\frac{ds}{dt} = 1$. That means $s = t + c$, here c is a constant. We have proven the property below definition 5.1.

Similar to the case of $\mathbf{R}^2$, we denote ∇ the usual flat, torsion-free affine connection of $\mathbf{R}^3$. Then we have the same structural equation. We denote $X_w(t)$ a variation $X(w,t) : (-\varepsilon, \varepsilon) \times I \to \mathbf{R}^3$ with $X(0,t) = X(t)$ and we have the similar notation.

$$\begin{aligned} W(\tfrac{ds}{dt}) &= \tfrac{1}{6}|X'(t), X''(t), X^{(3)}(t)|^{-\frac{5}{6}} W(|X'(t), X''(t), X^{(3)}(t)|) \\ &= -g\tfrac{ds}{dt}. \end{aligned}$$

Here we write $g = -\frac{1}{6}|X'(t), X''(t), X^{(3)}(t)|^{-1} W(|X'(t), X''(t), X^{(3)}(t)|)$ and same as in $\mathbf{R}^2$, we have

$$[W, X'(s)] = gX'(s).$$

But for parameter t, this relation becomes

$$[W, X'(t)] = 0.$$

The torsion-free and flatness property of the connection ∇ gives the relations

$$\nabla_W X'(t) = \nabla_{X'(t)} W, \quad \nabla_W X''(t) = \nabla^2_{X'(t)} W, \quad \nabla_W X^{(3)}(t) = \nabla^3_{X'(t)} W.$$

The main part of g becomes

$$\begin{aligned} &\tfrac{\partial}{\partial w}|X'(t), X''(t), X^{(3)}(t)| \\ &= |\nabla_W X', X'', X^{(3)}| + |X', \nabla_W X'', X^{(3)}| + |X', X'', \nabla_W X^{(3)}| \\ &= |\nabla_{X'} W, X'', X^{(3)}| + |X', \nabla^2_{X'} W, X^{(3)}| + |X', X'', \nabla^3_{X'} W|. \end{aligned}$$

We may assume $X(w,t)$ has a boundary condition such that $W(0,0) = W(0,L) = 0$, $\nabla_{X'} W(0,0) = \nabla_{X'} W(0,L) = 0$, $\nabla^2_{X'} W(0,0) = \nabla^2_{X'} W(0,L) = 0$. Then we obtain the first variation formula of the affine arclength.

$$\begin{aligned} \tfrac{d}{dw}\textstyle\int_I \ & |X', X'', X^{(3)}|^{\frac{1}{6}} dt|_{w=0} \\ &= \tfrac{1}{6}\textstyle\int_0^L [5|X'', W, X^{(4)}| + 2|X', W, X^{(5)}|] ds \\ &= \tfrac{1}{6}\textstyle\int_0^L [(2k' + 3\tau)|W, X', X''| + 2k|W, X', X^{(3)}|] ds. \end{aligned} \tag{5.5}$$

Therefore we have the Euler-Lagrange equation

$$2k' + 3\tau = 0, \qquad k = 0. \tag{5.6}$$

This means along the critical curve

$$k = 0, \qquad \tau = 0. \tag{5.7}$$

and finally the structure equation (5.2) becomes

$$X^{(4)}(s) = 0. \tag{5.8}$$

Integrating this equation, we have

$$X(s) = \frac{1}{6}s^3\mathbf{a} + \frac{1}{2}s^2\mathbf{b} + s\mathbf{c} + \mathbf{d}, \tag{5.9}$$

where $\mathbf{a}, \mathbf{b}, \mathbf{c}$ and $\mathbf{d}$ are constant vectors such that $|\mathbf{c}, \mathbf{b}, \mathbf{a}| = 1$. We change the coordinate system (x, y, z) such that

$$\mathbf{c} = (1,0,0)^T, \mathbf{b} = (0,1,0)^T, \mathbf{a} = (0,0,1)^T, \mathbf{d} = (0,0,0)^T.$$

We see that the equation of the curve is

$$x = s, y = \frac{1}{2}s^2, z = \frac{1}{6}s^3. \tag{5.10}$$

Similar to the straight line in Euclidean space, this is the simplest affine curve in the affine space $\mathbf{R}^3$. Since all quadric curves are plane curves, they are excluded from the three-dimensional affine curve theory.

We consider the second variation with the variational vector field $W = f(s)X''(s) + h(s)X^{(3)}(s)$. At $w = 0$, we have

$$\begin{aligned}
\nabla_{X'}W =& \; f'X'' + (f + h')X^{(3)} + hX^{(4)}, \\
\nabla^2_{X'}W =& \; f''X'' + (2f' + h'')X^{(3)} + (f + 2h')X^{(4)} + hX^{(5)}, \\
\nabla^3_{X'}W =& \; f^{(3)}X'' + (3f'' + h^{(3)})X^{(3)} + 3(f' + h'')X^{(4)} \\
& + (f + 3h')X^{(5)} + hX^{(6)}, \\
\nabla^4_{X'}W =& \; f^{(4)}X'' + (4f^{(3)} + h^{(4)})X^{(3)} + (6f'' + 4h^{(3)})X^{(4)} \\
& + (4f' + 6h'')X^{(5)} + (f + 4h')X^{(6)} + hX^{(7)}.
\end{aligned}$$

$$\begin{aligned}
g =& -\tfrac{1}{6}[|\nabla_{X'}W, X'', X^{(3)}| + |X', \nabla^2_{X'}W, X^{(3)}| + |X', X'', \nabla^3_{X'}W|] \\
=& -\tfrac{1}{6}[4f'' + h^{(3)} - (2f + 5h')k + 3h(\tau - k')].
\end{aligned}$$

We do one more derivative on the first variation formula and obtain

$$\frac{d^2}{dw^2}L(w)|_{w=0} = \frac{1}{6}\int_0^L [W(2k' + 3\tau)|W, X', X''| + 2W(k)|W, X', X^{(3)}|]ds$$

Now we only need to calculate $W(k)$, $W(k')$ and $W(\tau)$. Since

$$\begin{aligned}
\nabla_W X'(s) =& \; \nabla_{X'(s)}W + gX'(s) \\
\nabla_W X''(s) =& \; \nabla^2_{X'(s)}W + g_sX'(s) + 2gX''(s) \\
\nabla_W X^{(3)}(s) =& \; \nabla^3_{X'(s)}W + g_{ss}X'(s) + 3g_sX''(s) + 3gX^{(3)}(s) \\
\nabla_W X^{(4)}(s) =& \; \nabla^4_{X'(s)}W + g_{sss}X'(s) + 4g_{ss}X''(s) + 6g_sX^{(3)}(s) + 4gX^{(4)}(s).
\end{aligned}$$

We have the formula

$$\begin{aligned} W(k) = & \ |\nabla_{X'(s)}W, X^{(3)}(s), X^{(4)}(s)| + |X'(s), \nabla^3_{X'(s)}W, X^{(4)}| \\ & +|X'(s), X^{(3)}(s), \nabla^4_{X'(s)}W| - 4g_{ss} + 8gk \\ = & \ \tfrac{1}{3}(5f^{(4)} + 2h^{(5)}), \end{aligned} \tag{5.11}$$

$$\begin{aligned} W(\tau) = & \ |\nabla^2_{X'(s)}W, X^{(3)}(s), X^{(4)}(s)| + |X''(s), \nabla^3_{X'(s)}W, X^{(4)}| \\ & +|X''(s), X^{(3)}(s), \nabla^4_{X'(s)}W| + g_{sss} + 9g\tau + g_s k \\ = & \ -\tfrac{1}{6}(4f^{(5)} + h^{(6)}), \end{aligned} \tag{5.12}$$

$$W(k') = X'W(k) + gX'(k) = \frac{1}{3}(5f^{(5)} + 2h^{(6)}).$$

We transfer the initial condition to f and h and know

$$\begin{aligned} f(0) = & \ f(L) = f'(0) = f'(L) = f''(0) = f''(L), \\ h(0) = & \ h(L) = h'(0) = h'(L) = h''(0) = h''(L). \end{aligned}$$

Therefore we have the second variational formula of the total affine arclength

$$\begin{aligned} \tfrac{d^2}{dw^2}L(w)|_{w=0} = & \ -\tfrac{1}{36}\int_0^L [5(h^{(3)})^2 + 16h^{(3)}f^{(2)} + 20(f'')^2]ds \\ = & \ -\tfrac{1}{36}\int_0^L [5(h^{(3)} + \tfrac{8}{5}f^{(2)})^2 + \tfrac{36}{5}(f'')^2]ds \le 0. \end{aligned} \tag{5.13}$$

The above equality holds if and only if $f''(s) = 0$ and $h^{(3)}(s) = 0$. Since there is no second order or third order polynomial having eight zeros, we know the solution is stable. But similar to dimension 2, it obtains the maximum affine arclength.

5.2 Variation of the energy functional $\int_X (k^2 + \lambda)ds$ in the affine space $\mathbf{R}^3$

Now we consider the energy functional

$$\int_0^L (k^2 + \lambda)ds.$$

Here s is the affine arclength parameter of the curve $X_w(t) = X(w, t)$. Since we know

$$\begin{aligned} W(k) = & \ |\nabla_{X'(s)}W, X^{(3)}(s), X^{(4)}(s)| + |X'(s), \nabla^3_{X'(s)}W, X^{(4)}(s)| \\ & +|X'(s), X^{(3)}(s), \nabla^4_{X'(s)}W| + 8gk - 4g_{ss}. \end{aligned}$$

When we refine on the curve $X(t) = X(0,t)$, since $s = t$, we will drop t, s in $X'(t), X'(s)$ respectively. We consider the first variation of the energy functional $\int_X (k^2 + \lambda)ds$

$$
\begin{aligned}
& \tfrac{d}{dw} \int_0^L (k^2 + \lambda)ds|_{w=0} \\
= \; & \int_0^L [2k(|\nabla_{X'}W, X^{(3)}, X^{(4)}| + |X', \nabla^3_{X'}W, X^{(4)}| + |X', X^{(3)}, \nabla^4_{X'}W| \\
& +8gk - 4g_{ss}) - (k^2 + \lambda)g]ds.
\end{aligned}
$$

Now we give $X(w,t)$ a boundary condition such that $W(0,0) = W(0,L) = 0$, $\nabla_{X'}W(0,0) = \nabla_{X'}W(0,L) = 0$, $\nabla^2_{X'}W(0,0) = \nabla^2_{X'}W(0,L) = 0$, $\nabla^3_{X'}W(0,0) = \nabla^3_{X'}W(0,L) = 0$, $\nabla^4_{X'}W(0,0) = \nabla^4_{X'}W(0,L) = 0$. We calculate the integral part by part.

$$
\begin{aligned}
\int_0^L \; & 2k(|\nabla_{X'}W, X^{(3)}, X^{(4)}| + |X', \nabla^3_{X'}W, X^{(4)}| + |X', X^{(3)}, \nabla^4_{X'}W|)ds \\
= \; & 2\int_0^L [|\nabla_{X'}W, kX^{(3)}, X^{(4)}| - |\nabla^3_{X'}W, kX', X^{(4)}| + |\nabla^4_{X'}W, kX', X^{(3)}|]ds \\
= \; & 2\int_0^L [|\nabla_{X'}W, kX^{(3)}, X^{(4)}| - 2|\nabla^3_{X'}W, kX', X^{(4)}| \\
& -|\nabla^3_{X'}W, k'X', X^{(3)}| - |\nabla^3_{X'}W, kX'', X^{(3)}|]ds \\
= \; & 2\int_0^L [|\nabla_{X'}W, kX^{(3)}, X^{(4)}| + 3|\nabla^2_{X'}W, k'X', X^{(4)}| + 3|\nabla_{X'}W, kX'', X^{(4)}| \\
& +2|\nabla^2_{X'}W, kX', X^{(5)}| + 2|\nabla^2_{X'}W, k'X'', X^{(3)}| + |\nabla^2_{X'}W, k''X', X^{(3)}|]ds \\
= \; & -2\int_0^L [2|\nabla_{X'}W, kX^{(3)}, X^{(4)}| + 4|\nabla_{X'}W, k''X', X^{(4)}| \\
& +8|\nabla_{X'}W, kX'', X^{(4)}| + 5|\nabla_{X'}W, k'X', X^{(5)}| \\
& +5|\nabla_{X'}W, kX'', X^{(5)}| + 3|\nabla_{X'}W, k''X'', X^{(3)}| \\
& +2|\nabla_{X'}W, kX', X^{(6)}| + |\nabla_{X'}W, k^{(3)}X', X^{(3)}]ds \\
= \; & 2\int_0^L [10|W, k'X^{(3)}, X^{(4)}| + 7|W, kX^{(3)}, X^{(5)}| + 5|W, k^{(3)}X', X^{(4)}| \\
& +15|W, k''X'', X^{(4)}| + 9|W, k''X', X^{(5)}| + 18|W, k'X'', X^{(5)} \\
& +7|W, k'X', X^{(6)}| + 7|W, kX'', X^{(6)}| + 4|W, k^{(3)}X'', X^{(3)}| \\
& +2|W, kX', X^{(7)}| + |W, k^{(4)}X', X^{(3)}|]ds.
\end{aligned}
$$

From the definition of k, τ, we know

$$
\begin{aligned}
X^{(4)} = \; & \tau X' - kX'' \\
X^{(5)} = \; & \tau' X' + (\tau - k')X'' - kX^{(3)}, \\
X^{(6)} = \; & (\tau'' - k\tau)X' + (2\tau' - k'' + k^2)X'' + (\tau - 2k')X^{(3)} \\
X^{(7)} = \; & (\tau^{(3)} + \tau^2 - \tau' k - 3\tau k')X' + (3\tau'' - 2k\tau - k^{(3)} + 4kk')X'' \\
& +(3\tau' - 3k'' + k^2)X^{(3)}.
\end{aligned}
$$

Therefore we have

$$
\begin{aligned}
& \int_0^L k(|\nabla^2_{X'}W, X^{(3)}, X^{(4)}| + |X'', \nabla^3_{X'}W, X^{(4)}| + |X'', X^{(3)}, \nabla^4_{X'}W|)ds \\
= \; & \int_0^L [(4k^{(3)} - 15kk')|W, X'', X^{(3)}| \\
& +(-7kk^{(3)} - 6\tau k'' - 16k'k'' - 4k'\tau' + 15k^2k' - \tau'' k + 3\tau k^2)|W, X', X''| \\
& +(3\tau k' + k\tau' + 15kk'' + 14(k')^2 - 2k^3 - k^{(4)})|W, X^{(3)}, X'|]ds,
\end{aligned}
$$

$$\int_0^L [2k(8gk-4g'')-(k^2+\lambda)g]ds = -8g'k|_0^L + \int_0^L [8g'k' + (15k^2-\lambda)]ds$$
$$= 8gk'|_0^L + \int_0^L (-8k''+15k^2-\lambda)gds.$$

We denote $h = 15k^2 - 8k'' - \lambda$, then we have

$$\begin{aligned}
&\int_0^L (15k^2-8k''-\lambda)gds \\
= &-\tfrac{1}{6}\int_0^L h(|\nabla_{X'}W, X'', X^{(3)}| + |X', \nabla^2_{X'}W, X^{(3)}| + |X', X'', \nabla^3_{X'}W|)ds \\
= &-\tfrac{1}{6}\int_0^L (|\nabla_{X'}W, hX'', X^{(3)}| - |\nabla^2_{X'}W, hX', X^{(3)}| + |\nabla^3_{X'}W, hX', X''|)ds \\
= &-\tfrac{1}{6}\int_0^L [|\nabla_{X'}W, hX'', X^{(3)}| - 2|\nabla^2_{X'}W, hX', X^{(3)}| \\
&-|\nabla^2_{X'}W, h'X', X''|)ds \\
= &-\tfrac{1}{6}\int_0^L [3|\nabla_{X'}W, hX'', X^{(3)}| + 3|\nabla_{X'}W, h'X', X^{(3)}| \\
&+2|\nabla_{X'}W, hX', X^{(4)}| + |\nabla_{X'}W, h''X', X''|)ds \\
= &\tfrac{1}{6}\int_0^L [6|W, h'X'', X^{(3)}| + 5|W, hX'', X^{(4)}| + 4|W, h''X', X^{(3)}| \\
&+5|W, h'X', X^{(4)}| + 2|W, hX', X^{(5)}| + |W, h^{(3)}X', X''|)ds \\
= &\tfrac{1}{6}\int_0^L [(h^{(3)} - 5h'k - 3\tau h - 2hk')|W, X', X''| \\
&+6h'|W, X'', X^{(3)}| + (2hk - 4h'')|W, X^{(3)}, X'|]ds.
\end{aligned}$$

Since

$$\begin{aligned}
h' &= 30kk' - 8k^{(3)} \\
h'' &= 30(k')^2 + 30kk'' - 8k^{(4)} \\
h^{(3)} &= 90k'k'' + 30kk^{(3)} - 8k^{(5)}.
\end{aligned}$$

We have

$$\begin{aligned}
&\tfrac{d}{dw}\int_0^L (k^2+\lambda)ds|_{w=0} = \tfrac{1}{6}\int_0^L [(-8k^{(5)} - 14kk^{(3)} - 86k'k'' - 9\tau k^2 - 48\tau k'' \\
&-48k'\tau' - 12k\tau'' + (2k' + 3\tau)\lambda)|W, X', X''| \\
&+(6k^3 + 48(k')^2 + 44kk'' + 20k^{(4)} + 36\tau k' + 12k\tau' - 2\lambda k)|W, X', X''|]ds.
\end{aligned}$$

Then the Euler-Lagrange equation is

$$\begin{aligned}
&8k^{(5)} + 14kk^{(3)} + 86k'k'' + 9k^2\tau \\
&+48\tau k'' + 48k'\tau' + 12k\tau'' - (2k' + 3\tau)\lambda = 0, \\
&6k^3 + 48(k')^2 + 44kk'' + 20k^{(4)} + 36k'\tau + 12k\tau' - 2\lambda k = 0.
\end{aligned} \tag{5.14}$$

We subtract the derivative of the second equation from the first equation. Then the Euler equation becomes the following form

$$\begin{aligned}
-12k^{(5)} - 30kk^{(3)} - 54k'k'' + 9k^2\tau + 12k''\tau - 3\tau\lambda &= 0, \\
6k^3 + 48(k')^2 + 44kk'' + 20k^{(4)} + 36k'\tau + 12k\tau' - 2\lambda k &= 0,
\end{aligned} \tag{5.15}$$

These equation can be written as

$$\begin{aligned}&(4k''+3k^2-\lambda)^{(3)}+k(4k''+3k^2-\lambda)'-\tau(4k''+3k^2-\lambda)=0,\\&10(4k''+3k^2-\lambda)''+7(4k''+3k^2-\lambda)k\\&+3(12(k')^2+24k'\tau+8k\tau'-3k^3+\lambda k)=0.\end{aligned}\tag{5.16}$$

We consider the following equation whose solutions are also solutions of the above equation.

$$\begin{aligned}12(k')^2+24k'\tau+8k\tau'-3k^3+\lambda k=&\ 0,\\4k''+3k^2-\lambda=&\ 0.\end{aligned}\tag{5.17}$$

The second equation $4k''+3k^2-\lambda=0$ can be solved completely by the Jacobi elliptic functions as in chapter 2. The first integral $\tau=\frac{c}{k^3}-\frac{1}{2}k'$ gives the affine torsion. Then we can solve $X'(s)$ from the linear equation

$$X^{(4)}(s)+k(s)X''(s)-\tau(s)X'(s)=0,\tag{5.18}$$

as in chapter 3. One more integration, we obtain $X(s)$, Therefore we partially solve the critical curve.

From the above analysis, even if the Lagrange is a nice function, it is still difficult to solve the motion equations. In chapter 4, we completely solve the Euler-Lagrange equation by Jacobi elliptic functions for the energy functional which is the integral of a linear function of the affine curvature. But in three-dimensional case, the Euler-lagrange equation for the energy functional $p(k)=k+\lambda$ is

$$\begin{aligned}&k^{(3)}+kk'-(k-\lambda)\tau=0,\\&4k''+k^2-\lambda k+3\tau'=0.\end{aligned}\tag{5.19}$$

It seems to be difficult to solve this system completely.

At last, we list all affine curves with constant affine curvature and affine torsion to finish this chapter. This could be done by the similar work in chapter 3.

The characteristic equation of differential equation (5.18) for $X'(s)$ is

$$r^3+kr-\tau=0.\tag{5.20}$$

We divide it into the following cases.

I. $\tau=0$. At least one characteristic root is zero.

(1). $k = 0$. The curve is the twisted cubic (5.9).

$$X(s) = \frac{1}{6}s^3\mathbf{a} + \frac{1}{2}s^2\mathbf{b} + s\mathbf{c} + \mathbf{d}, \tag{5.21}$$

where $\mathbf{a}, \mathbf{b}, \mathbf{c}$ and $\mathbf{d}$ are constant vectors such that $|\mathbf{c}, \mathbf{b}, \mathbf{a}| = 1$.

(2). $k = a^2 > 0$.

$$X(s) = \cos(as)\mathbf{a} + \sin(as)\mathbf{b} + \mathbf{c}s + \mathbf{d}, \tag{5.22}$$

where $\mathbf{a}$, $\mathbf{b}$, $\mathbf{c}$ and $\mathbf{d}$ are constant vectors such that $|\mathbf{a}, \mathbf{b}, \mathbf{c}| = -1/a^5$.

(3). $k = -a^2 < 0$. A particular solution is

$$X(s) = (\frac{1}{a^2}\cosh(as), \frac{1}{a^2}\sinh(as), \frac{1}{a}s)^T. \tag{5.23}$$

II. $\tau \neq 0$.

(4). If the characteristic equation (5.20) has three distinct real roots α_1, α_2 and α_3, then $\alpha_1 + \alpha_2 + \alpha_3 = 0$ and $-4k^3 - 27\tau^2 > 0$. A particular solution of the structure equation (5.18) is

$$X(s) = (d_1 e^{\alpha_1 s}, d_2 e^{\alpha_2 s}, d_1 e^{\alpha_3 s})^T \tag{5.24}$$

where d_1, d_2 and d_3 are constants and satisfy $d_1 d_2 d_3 \tau\sqrt{-4k^3 - 27\tau^2} = 1$. This curve is on the cubic surface

$$xyz = d_1 d_2 d_3 = \frac{1}{\tau\sqrt{-4k^3 - 27\tau^2}}.$$

(5). If the characteristic equation (5.20) has a double root $\alpha_1 = \alpha_2 = \alpha$, then $\alpha_3 = -2\alpha$ and $k = -3\alpha^2$, $\tau = -2\alpha^3$. The general solution of the structure equation (5.18) is

$$X(s) = \mathbf{a}e^{\alpha s} + \mathbf{b}se^{\alpha s} + \mathbf{c}e^{-2\alpha s} + \mathbf{d}, \tag{5.25}$$

where $\mathbf{a}$, $\mathbf{b}$, $\mathbf{c}$ and $\mathbf{d}$ are constant vectors such that $3k\tau|\mathbf{a}, \mathbf{b}, \mathbf{c}| = -1$.

(6). If the characteristic equation (3.16) has only one real root α_1 and the other two roots are

$$\alpha_2 = a + bi, \qquad \alpha_3 = a - bi, \qquad b \neq 0.$$

Then $\alpha_1 = -2a$. A particular solution of the structure equation (3.6) is

$$X(s) = (d_1 e^{-2as}, d_2 e^{as}\cos(bs), d_3 e^{as}\sin(bs))^T, \tag{3.27}$$

where d_1, d_2 and d_3 are constants and satisfy $2d_1 d_2 d_3 ab(a^2 + b^2)(9a^2 + b^2) = -1$.

Bibliography

[1] M. Ablowitz and P. Clarkson, *Solitons, nonlinear evolution equations and inverse scattering*, Cambridge University press, 1991.

[2] S. Angenent, G. Sapiro and A. Tannenbaum, On the affine heat equation for non-convex curves, *Journal of the American Mathematical Society*, 11(1998), 601-634.

[3] F. Arscott, *Periodic differential equations*, Pergamon, 1964.

[4] W. Blaschke, *Vorlesungen über Differentialgeometrie. I: Elementare Differentialgeomrtrie.* Springer, Berlin, 1923.

[5] W. Blaschke, *Vorlesungen über Differentialgeometrie. II: Affine Differentialgeomrtrie.* Springer, Berlin, 1923.

[6] G. Boole, *A treatise on differential equations*, G. E. Stechert & co., New York, 1931.

[7] R. Bryant and P. Griffiths, Reduction for constrained variational problem and $\int k^2/2ds$, *American Journal of Mathematics*, 108(1986), 525-570.

[8] P. Byrd and M. Friedman, *Handbok of elliptic integrals for engineers and physicists*, Springer, Berlin, 1954.

[9] P. Drazin and R. Johnson, Solitons: an introduction, Cambridge University Press, 1989.

[10] A. Ferrández, J. Guerrero, M. Javaloyes and P. Lucas, Particles with curvature and torsion in three-dimensional pseudo-Riemannian space forms, *J. of Geometry and Physics*, 56(2006), 1666-1687.

[11] T. Flash and A. Handzel, Affine differential geometry analysis of human arm movements, *Biological. Cybernetics*, 96(2007), 577-601.

[12] C. Fraser, Mathematical technique and physical conception in Euler's invention of the elastica, *Centaurus*, 34(1991), 211-246.

[13] C. Gardner, J. Greene, M. Kruskal and R. Miura, Method for solving the Korteweg-de Vries equation, *Physical Review Letter*, 1967, 1095-1097.

[14] H. Goldstine, *A history of the calculus of variations from the 17th through the 19th century*, Springer-Verlag, New York, 1980.

[15] I. Gradsbteyn and L. Ryzbik, *Table of integrals, series and products*, Elsevier, Singapore, 2004.

[16] H. Guggenheimer, *Differential Geometry*, Dover Publications Inc., New York, 1977.

[17] H. Hasimoto, A soliton on a vortex filament, *J. Fluid Mech.*, 51(3)(1972), 477-486.

[18] R.Huang, *Affine and subaffine elastic Curves in* $\mathbf{R}^2$ *and* $\mathbf{R}^3$, thesis, Case Western Reserve University, 2000.

[19] R. Huang and D. Singer, A new flow on starlike curves in $\mathbf{R}^3$, *Proc. of the American Mathematical Society*, 130(9)(2002), 2725–2735.

[20] R. Huang, Starlike affine elastic curves on $\mathbf{R}^2$, *Acta Math. Scientia*, A23(4)(2003), 410-418.

[21] R. Huang, A note on the generalized centroaffine elastica in $\mathbf{R}^2$, *Chin. Quart. J. of Math.*, 18(1)(2003), 88-92.

[22] R. Huang, A note on the p-elastica in a constant sectional curvature manifold, *J. of Geometry and Physics*, 49(3-4)(2004), 343-349.

[23] M. Kline, *Mathematical thought from ancient to modern times*, Volume 2, Oxford University press, 1972.

[24] G. Lamb, *Elements of Soliton Theory*, Wiley Interscience, New York, 1980

[25] J. Langer and R. Perline, Poisson geometry of the filament equation, *J. Nonlinear Science*, 1(1991), 71-93.

[26] J. Langer and R. Perline, Local geometric invariants of integrable evolution equations, *J. Math. Phys.*, 35(4)1994, 1732-1737.

[27] J. Langer and D. Singer, The total squared curvature of closed curves, *Journal of Differential Geometry*, 20(1984), 1-22.

[28] J. Langer and D. Singer, Curve straightening and a minimax argument for closed elastic curves, *Topology*, 24(1) (1985), 75-88.

[29] J. Langer and D. Singer, Lagrangian aspects of the Kirchhoff elastic rod, *SIAM Review*, 38(4) (1996), 605-618.

[30] A. Linner, Explicit Elastic Curves, *Annals of Global Analysis and Geometry*, 16(1998), 445–475.

[31] A.Love, *A treatise on the mathematical theory of elasticity*, Cambridge University press, 1927.

[32] E. Musso, Variational problems for plane curves in centroaffine geometry, *J. Phys.*, A43(2010), 305206.

[33] K.Nomizu and T.Sasaki, *Affine differential geometry*, Cambridge University press, 1994.

[34] S.Novikov, *Solitons and geometry*, Accademia Nazionale Deilincei, 1992.

[35] U. Pinkall, Hamiltonian flows on the space of star-shaped curves, *Results in Mathematics*, 27(1995), 328–332.

[36] G. Sapiro and A. Tannenbaum, On affine plane curve evolution, *Journal of Functional Analysis*, 119(1994), 79-120.

[37] Y. Shi and J. Hearst, The Kirchhoff elastic rod, the nonlinear Schrodinger equation, and DNA supercoiling, *Journal of Chemical Physics*, 101(6)(1994), 5186–5200.

[38] B. Su, *Affine differential geometry*, Beijing, 1983.

[39] C. Truesdell, The influence of elasticity on analysis: The classic heritage, *Bulletin of the American Mathematical Society*, 9(3)(1983), 293–310.

[40] S. Verpoort, Curvature functionals for curves in the equiaffine plane, *Czech. Math. J.*, 61(2011), 419-435.

Index

◎ 编辑手记

本书是一部英文版的数学专著，中文书名可译为《$\mathbf{R}^2$ 和 $\mathbf{R}^3$ 中的仿射弹性曲线：概念与方法》.

本书作者黄荣培，华东师范大学助教. 2020 年，他在凯斯西储大学获得了博士学位. 他的研究方向是几何学，特别是黎曼几何、积分几何和变分问题.

作者在前言中介绍道：

> 本书致力于研究在 $\mathbf{R}^2$ 和 $\mathbf{R}^3$ 中的中心仿射变换或等仿射变换下的仿射曲线不

变的变分性质.它可以被认为是经典欧几里得弹性曲线研究的对应物.本书内容以作者在凯斯西储大学的博士学位论文以及作者对该主题的进一步研究为基础.

感谢我的导师David Singer教授的指导和支持,他将我带入了这一领域,是他让我意识到如何从“简单的”离散对象中找到一些研究思路.这改变了我过去学习数学的方式.我还要感谢Joel Langer教授的协助和他提供给我的大量的技术评论.

本书目录翻译如下:

本书的第一章第一节是讲弹性线的.

中国石油大学(华东)工程力学系的刘建林教授,2013 年曾撰文介绍过《细长杆弹性线模型的发展历史》.①

细长弹性杆(slender & elastic rod)模型广泛存在于自然界和工程实际中.例如,海底电缆、高压输电线、柔性绳索、弹簧、石油工程中的钻杆和抽油杆、纳米纤维和纳米管、DNA 和大分子聚合物、攀缘类植物的茎

① 摘自《自然杂志》,2013 年第 36 卷第 5 期.

等,都可以简化成细长弹性杆的模型来进行力学分析[①]. 这种三维模型的一个显著特点就是一个方向上的尺寸远大于其他两个方向上的尺寸. 在外力作用下,细长杆容易发生弯曲、剪切、扭转甚至打结等复杂的变形,因而细长弹性杆的变形具有独特的力学特点,它往往伴随着结构很强的几何非线性. 在力学上,细长弹性杆的形貌往往被称为"弹性线"(elastica),这个单词实际上来源于拉丁文,其意思是弹性薄片.

1. 平面弹性线

根据文献记载,弹性线问题最早由13世纪的数学家Jordanus de Nemore提出[②],正是他第一次从数学曲线的角度研究了弹性线的形状. 他研究了一根细杆,指出"如果快速握住其中部,杆的两端弯曲的程度更大",但是他错误地认为弹性线的形状是一个圆,而实际上圆仅仅是弹性线的一个特殊解.

此后关于弹性线的研究则没有太大进展,但是一些相关的实验和理论为取得弹性线的新结果奠定了基础. 例如,Leonardo da Vinci 对两端铰支梁和悬臂梁的

---

① 刘延柱. 弹性细杆的非线性力学——DNA力学模型的理论基础. 北京:清华大学出版社,2006.

② RAPH L. The elastica: a mathematical history. Technical Report No. UCB/EECS-2008-103. [2012-12-01]. http://www.eecs.berkeley.edu/Pubs/TechRpts/2008/EECS-2008-103.html.

强度进行了实验研究,得到了结构强度与梁的长度成反比的结论[1]. Galileo Galilei 在 1638 年开展了一个实验[2],将一根悬臂梁的一端插入到一堵墙里面,另外一端悬挂着一个重物,具体研究多大的重物能够使梁破坏.尽管这个实验比较粗糙而且装置简单,但是 Galileo 在此基础上提出了脆性材料破坏的第一强度理论.他从弯矩的角度对梁进行了受力分析,但是他没有考虑梁的横向位移,这是由于梁是脆性材料,因而其破坏之前的变形较小.与 Galileo 几乎同时代的一位学者 Ignace-Gaston Pardies 在 1673 年给出了弹性线问题的一个可能解答,认为 Galileo 实验中梁的轴线变形后为抛物线,但是此结论后来被证明是错误的[3].

这一阶段的相关工作还有 Robert Hooke 在 1678 年发表的胡克定律(东汉郑玄(127—200)发现此定律比 Hooke 早 1 500 年,故也称为郑玄 - Hooke 定律[4]),提出了在弹性范围内,结构的外力与其变形成正比.同

---

① 铁摩辛柯.材料力学史.常振楫,译.上海:上海科学技术出版社,1961.

② 老亮.中国古代材料力学史.长沙:国防科技大学出版社,1991.

③ RAPH L. The elastica: a mathematical history. Technical Report No. UCB/EECS-2008-103. [2012-12-01]. http://www.eecs.berkeley.edu/Pubs/TechRpts/2008/EECS-2008-103.html.

④ 老亮.中国古代材料力学史.长沙:国防科技大学出版社,1991.

时 Hooke 也指出梁发生弯曲变形时，其横截面应该是一部分受压，另一部分受拉，但他并没有确定中性轴的具体位置. 1687 年，Newton 的《自然哲学的数学原理》也发表了，这标志着微积分宏伟精巧的大厦已经建立. 牛顿给出了在直角坐标系 xOy 中平面曲线曲率的表达式

$$\frac{1}{\rho}=\frac{y''}{(1+y'^2)^{3/2}} \tag{1}$$

其中 ρ 为曲线上任意一点的曲率半径. 这些工作为弹性线的进一步研究做好了铺垫.

站在这些巨人的肩膀上，James Bernoulli（也称为 Jacob Bernoulli）于 1691 年对弹性线问题进行了深入思考，并给出了弹性线的一个解答，称为“矩形弹性线”，但是这个解答并不足够准确①. 于是他在 1694 年提出了他自己称之为“黄金定理（Golden Theorem）”的公式对弹性线进行进一步的阐释，其基本思想为：弹性杆曲线上任一点的曲率与弯矩成正比. 正是这项工作初步描绘出了弹性线基本方程的雏形. 在同一时期，荷兰力学家 Christiaan Huygens 研究了悬索线问题，即两

---

① RAPH L. The elastica: a mathematical history. Technical Report No. UCB/EECS-2008-103. [2012-12-01]. http://www.eecs.berkeley.edu/Pubs/TechRpts/2008/EECS-2008-103.html.

端固定的柔性绳索在自身重力作用下的沉降量. Huygens 对细长杆的弹性线问题也进行了研究,并指出其他几种弹性线的形状也有可能存在①.

关于弹性线理论发展的一个里程碑式的标志就是 Leonhard Euler 参与该问题的研究. 对我们来说, Euler 绝不是一个陌生的名字. 实际上,他的研究领域遍布数学、物理、力学和电学的各个角落,单以 Euler 命名的公式和定理就多得让我们数不清. 值得一提的是, Euler 正是 Bernoulli 家族的杰出的学生. 1742 年, James Bernoulli 的侄子 Daniel Bernoulli 在写给 Euler 的信中提出,能否采用能量极值的方法来求解弹性线问题的一般解. 实际上弹性结构的平衡构型与其总势能的极小值是一一对应的,这个原理就是最小势能原理. 这是最小作用量原理的一种特殊情况. 关于为什么自然界中最小作用量原理是一个普适的规律,用 Galileo 的一句名言来形容比较恰当:"自然界总是习惯于使用最简单和最容易的手段行事."因此,在 Bernoulli 的启发下, Euler 发展了变分法,提出了弹性线的平衡形状对

---

① RAPH L. The elastica: a mathematical history. Technical Report No. UCB/EECS-2008-103. [2012-12-01]. http://www.eecs.berkeley.edu/Pubs/TechRpts/2008/EECS-2008-103.html.

应于应变能的极小值,而应变能可以由下式来度量

$$\int_0^l \frac{1}{\rho^2}\mathrm{d}s \tag{2}$$

其中 l 为细长杆的长度,s 为弹性线的弧长坐标. 对上述应变能进行变分运算,Euler 推出了在直角坐标系中弹性线的平衡方程

$$B\frac{y''}{(1+y'^2)^{3/2}}=Px \tag{3}$$

其中 B 为杆的抗弯刚度. Euler 把常数 B 称为“绝对弹性”,但是他并没有讨论 B 的物理意义,仅仅说明它与材料的弹性有关. Euler 还认为对于矩形截面梁,抗弯刚度 B 正比于梁宽,并与梁高的平方成正比. 我们现在知道这个结论是错误的,因为抗弯刚度应该与梁高的三次方成正比①.

Euler 考察了如图 1 所示的各种弯曲情况,并根据外载荷的作用方向与载荷作用点的切线之间的夹角大小,把相应的弹性曲线分为多种类型. 根据这些弹性线,Euler 还进一步研究了细长杆的屈曲(buckling)问题. 他求得了一根一端固定、一端自由的梁在轴线压缩载荷 P 作用下发生屈曲的临界载荷

① 铁摩辛柯. 材料力学史. 常振楫,译. 上海:上海科学技术出版社,1961.

$$P_{cr} = \frac{B\pi^2}{4l^2} \tag{4}$$

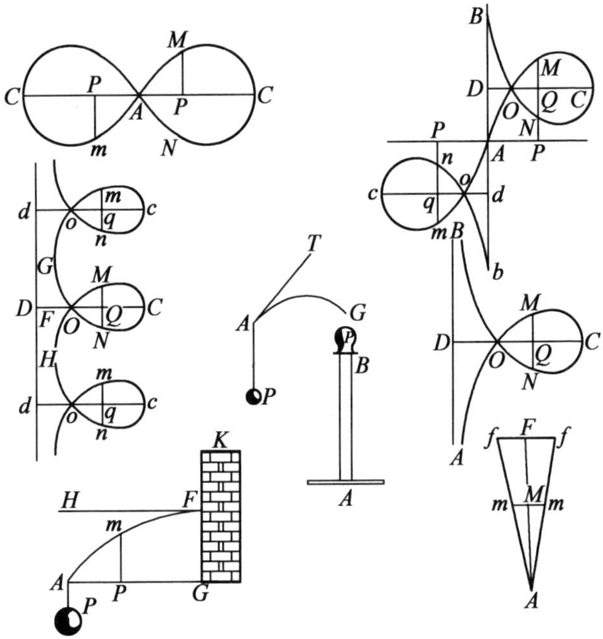

图 1　Euler 分类的各种弹性线形状

但是在 Euler 的时代，这个公式没有得到足够的重视. 这是因为当时主要的建筑材料，例如木头、石头和铸铁大多是脆性的. 这个公式直到后来低碳钢得到大量应用时才重新得到重视，并在工业设计里面起到举足轻重的作用. 实际上早在 1729 年,荷兰力学家 Pieter van Musschenbroek 就通过实验研究了矩形截面梁柱发生屈曲的临界载荷,并指出临界载荷与杆长的平方成反比[①]. 在 1811 年,A. J. C. B. Duleau 做了 105 个实验,其中包括通过设计准静态试验研究梁柱的屈曲. 但是他的实验结果与 Euler 的理论计算有所出入,主要还是因为当时的实验设备精度不够[②].

比 Euler 年轻的 Joseph-Louis Lagrange 是 Euler 狂热的追随者,他跟踪研究了 Euler 的很多工作. 他与 Euler 先后研究了小变形情况下的弹性线问题,并于 1770 年针对细长杆发生屈曲的微分方程

$$Py + By'' = 0 \tag{5}$$

得到了临界载荷的表达式

$$P_{\mathrm{cr}} = B\left(\frac{n\pi}{l}\right)^2 \quad (n = 0, 1, 2, 3, \cdots) \tag{6}$$

这正是我们今天材料力学课本中的常见表述. Lagrange 针对方程(3)进行级数展开,得到了外载荷和梁最大

---

① RAPH L. The elastica: a mathematical history. Technical Report No. UCB/EECS-2008-103. [2012-12-01]. http://www. eecs. berkeley. edu/Pubs/TechRpts/2008/EECS-2008-103. html.

② 铁摩辛柯. 材料力学史. 常振楫,译. 上海:上海科学技术出版社,1961.

挠度之间的关系式[1].

Euler 还综合考虑了细长杆以及悬索线的受力平衡,给出了这些弹性结构的一般方程

$$\frac{\mathrm{d}N}{\mathrm{d}s}+Q\frac{\mathrm{d}\theta}{\mathrm{d}s}=-\boldsymbol{f}_{\mathrm{t}} \tag{7}$$

$$\frac{\mathrm{d}Q}{\mathrm{d}s}-N\frac{\mathrm{d}\theta}{\mathrm{d}s}=-\boldsymbol{f}_{\mathrm{n}} \tag{8}$$

$$\frac{\mathrm{d}M}{\mathrm{d}s}-Q=0 \tag{9}$$

其中 Q 为剪力,N 为轴力,M 为弯矩,θ 为杆上任意一点的倾角,$\boldsymbol{f}_{\mathrm{t}}$ 和 $\boldsymbol{f}_{\mathrm{n}}$ 分别为体力沿着杆截面切向和法向的分量. 这组方程不依赖于材料的性质,因为它将具有刚度的弹性杆件和不具有刚度的柔索都囊括在内.

尽管弹性线的方程已经建立,但是弹性线的形状直到 1906 年才由诺贝尔奖得主 Max Born 精确绘出. Born 在哥廷根大学获得了博士学位,其博士论文的题目为《平面与空间弹性线稳定性的研究》. 他设计了一套实验装置,主要是利用重物和标度盘,通过不同的边界条件对细长弹性薄片进行加载,并对其变形形貌进行拍摄. Born 的数学功底比较深厚,他利用现代数学的技巧对弹性线问题进行了表述,并将其推广到三维

① 铁摩辛柯. 材料力学史. 常振楫,译. 上海:上海科学技术出版社,1961.

空间问题中[①].

因为弹性线方程是一个很强的非线性方程,所以其解析解只有在一些特殊情况下才能求得. 1945 年, Bisshopp 和 Drucker 给出了以椭圆积分表示的、自由端承受集中力的悬臂梁的大位移解答[②]. 具有不同边界条件的弹性线解答对于工程应用,例如纳米材料的制备等,有着非常重要的意义. 在碳纳米管阵列的制备过程中,由于水分的蒸发会导致相邻的纳米管发生毛细粘附,从而结构产生较强的几何大变形[③]. 刘建林[④]采用弹性线模型求解了两根纳米管的毛细粘附临界参数,这对于如何避免碳纳米管阵列的粘附有一定的参考价值.

2. 空间弹性线

从平面弹性线转变到空间弹性线的研究,必须对

---

① RAPH L. The elastica: a mathematical history. Technical Report No. UCB/EECS-2008-103. [2012-12-01]. http://www.eecs.berkeley.edu/Pubs/TechRpts/2008/EECS-2008-103.html.

② BISSHOPP K E, DRUCKER D C. Large deflection of cantilever beams. Quarterly of Applied Mathematics, 1945,3:272-275.

③ JOURNET C, MOULINET S, YBERT C, et al. Contact angle measurements on superhydrophobic carbon nanotube forests: effect of fluid pressure. Europhysics Letters, 2005,71:104-109.

④ LIU J L, FENG X Q. Capillary adhesion of micro-beams: finite deformation analyses. Chinese Physics Letters, 2007,8:2349-2352.

柱体的扭转进行研究. 1784 年, Charles Augustin de Coulomb 发表了关于扭转的研究报告,他给出了受扭转的金属丝的扭矩与扭转角成正比的结论,并设计了库仑扭秤的装置[①]. 此后,法国工业学院于 1795 年成立,该校涌现出了一批著名的力学家,如 Cauchy, Lame, Poisson, Saint-Venant, Duleau, Navier 等. 正是由于这些科学家的努力,才逐渐完善了弹性理论. 1853 年,Saint-Venant 向法国科学院提交了关于柱体扭转的研究报告. 他引入了柱体截面的翘曲函数,并提出了"半逆解法",从而顺利地解决了该问题. 在研究柱体扭转的过程中,Saint-Venant 也提出了著名的"Saint-Venant 原理",即静力等效的载荷引起的差异只在加载位置附近有所差异,所以对离开柱体两端较远处的柱体中的应力,他的解答是足够精确的.

Augustus Edward Hough Love 在 Saint-Venant 的基础上发展了扭转理论,将平面梁的 Euler-Bernoulli 假设推广到了三维构型. 他指出,细长杆的总扭转包括两部分,实际上代表了截面的绝对角速度等于相对角速度

---

① 铁摩辛柯. 材料力学史. 常振楫,译. 上海:上海科学技术出版社,1961.

与牵连角速度之和[①].

尽管弹性理论的大厦逐渐夯实,但是细长杆空间弹性线还不同于经典的弹性力学,这是由于弹性杆在变形过程中往往伴随着非常强的几何非线性,因而必须发展独特的理论才能加以分析.关于弹性线发展史的另一个里程碑式的贡献,则是1859年Gustav Robert Kirchhoff对空间弹性线的研究[②].他从三维弹性理论的角度出发,认为杆是由许多小柱体组成的,每个小柱体与周围的柱体通过力和力偶发生接触作用.为了说明这许多小柱体能够相互配合和协调,Kirchhoff提出了著名的以他的名字命名的假设:①杆不可伸长,不受剪力,中心线在变形前后均为2阶以上光滑曲线;②杆的长度和曲率半径远大于横截面尺寸;③横截面为刚性平面;④忽略弯曲引起的剪切变形,横截面与中心线正交;⑤忽略中心线的拉伸变形,任意两截面沿中心线的距离不变;⑥相邻截面可绕中心线作相对扭转,扭角为弧长的连续函数.

---

① RAPH L. The elastica: a mathematical history. Technical Report No. UCB/EECS-2008-103. [2012-12-01]. http://www.eecs.berkeley.edu/Pubs/TechRpts/2008/EECS-2008-103.html.

② 刘延柱.弹性细杆的非线性力学——DNA力学模型的理论基础.北京:清华大学出版社,2006.

继而，Cosserat 兄弟（Eugene Cosserat 和 Francois Cosserat）在 Kirchhoff 杆的基础上，进一步考虑了杆的轴向线应变和弯曲剪应变等因素，提出了采用方向矢量（director）沿中心线的运动来描述弹性杆的模型. 如图 2 所示，他们把杆看作是一种由物质点组成的数学曲线，而其中的方向矢量即与截面固结的沿形心主轴的 3 个单位向量形成的右手系. 这样，Cosserat 模型便可以同时考虑细长杆的拉压、剪切、扭转和弯曲变形[①②].

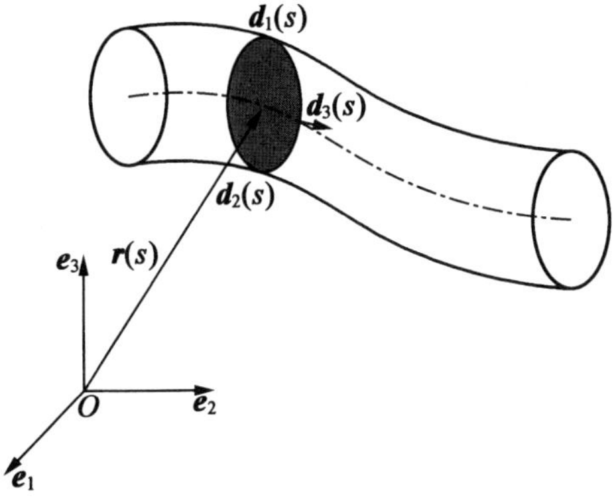

图 2　Cosserat 杆示意图

---

① VAN DER HEIJDEN G H M, CHAMPNEYS A R, THOMPSON JMT. Spatially complex localization in twisted elastic rods constrained to lie in the plane. Journal of the Mechanics and Physics of Solids, 1999, 47: 59-79.

② 薛纭，王鹏. Cosserat 弹性杆动力学普遍定理的守恒量问题. 物理学报，2011, 60: 114501.

3. 弹性线与其他物理现象的比拟

在研究弹性线的过程中，也有很多科学家注意到了弹性线的性质与其他一些不同的物理现象存在很多相似性. 早在 1807 年，Pierre Simon Laplace 就建立了毛细管中弯液面的形状方程（图 3），即液体的表现张力 γ 与弯液面的内外压差 Δp 之间存在通过曲率相互关联的方程[①]

$$\gamma \frac{y''}{(1+y'^2)^{3/2}}=\Delta p \qquad (10)$$

将方程(10)与弹性线的控制方程(3)相比较，可以发现两者之间具有惊人的相似性. 这一性质后来被 James Clerk Maxwell[②] 发现，并指出毛细效应引起的弯液面与弹性薄片的变形之间具有相似性. 2002 年，法国国家科学中心的 David Quere[③] 对两者之间的相似性进行了详尽的实验研究. 刘建林[④]对上述两个方程(3)和(10)

---

① LAPLACE P S. Euvres completes de Laplace, volume 4. Paris: Gauthier-Villars, 1880.

② MAXWELL J C. Capillary action. 9th ed. New York: Encyclopedia Britannica, 1875.

③ CLANET C, QUERE D. Onset of menisci. Journal of Fluid Mechanics, 2002, 460: 131-149.

④ LIU J L. Analogies between a meniscus and a cantilever. Chinese Physics Letters, 2009, 26: 116803.

进行了坐标变换,发现两者可以写成统一的形式,并详细探讨了两者之间参数的相似性. 例如,对于弹性杆而言,其刚度用 B 来表示,而对于液体而言,其刚度可以用表面张力来度量. 对于曲率、应变能、载荷等,两个系统均有对应的比拟关系. 这一研究将为我们设计新型的比拟实验提供理论上的依据.

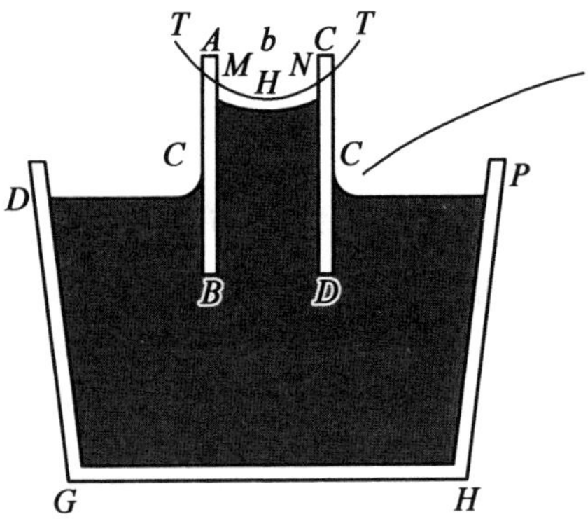

图 3　Laplace 研究的毛细效应

另外,1859 年 Kirchhoff 提出了著名的静力学—动力学比拟,指出弹性杆的平衡微分方程与经典力学中刚体定点转动微分方程之间具有相似性[①]. 如果明确了

① 刘延柱. 弹性细杆的非线性力学——DNA 力学模型的理论基础. 北京:清华大学出版社,2006.

两个系统之间参数的比拟关系,就可以把我们熟悉的刚体运动规律借用来分析弹性杆的性质. 这两个不同系统之间的参数比较如表 1 所示.

表 1　弹性杆与刚体定点转动之间的比拟

超细长弹性杆	刚体定点转动
研究对象为杆横截面	研究对象为定点转动刚体
自变量为弧长	自变量为时间
截面的弧坐标历程	刚体运动时间历程
截面上的主矢分量	刚体的重力矩
截面姿态的 Euler 角	刚体姿态的 Euler 角
截面形心主轴的抗弯(扭)刚度	刚体对连体主轴的转动惯量
截面的弯(扭)度	刚体的角速度
弹性应变能	刚体的动能
杆的平衡构型	刚体的运动过程
无扭直杆	静止刚体
受扭直杆	定轴转动刚体
螺旋杆	规则进动刚体

总之,从弹性线概念的提出,到平面弹性线和空间弹性线模型的逐渐成熟,是众多力学家和工程师携手并进、共同研究的结果. 从建立弯矩与曲率之间的关

系，到 Kirchhoff-Cosserat 杆模型建立的过程，经历了众多力学家之间的辩论和竞争，从而使该理论不断趋于完善. 在这个过程中，需要引入符合工程实际的假设，需要精密实验的验证，也需要高深数学技巧的推导. 弹性线理论的建立也证明了最小势能原理与弹性结构的平衡构形是一一对应的. 弹性线的发展历程告诉我们，只有将工程实践与理论分析紧密结合起来，才能使力学发挥出惊人的威力. 只有精深的理论而不关心工程的需求，只能是纸上谈兵；只有工程经验而不探究力学的规律，也无异于合眼摸象. 这种理论结合实践的理论与钱学森先生所提倡的“技术科学(Engineering Science)”思想实际上是一脉相承的. 他认为，技术科学不限于是自然科学与工程技术间的桥梁，它同时也是人类认识的源泉，应该从工程实践中提取力学模型，上升到理论高度，然后反过来指导工程应用. 弹性线的发展历程正是践行技术科学思想的一个范例.

超细长弹性杆作为 DNA 等生物大分子链的力学模型，其平衡和稳定性问题已成为力学与分子生物学交叉的研究热点. 虽然在 Kirchhoff 动力学比拟的基础上，用分析力学方法讨论弹性杆的文章已见诸文献，但尚未形成弹性杆分析力学的严格理论.

上海大学、上海市应用数学和力学研究所的薛纭，上海应用技术学院机械与自动化工程学院的刘延柱，上海交通大学工程力学系的陈立群三位教授2005年研究了《超细长弹性杆的分析力学问题》，对杆截面的自由度、虚位移、约束方程及约束力等基本概念给出严格的定义和表达式；建立弹性杆平衡的D'Alembert-Lagrange原理，Jourdain原理和Gauss原理；从D'Alembert-Lagrange原理导出Hamilton原理，从变分原理出发导出Lagrange方程，Nielsen方程，Appell方程和Hamilton正则方程；对于受约束的弹性杆，导出了带乘子的Lagrange方程；讨论了Lagrange方程的首次积分；对于杆中心线存在尖点的情形，导出了微段杆平衡的近似方程[①].

细长弹性杆的研究始于Daniel Bernoulli和Euler(1730). Kirchhoff(1859)在若干假定下建立了弹性细杆静力学的基本方程. 由于在形式上与定点转动刚体的动力学方程的一致性而形成Kirchhoff的动力学比拟理论[②③]. 弹性杆作为电缆、钻杆、纤维等的力学模

---

① 摘自《力学学报》,2005年,第37卷,第4期.

② LOVE A E H. A treatise on mathematical theory of elasticity. 4th. ed. New York: Dover, 1927.

③ 陈至达. 杆板壳大变形理论. 北京:科学出版社,1994:45-85.

型，其平衡和稳定性问题有着广泛的工程背景．20 世纪 70 年代以来，由于超细长弹性杆模型在以 DNA 为代表的生物大分子链的研究工作中得到应用[1]，使这一经典力学问题重新受到关注[2][3][4][5][6][7][8][9]．在研究受约束弹性杆的力学问题时，分析力学方法更具有优越性[10][11][12]．虽然直接应用 Hamilton 原理和 Lagrange 方程

---

① FULLER F B. The writhing number of a space curve. Proc. Natl. Acad. Sci.，1971，68(4)：815-819.

② 刘延柱．弹性杆基因模型的力学问题．力学与实践，2003，25(1)：1-5.

③ 刘延柱．具有初曲率和初挠率弹性直杆的平衡稳定性．上海交通大学学报，2002，36(11)：1587-1590.

④ TANAKA F, TAKAHASHI H. Elastic theory of supercoiled DNA. J. Chem. Physics，1985，83：6017-6026

⑤ VAN DER HEIJDEN, CHAMPNEYS A R, THOMPSEN J M T. Spatially complex localisation in twisted elastic rods constrained to a cylinder. Solids and Structures，2002，39：1863-1883

⑥ 刘延柱，薛纭，陈立群．弹性细杆平衡的动态稳定性．物理学报，2004，53(8)：2424-2428.

⑦ XUE YUN, LIU YANZHU, CHEN L Q. The Schrödinger equation for a Kirchhoff elastic rod with noncircular cross section. Chinese Physics，2004，13(6)：794-797

⑧ 薛纭，陈立群，刘延柱．受曲面约束弹性细杆的平衡问题．物理学报，2004，53(7)：2040-2045.

⑨ 薛纭，陈立群，刘延柱．Kirchhoff 方程的相对常值特解及其 Lyapunov 稳定性．物理学报，2004，53(12)：4029-4036.

⑩ 刘延柱．高等动力学．北京：高等教育出版社，2001.

⑪ 陈滨．分析动力学．北京：北京大学出版社，1987.

⑫ 梅凤翔，刘端，罗勇．高等分析力学．北京：北京理工大学出版社，1991.

的讨论已见诸文献[①②],但目前尚未形成弹性杆分析力学的严格理论.因此建立严格的弹性杆分析力学的基本概念和方法,对于完善和发展弹性细杆平衡问题的研究具有重要的理论和实际意义.

1. 连续杆的离散化

Kirchhoff 假定为连续杆的离散化提供依据,即将杆视为刚性截面沿中心线以"单位速度"运动的轨迹.除端部以外不受约束作用的杆为自由杆,则截面的位形需要由 6 个坐标确定,即形心和相对形心的姿态各 3 个坐标.

建立惯性坐标系 $O\text{-}\xi\eta\zeta$ 和与截面固结的形心主轴坐标系 $p\text{-}xyz$,沿坐标轴的单位基矢量分别为 $\boldsymbol{e}_\xi,\boldsymbol{e}_\eta,\boldsymbol{e}_\zeta$ 和 $\boldsymbol{e}_1(s),\boldsymbol{e}_2(s),\boldsymbol{e}_3(s)$,其中 s 为中心线的弧坐标,$\boldsymbol{e}_3$ 为切向基矢量,指向弧坐标增加方向,外法矢与 $\boldsymbol{e}_3$ 一致的截面记为 s^+,否则记为 s^-. 用 Euler 角 $\psi(s)$, $\vartheta(s)$, $\varphi(s)$ 描述主轴坐标系 $p\text{-}xyz$ 相对惯性坐标系 $O\text{-}\xi\eta\zeta$ 的姿态. 杆截面的位形可表示为

$$\begin{cases}\xi=\xi(s),\eta=\eta(s),\zeta=\zeta(s)\\ \psi=\psi(s),\vartheta=\vartheta(s),\varphi=\varphi(s)\end{cases}\tag{1}$$

---

① LANGER J, SINER D A. Lagrangian aspects of the Kirchhoff elastic rod. SIAM Rev, 1996,38(4):605-618.

② POZO CORONADO L M. Hamilton equations for elasticae in the Euclidean 3 space. Physica D,2000,141:248-260.

设式(1)关于弧坐标 s 为二阶连续可微. 这 6 个广义坐标为独立变量,但需满足方程

$$\dot{\boldsymbol{r}} = \boldsymbol{e}_3 \tag{2}$$

因此对 s 的导数不独立,式中 $\boldsymbol{r}$ 为中心线的矢径,顶部点号表示对 s 的导数,投影式为

$$\dot{\xi} = \sin\psi\sin\vartheta, \dot{\eta} = -\cos\psi\sin\vartheta, \dot{\zeta} = \cos\vartheta \tag{3}$$

方程(3)是不可积的,构成对截面状态的非完整约束,使截面的自由度减为 3. 从而表明自由 Kirchhoff 杆为非完整系统. 只要变形前后的挠性线满足光滑条件,约束(3)就能自动实现. 因此这种特殊约束形式是无须外界约束力作用的伪非完整约束.

截面的姿态相对弧坐标的变化率称为截面的弯扭度,记作 $\boldsymbol{\omega}(s)$,其沿主轴的分量为

$$\begin{cases} \omega_1 = \dot{\vartheta}\cos\varphi + \dot{\psi}\sin\vartheta\sin\varphi \\ \omega_2 = -\dot{\vartheta}\sin\varphi + \dot{\psi}\cos\varphi\sin\vartheta \\ \omega_3 = \dot{\varphi} + \dot{\psi}\cos\vartheta \end{cases} \tag{4}$$

弯扭度也可理解为截面沿中心线正向以单位速度运动时相对惯性坐标系的角速度.

2. 约束、约束方程和约束力

设惯性空间中的曲面约束由方程

$$g(\xi_c, \eta_c, \zeta_c) = 0 \tag{5}$$

描述,设约束曲面为小曲率曲面,且为连续、光滑和有向;并设约束为刚性和双面的,且不计摩擦和约束力对杆截面形状的影响. 设截面的边界上有且只有一点与约束曲面接触,即满足

$$\boldsymbol{\rho}=\boldsymbol{r}+\boldsymbol{b} \tag{6}$$

式中 $\boldsymbol{\rho}$ 为约束曲面上的点;$\boldsymbol{r}=\xi\boldsymbol{e}_\xi+\eta\boldsymbol{e}_\eta+\zeta\boldsymbol{e}_\zeta$ 为截面形心的矢径;$\boldsymbol{b}(s,t)$ 为 s 截面的边界曲线方程,t 为参数,设 $\boldsymbol{b}$ 关于 s,t 具有二阶连续偏导数,且 $\boldsymbol{b}\cdot\boldsymbol{e}_3=0$,在 p-xy 平面中围成的区域是凸的. 对于给定的 s,存在唯一的一点 $t=t(s)$ 满足式(6). 由此得

$$\xi_c=\xi+b_\xi,\eta_c=\eta+b_\eta,\zeta_c=\zeta+b_\zeta \tag{7}$$

式中 b_ξ,b_η,b_ζ 为 $\boldsymbol{b}$ 在轴 ξ,η,ζ 上的投影. 导出对杆截面位形的约束方程

$$g(\xi+b_\xi,\eta+b_\eta,\zeta+b_\zeta)=0 \tag{8}$$

一般情况下,式(8)显含 s 因而是非定常的,其对弧坐标的导数可化为

$$\dot{g}=\boldsymbol{n}_c\cdot\dot{\boldsymbol{r}}+\boldsymbol{n}_c\cdot\mathring{\boldsymbol{b}}+\boldsymbol{\omega}\cdot(\boldsymbol{b}\cdot\boldsymbol{n}_c)=0 \tag{9}$$

式中 $\boldsymbol{n}_c$ 为约束曲面在接触点的法向量;顶部“。”号表示是在主轴坐标系 p-xyz 中对 s 求导. 式(8)和式(9)使截面的广义坐标数和自由度分别减为 5 和 2. 将线分布约束力向截面形心简化,得

$$\boldsymbol{f}^C=\lambda\boldsymbol{n}_c,\boldsymbol{m}^C=\boldsymbol{b}\times\boldsymbol{f}^C \tag{10}$$

其中 λ 为不定乘子，对于圆形等截面杆，有关系：$f_3^C = \boldsymbol{f}^C \cdot \boldsymbol{e}_3 = 0$ 和 $\boldsymbol{m}^C = \boldsymbol{0}$. 如果约束为单面，则截面的受约束条件为

$$\lambda = \boldsymbol{f}^C \cdot \boldsymbol{n}_c \geqslant 0, \text{或 } \lambda = \boldsymbol{f}^C \cdot \boldsymbol{n}_c \leqslant 0 \tag{11}$$

式中等号为临界情形. 以上讨论容易推广到受两个曲面约束的情形.

给定 $O\text{-}\xi\eta\zeta$ 中的一个单位常矢量 $\boldsymbol{h}$，限制截面的弯扭度方向构成非完整约束. 非完整约束仅使自由度减少. 例如：

(1) 截面弯扭度与给定方向 $\boldsymbol{h}$ 垂直：$\boldsymbol{\omega} \cdot \boldsymbol{h} = 0$. 显然其坐标式是不可积的. 在此约束下，杆截面的自由度减为 2.

(2) 截面弯扭度与给定方向 $\boldsymbol{h}$ 平行：$\boldsymbol{\omega} \times \boldsymbol{h} = 0$，其分量形式也是不可积的. 截面的自由度为 0，约束方程已完全规定了截面的位形，Kirchhoff 方程寻求实现这一约束力的作用方式.

(3) 给定截面弯扭度的模平方变化规律：$\boldsymbol{\omega}^2 = u(s)$，在 s 的定义域内 $u(s) \geqslant 0$. 用 Euler 角表示为 $\dot{\psi}^2 + \dot{\vartheta}^2 + \dot{\varphi}^2 + 2\cos\vartheta\dot{\psi}\dot{\varphi} = u(s)$，它构成对截面姿态的非线性非完整约束.

3. 虚位移及其限制方程

弹性理论表明，直杆的弯扭变形（小应变）不改变

截面上的点在 $p\text{-}xy$ 平面上投影的位置. 因此,横截面上的点可表为:$\boldsymbol{R}_i(s)=\boldsymbol{r}(s)+\boldsymbol{b}_i$,其中 $\boldsymbol{b}_i=x_i\boldsymbol{e}_1(s)+y_i\boldsymbol{e}_2(s)$,$i$ 为 s 截面上点的标号. 称同时满足约束条件和平衡条件的位形为截面的实际平衡状态,只满足约束条件的位形称为截面的可能平衡状态.

定义 1(点的真实位移) 在实际平衡状态 $\boldsymbol{R}_i$ 下,s 截面上一点的真实位移定义为

$$\Delta\boldsymbol{R}_i=\boldsymbol{R}_i(s+\Delta s)-\boldsymbol{R}_i(s) \tag{12}$$

定义 2(点的可能位移) 在可能平衡状态 $\overline{R}_i$ 下,s 截面上一点的可能位移定义为

$$\Delta\overline{\boldsymbol{R}}_i=\overline{\boldsymbol{R}}_i(s+\Delta s)-\overline{\boldsymbol{R}}_i(s) \tag{13}$$

显然这里的"位移"不是运动学意义上的,而是不同两点的位置矢径之差.

定义 3(点的虚位移) 在给定位形下发生的、约束所允许的、假想的、与弧坐标无关的截面上一点的无限小位移称为点的虚位移,记为 $\delta\boldsymbol{R}_i$,称 δ 为等弧长变分.

显然,一点的可能位移是虚位移中的一个. 两者的差别在于可能位移源于弧坐标的变化,而虚位移是运动学意义上的,与弧坐标无关. 根据 Kirchhoff 刚性截面假定,同一截面上不同点的虚位移形成截面的虚角位移.

定义 4(截面虚角位移)　约束所允许的、与弧坐标变化无关的、假想的截面无限小角位移定义为虚角位移，记为 $\delta\boldsymbol{\Phi}$.

截面的虚角位移导致杆的虚变形，它也是杆的可能平衡状态. 因此可以计算点的虚位移和截面的虚角位移对弧坐标的导数. 设微分和变分服从交换关系

$$\frac{\mathrm{d}}{\mathrm{d}s}\delta(\ ) = \delta\frac{\mathrm{d}}{\mathrm{d}s}(\ ) \tag{14}$$

则有

$$\begin{cases} \dfrac{\mathrm{d}}{\mathrm{d}s}(\delta \boldsymbol{R}_i) = \delta \boldsymbol{v}_i \\ \dfrac{\mathrm{d}}{\mathrm{d}s}\delta\boldsymbol{\Phi} = \tilde{\delta}\boldsymbol{\omega} \\ \dfrac{\tilde{\mathrm{d}}}{\mathrm{d}s}\delta\boldsymbol{\Phi} = \delta\boldsymbol{\omega} \end{cases} \tag{15}$$

其中 $\boldsymbol{v}_i = \mathrm{d}\boldsymbol{R}_i/\mathrm{d}s$. d,$\delta$ 上的波浪号表示运算是相对主轴坐标系的. 给定截面 $\boldsymbol{r}(s)$ 上的一点 $\boldsymbol{R}_i(s)$，有如下关系

$$\delta\boldsymbol{R}_i = \delta\boldsymbol{r} + \delta\boldsymbol{b}_i,\quad \delta\boldsymbol{b}_i = \delta\boldsymbol{\Phi}\times\boldsymbol{b}_i \tag{16}$$

选定截面的姿态坐标 $q_i(i=1,2,3)$ 后，截面的弯扭度和虚角位移可表示为

$$\boldsymbol{\omega} = \sum_{i=1}^{3}\boldsymbol{\omega}_{\dot{q}_i}\dot{q}_i \tag{17a}$$

$$\delta\boldsymbol{\Phi} = \sum_{i=1}^{3}\boldsymbol{\omega}_{\dot{q}_i}\delta q_i \tag{17b}$$

其中 $\boldsymbol{\omega}_{\dot{q}_i}$ 为姿态坐标 $q_i(i=1,2,3)$ 的可微矢值函数.

设横截面受有一般的几何约束

$$g_i(q_1,q_2,q_3,q_4,q_5,q_6,s)=0 \quad (i=1 \text{ 或 } 1,2) \tag{18}$$

其中 $q_4=\xi,q_5=\eta,q_6=\zeta$. 约束(18)加在虚位移上的限制方程为

$$\delta g_i=\sum_{j=1}^{6}\frac{\partial g_i}{\partial q_j}\delta q_j=0 \quad (i=1 \text{ 或 } 1,2) \tag{19}$$

非完整约束的一般形式由形如

$$h_i(q_1,\cdots,q_6,\dot{q}_1,\cdots,\dot{q}_6,s)=0 \quad (i=1 \text{ 或 } 1,2) \tag{20}$$

的不可积约束方程给出. 其加在虚位移上的限制方程按 Appell-Chetaev 定义为

$$\delta h_i=\sum_{j=1}^{6}\frac{\partial h_i}{\partial \dot{q}_j}\alpha q_j=0 \tag{21}$$

由于虚位移的任意性,不能将式(3)预先嵌入约束方程(20). Appell-Chetaev 定义是对非完整约束条件(20)实现方式的一个选择①②.

曲面约束(9)加在坐标空间虚位移上的限制可化

---

① 梅凤翔,陈滨. 关于分析力学的学科发展问题. //黄文虎,陈滨,王照林. 一般力学(动力学、振动与控制)最新进展. 北京:科学出版社,1994:37-45.

② 薛纭. 约束条件的实现与非完整非完备力学系统的数学模型. //陈滨. 动力学、振动与控制的研究. 北京:北京大学出版社,1994:5-8.

为

$$\delta g = \boldsymbol{n}_c \cdot \delta \boldsymbol{r} + (\boldsymbol{b} \times \boldsymbol{n}_c) \cdot \delta \boldsymbol{\Phi} = 0 \tag{22}$$

其中函数的变分源于姿态坐标的变分,它与约束力的虚功相关

$$\lambda \delta g = \boldsymbol{f}^C \cdot \delta \boldsymbol{r} + \boldsymbol{m}^C \cdot \delta \boldsymbol{\Phi} \tag{23}$$

若约束力在所有虚位移上所作的功为零,则称这种约束为理想约束. 值得注意的是,由于式(3)是伪非完整约束,它仅限制截面的自由度而不构成对虚位移的限制.

4. 微分变分原理

建立截面平衡的微分变分原理. 杆 s^- 截面上的点 m_i 相对主轴坐标系的矢径为 $\boldsymbol{b}_i$,围绕 m_i 点取面积微元 A_i,其上的应力为 $-\boldsymbol{p}_i$. 当截面 s 沿中心线"移动"到 $s+\Delta s$ 时,m_i"移动"到点 m'_i,矢径为 $\boldsymbol{b}_i+\Delta \boldsymbol{b}_i$,应力为 $\boldsymbol{p}_i+\Delta \boldsymbol{p}_i$. 面积微元的"移动"形成微元体 V_i. 此微元体的虚位移由 $\delta \boldsymbol{r}$ 和 $\delta \boldsymbol{\Phi}$ 给出. 略去二阶微量,作用在 V_i 上的所有力的虚功之和为

$$\delta W_D = \sum_i \left[(A_i \Delta \boldsymbol{p}_i + \boldsymbol{f}_i^G \Delta s + \boldsymbol{f}_i^C \Delta s) \cdot \delta \boldsymbol{R}_i + (\Delta \boldsymbol{R}_i \times A_i \boldsymbol{p}_i + \boldsymbol{m}_i^G \Delta s) \cdot \delta \boldsymbol{\Phi} \right] \tag{24}$$

其中 $\boldsymbol{f}_i^G$,$\boldsymbol{m}_i^G$ 为主动力和力偶关于弧坐标 s 的集度,$\boldsymbol{f}_i^C$ 为侧面约束力关于弧坐标 s 的集度. 将上式各项除以 Δs,并令 $\Delta s \to 0$,记为 W_D^*,简化后导出

$$W_D^* = (\dot{\boldsymbol{F}} + \boldsymbol{f}^G) \cdot \delta \boldsymbol{r} + \left(\frac{\mathrm{d}\boldsymbol{M}}{\mathrm{d}s} + \boldsymbol{e}_3 \times \boldsymbol{F} + \boldsymbol{m}^G \right) \cdot \delta \boldsymbol{\Phi} \tag{25a}$$

$$\boldsymbol{f}^C \cdot \delta \boldsymbol{r} + \boldsymbol{m}^C \cdot \delta \boldsymbol{\Phi} = 0 \tag{25b}$$

其中

$$\boldsymbol{F} = \sum_i A_i \boldsymbol{p}_i, \boldsymbol{M} = \sum_i \boldsymbol{b}_i \times A_i \boldsymbol{p}_i$$

$$\boldsymbol{f}^G = \sum_i \boldsymbol{f}_i^G, \boldsymbol{m}^G = \sum_i \boldsymbol{b}_i \times \boldsymbol{f}_i^G$$

$$\boldsymbol{f}^C = \sum_i \boldsymbol{f}_i^C, \boldsymbol{m}^C = \sum_i (\boldsymbol{b}_i \times \boldsymbol{f}_i^C)$$

式(25b)为理想约束条件,对于自由弹性杆自然满足.

上述过程并非简单地将动力学普遍方程中的时间变量替换成弧坐标,两者的不同在于弧坐标既是空间变量,按 Kirchhoff 动力学比拟,又充当“时间变量”.

D’Alembert-Lagrange 原理　受有理想双面约束的 Kirchhoff 杆平衡时,对满足理想约束条件(25b)的任意虚位移,有

$$W_D^* = (\dot{\boldsymbol{F}} + \boldsymbol{f}^G) \cdot \delta \boldsymbol{r} + \left(\frac{\mathrm{d}\boldsymbol{M}}{\mathrm{d}s} + \boldsymbol{e}_3 \times \boldsymbol{F} + \boldsymbol{m}^G \right) \cdot \delta \boldsymbol{\Phi} = 0 \tag{26}$$

式(26)可以从平衡方程导出. 微元体 V_i 的力和力矩平衡方程为

$$\begin{cases} A_i \Delta \boldsymbol{p}_i + \boldsymbol{f}_i^G \Delta s + \boldsymbol{f}_i^C \Delta s = \boldsymbol{0} \\ \Delta \boldsymbol{R}_i \times A_i \boldsymbol{p}_i + \boldsymbol{m}_i^G \Delta s + \boldsymbol{m}_i^C \Delta s = \boldsymbol{0} \end{cases} \tag{27}$$

用 Δs 去除上式，并令 $\Delta s \to 0$，导出微元体 V_i 当 $\Delta s \to 0$ 时的平衡微分方程

$$\begin{cases} A_i \dot{\boldsymbol{p}}_i + \boldsymbol{f}_i^G + \boldsymbol{f}_i^C = \boldsymbol{0} \\ \dot{\boldsymbol{R}}_i \times A_i \boldsymbol{p}_i + \boldsymbol{m}_i^G + \boldsymbol{m}_i^C = \boldsymbol{0} \end{cases} \quad (i = 1, 2, \cdots) \tag{28}$$

将 $\delta \boldsymbol{R}_i$ 和 $\delta \boldsymbol{\Phi}$ 分别点乘式(28)两式后相加并对 i 求和，利用式(25b)即得式(26).

原理(26)不涉及杆的本构关系. 设杆服从线性本构关系

$$\begin{cases} M_1 = A(\omega_1 - \omega_1^0) \\ M_2 = B(\omega_2 - \omega_2^0) \\ M_3 = C(\omega_3 - \omega_3^0) \end{cases} \tag{29}$$

式中 A, B 为关于 x, y 轴的抗弯刚度，C 为关于 z 轴的抗扭刚度；$\omega_i^0 = \omega_i^0(s)\ (i = 1, 2, 3)$ 为原始弯扭度分量. 用 $q_i\ (i = 1, 2, 3)$ 表示截面的 3 个 Euler 角，存在如下关系①

$$\frac{\tilde{\mathrm{d}}}{\mathrm{d}s} \frac{\partial \boldsymbol{\omega}}{\partial \dot{q}_i} - \frac{\tilde{\partial} \boldsymbol{\omega}}{\partial q_i} = \frac{\partial \boldsymbol{\omega}}{\partial \dot{q}_i} \times \boldsymbol{\omega} \quad (i = 1, 2, 3) \tag{30}$$

---

① WESTCOTT T P, TOBIAS I, OLSON W K. Elasticity theory and numerical analysis of DNA supercoiling: An application to DNA looping. J. Phys. Chemistry, 1995, 99: 17926-17935.

式中波浪号表示相对主轴坐标系的导数. 导出

$$\left(\frac{\tilde{\mathrm{d}}\boldsymbol{M}}{\mathrm{d}s}+\boldsymbol{\omega}\times\boldsymbol{M}\right)\cdot\frac{\partial\boldsymbol{\omega}}{\partial\dot{q}_i}=\frac{\mathrm{d}}{\mathrm{d}s}\frac{\partial T}{\partial\dot{q}_i}-\frac{\partial T}{\partial q_i}\quad(i=1,2,3)\tag{31}$$

式中 $T=\frac{1}{2}[A(\omega_1-\omega_1^0)^2+B(\omega_2-\omega_2^0)^2+C(\omega_3-\omega_1^0)^2]$ 为 s 截面的弹性应变势能. 记 $\boldsymbol{m}^F=\boldsymbol{e}_3\times\boldsymbol{F}$, 原理(26)可表示为 Euler-Lagrange 形式

$$\begin{aligned}W_D^*=&(\dot{\boldsymbol{F}}+\boldsymbol{f}^G)\cdot\delta\boldsymbol{r}+\\&\sum_{i=1}^{3}\left(\frac{\mathrm{d}}{\mathrm{d}s}\frac{\partial T}{\partial\dot{q}_i}-\frac{\partial T}{\partial q_i}+m_{\dot{q}_i}^F+m_{\dot{q}_i}^G\right)\delta q_i=0\end{aligned}\tag{32}$$

式中 $m_{\dot{q}_i}^F=\boldsymbol{m}^F\cdot\partial\boldsymbol{\omega}/\partial\dot{q}_i$, $\boldsymbol{m}_{\dot{q}_i}^G=\boldsymbol{m}^G\cdot\partial\boldsymbol{\omega}/\partial\dot{q}_i$. 对于不受主动力和侧面约束力作用的特殊情形, 取虚位移为平移, 推知 $\boldsymbol{F}$ 在惯性空间中为常矢量, 设与 ζ 轴平行, 可写为

$$\boldsymbol{F}=F(\sin\vartheta\sin\varphi\boldsymbol{e}_1+\sin\vartheta\cos\varphi\boldsymbol{e}_2+\cos\vartheta\boldsymbol{e}_3)\tag{33}$$

存在关系

$$\left(\frac{\partial\boldsymbol{\omega}}{\partial\dot{q}_i}\times\boldsymbol{e}_3\right)\cdot\boldsymbol{F}=\frac{\partial}{\partial q_i}(\boldsymbol{F}\cdot\boldsymbol{e}_3)\tag{34}$$

引进新函数

$$\varGamma=T+V\tag{35}$$

其中 $V=-\boldsymbol{F}\cdot\boldsymbol{e}_3$. 式(32)可写作

$$W_D^* = \sum_{i=1}^{3}\left(\frac{\mathrm{d}}{\mathrm{d}s}\frac{\partial\Gamma}{\partial\dot{q}_i}-\frac{\partial\Gamma}{\partial q_i}+m_{\dot{q}_i}^G\right)\delta q_i=0 \tag{36}$$

原理(36)也可化作 Nielsen 形式

$$\sum_{i=1}^{3}\left(\frac{\partial\dot{\Gamma}}{\partial\ddot{q}_i}-2\frac{\partial\Gamma}{\partial q_i}+m_{\dot{q}_i}^G\right)\delta q_i=0 \tag{37}$$

和 Appell 形式

$$\sum_{i=1}^{3}\left(\frac{\partial\Pi}{\partial\ddot{q}_i}+m_{\dot{q}_i}^G\right)\delta q_i=0 \tag{38}$$

其中函数 Π 定义为

$$\Pi=\frac{1}{2}[A(\dot{\omega}_1-\dot{\omega}_1^0)^2+B(\dot{\omega}_2-\dot{\omega}_2^0)^2+C(\dot{\omega}_3-\dot{\omega}_3^0)^2]+[\boldsymbol{\omega}\times\boldsymbol{M}]+(\boldsymbol{e}_3\times\boldsymbol{F})]\cdot\dot{\boldsymbol{\omega}} \tag{39}$$

5. Hamilton **原理和** Hamilton **正则方程**

将式(26)乘 ds 后对 s 沿杆长积分,化作积分变分原理

$$\int_0^l W_D^*\,\mathrm{d}s=-\int_0^l(\boldsymbol{M}\cdot\tilde{\delta}\boldsymbol{\omega}+\boldsymbol{f}^G\cdot\delta\boldsymbol{r}+\boldsymbol{m}^G\cdot\delta\boldsymbol{\Phi})\,\mathrm{d}s-(\boldsymbol{F}\cdot\delta\boldsymbol{r}+\boldsymbol{M}\cdot\delta\boldsymbol{\Phi})|_{s=0}+(\boldsymbol{F}\cdot\delta\boldsymbol{r}+\boldsymbol{M}\cdot\delta\boldsymbol{\Phi})|_{s=l}=0 \tag{40}$$

在忽略体积力和接触力的条件下,截面主矢为常矢量.原理(40)化为

$$\int_0^l W_D^* \mathrm{d}s = -\int_0^l [\boldsymbol{M} \cdot \tilde{\delta}\boldsymbol{\omega} - \delta(\boldsymbol{F} \cdot \boldsymbol{e}_3)]\mathrm{d}s - (\boldsymbol{M} \cdot \delta\boldsymbol{\Phi})_{s=0} + (\boldsymbol{M} \cdot \delta\boldsymbol{\Phi})_{s=l} = 0 \tag{41}$$

如果杆服从线性本构关系(29),则式(41)化为

$$\int_0^l W_D^* \mathrm{d}s = -\int_0^l (\delta\Gamma)\mathrm{d}s - (\boldsymbol{M} \cdot \delta\boldsymbol{\Phi})_{s=0} + (\boldsymbol{M} \cdot \delta\boldsymbol{\Phi})_{s=l} = 0 \tag{42}$$

直接计算变分,注意到微分和变分的交换关系,式(42)化作Euler-Lagrange形式

$$\int_0^1 W_D^* \mathrm{d}s = \int_0^l \sum_{i=1}^3 \left\{\left[\frac{\mathrm{d}}{\mathrm{d}s}\left(\frac{\partial \Gamma}{\partial \dot{q}_i}\right) - \frac{\partial \Gamma}{\partial q_i}\right]\delta q_i\right\}\mathrm{d}s + \left[\sum_{i=1}^3 \left(\frac{\partial \Gamma}{\partial \dot{q}_i} - \boldsymbol{M} \cdot \boldsymbol{\omega}_i\right)\delta q_i\right]_{s=0} + \left[\sum_{i=1}^3 \left(-\frac{\partial \Gamma}{\partial \dot{q}_i} + \boldsymbol{M} \cdot \boldsymbol{\omega}_i\right)\delta q_i\right]_{s=l} = 0 \tag{43}$$

考虑到虚角位移 δq_i 的任意性和端点坐标变分 $(\delta q_i)_{s=0}$,$(\delta q_i)_{s=l}$的独立性,从式(43)导出

$$\begin{cases} \dfrac{\mathrm{d}}{\mathrm{d}s}\left(\dfrac{\partial \Gamma}{\partial \dot{q}_i}\right) - \dfrac{\partial \Gamma}{\partial q_i} = 0 \\ \left(\dfrac{\partial \Gamma}{\partial \dot{q}_i} - \boldsymbol{M} \cdot \boldsymbol{\omega}_i\right)_{s=0} = 0 \quad (i = 1,2,3) \\ \left(-\dfrac{\partial \Gamma}{\partial \dot{q}_i} + \boldsymbol{M} \cdot \boldsymbol{\omega}_i\right)_{s=l} = 0 \end{cases} \tag{44}$$

上述第 1 组方程是杆的 Euler-Lagrange 形式的平衡微

分方程,它等同于动力学中的第 2 类 Lagrange 方程,只要将 Γ 比作 Lagrange 函数,即将弹性应变势能 T 比作动能,$\boldsymbol{F}\cdot\boldsymbol{e}_3$ 比作势能. 第 2,3 组方程是弧坐标端面的边界条件. 由此表明,式(44)中杆两端力偶的虚功仅与边界条件相关. 形式上可以把变分问题(42)化为泛函

$$S=\int_0^l \Gamma \mathrm{d}s \tag{45}$$

在端点变分条件 $(\delta q_i)_{s=0}=(\delta q_i)_{s=l}=0\ (i=1,2,3)$ 下的驻值问题. 称 S 为弹性杆的 Hamilton 作用量,则有:

Hamilton **原理** 除端点外不受力作用的弹性杆的实际平衡状态与可能平衡状态的区别在于前者使弹性杆的 Hamilton 作用量取驻值,即

$$\delta S=0 \tag{46}$$

原理(46)即弹性力学中的最小势能原理. 进一步推知,稳定平衡位形使杆的总势能,即 Hamilton 作用量取极小值.

引入正则变量:$q_i, p_i=\partial\Gamma/\partial\dot{q}_i\ (i=1,2,3)$,定义 Hamilton 函数

$$H(q_1,q_2,q_3,p_1,p_2,p_3)=$$

$$\left(\sum_{i=1}^{3} p_i\dot{q}_i-\Gamma\right)\Big|_{\dot{q}_j=\dot{q}_j(q_1,q_2,q_3,p_1,p_2,p_3)}=T-V \tag{47}$$

直接计算偏导数,导出

$$\dot{q}_i = \frac{\partial H}{\partial p_i}, \dot{p}_i = -\frac{\partial H}{\partial q_i} \quad (i = 1,2,3) \tag{48}$$

此即描述 Kirchhoff 杆平衡的 Hamilton 正则方程.

6. Jourdain **变分与虚功率原理**

在位形保持不变的前提下,约束所允许的、假想的截面内一点的弧坐标速度 $\boldsymbol{v}_i = \mathrm{d}\boldsymbol{r}_i/\mathrm{d}s$ 的变更称为虚速度,记为 $\delta_J \boldsymbol{v}_i$. 称 δ_J 为 Kirchhoff 杆的 Jourdain 变分. 此处的速度是指点的矢径相对弧坐标的变化率. 同一截面上不同点的虚速度形成截面的虚弯扭度.

定义 5(截面虚弯扭度) 在保持截面位形的前提下,约束所允许的、与弧坐标变化无关的、假想的截面弯扭度的变更定义为虚弯扭度,记为 $\delta_J\omega$.

由式(17a)导出虚弯扭度为

$$\delta_J \boldsymbol{\omega} = \sum_{i=1}^{3} \boldsymbol{\omega}_{\dot{q}_i} \delta_J \dot{q}_i \tag{49}$$

给定点的虚速度和虚弯扭度,存在如下关系

$$\delta_J \dot{\boldsymbol{R}}_i = \delta_J \dot{\boldsymbol{r}} + \delta_J \dot{\boldsymbol{b}}_i, \delta_J \dot{\boldsymbol{b}}_i = \delta_J \boldsymbol{\omega} \times \dot{\boldsymbol{b}}_i \tag{50}$$

约束(18)和(20)加在虚弯扭度上的限制方程分别为

$$\begin{cases} \delta_J \dot{g}_i = \sum_{j=1}^{6} \dfrac{\partial g_i}{\partial q_j} \delta_J \dot{q}_j = 0 \\ \delta_J h_i = \sum_{j=1}^{6} \dfrac{\partial h_i}{\partial \dot{q}_j} \delta_J \dot{q}_j = 0 \end{cases} \tag{51}$$

作用在 V_i 上的所有力在虚速度上所作的虚功率 W_J^* 为

$$W_J^* = (\dot{\boldsymbol{F}} + \boldsymbol{f}^G) \cdot \delta_J \dot{\boldsymbol{r}} + \left(\frac{\mathrm{d}\boldsymbol{M}}{\mathrm{d}s} + \boldsymbol{e}_3 \times \boldsymbol{F} + \boldsymbol{m}^G\right) \cdot \delta_J \boldsymbol{\omega} \tag{52}$$

上式的推导中用到理想约束条件

$$\boldsymbol{f}^C \cdot \delta_J \dot{\boldsymbol{r}} + \boldsymbol{m}^C \cdot \delta_J \boldsymbol{\omega} = 0 \tag{53}$$

Jourdain **原理** 受有理想双面约束的 Kirchhoff 杆平衡时,对满足理想约束条件(53)的任意虚速度,截面作用力的虚功率为零,即

$$W_J^* = (\dot{\boldsymbol{F}} + \boldsymbol{f}^G) \cdot \delta_J \dot{\boldsymbol{r}} + \left(\frac{\mathrm{d}\boldsymbol{M}}{\mathrm{d}s} + \boldsymbol{e}_3 \times \boldsymbol{F} + \boldsymbol{m}^G\right) \cdot \delta_J \boldsymbol{\omega} = 0 \tag{54}$$

可用 Euler 角表示为

$$W_J^* = (\dot{\boldsymbol{F}} + \boldsymbol{f}^G) \cdot \delta_J \dot{\boldsymbol{r}} + \sum_{i=1}^{3} \left[\frac{\mathrm{d}}{\mathrm{d}s}\left(\boldsymbol{M} \cdot \frac{\partial \boldsymbol{\omega}}{\partial \dot{q}_i}\right) - \boldsymbol{M} \cdot \frac{\partial \boldsymbol{\omega}}{\partial q_i} + \boldsymbol{F} \cdot \left(\frac{\partial \boldsymbol{\omega}}{\partial \dot{q}_i} \times \boldsymbol{e}_3\right) + \boldsymbol{m}^G \cdot \frac{\partial \boldsymbol{\omega}}{\partial \dot{q}_i}\right] \delta_J \dot{q}_i = 0 \tag{55}$$

Jourdain 原理指出杆截面的实际平衡状态与位置相同,但弯扭度不同的可能平衡状态的区别. 如果截面主矢 $\boldsymbol{F}$ 为常矢量,注意到式(34),式(55)化作

$$W_J^* = \sum_{i=1}^{3} \left[\frac{\mathrm{d}}{\mathrm{d}s}\left(\boldsymbol{M} \cdot \frac{\partial \boldsymbol{\omega}}{\partial \dot{q}_i}\right) - \boldsymbol{M} \cdot \frac{\partial \boldsymbol{\omega}}{\partial q_i} + \frac{\partial}{\partial q_i}(\boldsymbol{F} \cdot \boldsymbol{e}_3) + m_{\dot{q}_i}^G \right] \delta_J \dot{q}_i = 0 \tag{56}$$

满足线性本构关系(29)时,式(56)化作 Euler-Lagrange 形式

$$W_J^* = \sum_{i=1}^{3} \left(\frac{\mathrm{d}}{\mathrm{d}s} \frac{\partial \Gamma}{\partial \dot{q}_i} - \frac{\partial \Gamma}{\partial q_i} + m_{\dot{q}_i}^G \right) \delta_J \dot{q}_i = 0 \tag{57}$$

同理也可化作 Nielsen 形式和 Appell 形式.

7. Gauss **变分与** Gauss **原理、**Gauss **最小拘束原理**

在保持点的位形和弧坐标速度不变的前提下,约束所允许的、假想的截面内一点的弧坐标加速度 $\boldsymbol{a}_i = \mathrm{d}\boldsymbol{v}_i/\mathrm{d}s$ 的变更,称为点的虚加速度,记为 $\delta_G \boldsymbol{a}_i$,称 δ_G 为 Gauss 变分. 在 Kirchhoff 假定下,同一截面上不同点的虚加速度形成截面的虚角加速度 $\delta_G \boldsymbol{\alpha}$,其中 $\boldsymbol{\alpha} = \mathrm{d}\boldsymbol{\omega}/\mathrm{d}s$.

定义 6(截面虚角加速度) 在保持截面位形和弯扭度的前提下,约束所允许的、与弧坐标变化无关的、假想的截面角加速度变更定义为虚角加速度,记为 $\delta_G \alpha$.

由式(17a)和角加速度的定义,角加速度的 Gauss 变分为

$$\delta_G \boldsymbol{\alpha} = \sum_{i=1}^{3} \boldsymbol{\omega}_{\dot{q}_i} \delta_G \ddot{q}_i \tag{58}$$

给定点的虚加速度和虚角加速度，存在如下关系

$$\delta_G \ddot{\boldsymbol{R}}_i = \delta_G \ddot{\boldsymbol{r}} + \delta_G \ddot{\boldsymbol{b}}_i, \quad \delta_G \ddot{\boldsymbol{b}}_i = \delta_G \boldsymbol{\alpha} \times \ddot{\boldsymbol{b}}_i \tag{59}$$

约束(18)和(20)加在 Euler 虚角加速度上的限制方程分别为

$$\begin{cases} \delta_G \bar{g}_i = \sum_{j=1}^{6} \frac{\partial g_i}{\partial q_j} \delta_G \bar{q}_j = 0 \\ \delta_G \dot{h}_i = \sum_{j=1}^{6} \frac{\partial h_i}{\partial \dot{q}_j} \delta_G \ddot{q}_j = 0 \end{cases} \tag{60}$$

V_i 上的作用力与虚加速度的数量积 W_G^* 在理想约束条件

$$\boldsymbol{f}^C \cdot \delta_G \ddot{\boldsymbol{r}} + \boldsymbol{m}^C \cdot \delta_G \boldsymbol{\alpha} = 0 \tag{61}$$

下为

$$W_G^* = (\dot{\boldsymbol{F}} + \boldsymbol{f}^G) \cdot \delta_G \ddot{\boldsymbol{r}} + \left(\frac{\mathrm{d}\boldsymbol{M}}{\mathrm{d}s} + \boldsymbol{e}_3 \times \boldsymbol{F} + \boldsymbol{m}^G \right) \cdot \delta_G \boldsymbol{\alpha} \tag{62}$$

Gauss **原理** 受有理想双面约束的 Kirchhoff 杆平衡时，对满足理想约束条件(61)的任意虚加速度，有

$$W_G^* = (\dot{\boldsymbol{F}} + \boldsymbol{f}^G) \cdot \delta_G \ddot{\boldsymbol{r}} + \left(\frac{\mathrm{d}\boldsymbol{M}}{\mathrm{d}s} + \boldsymbol{e}_3 \times \boldsymbol{F} + \boldsymbol{m}^G \right) \cdot \delta_G \boldsymbol{\alpha} = 0 \tag{63}$$

或用 Euler 角表示为

$$W_G^* = (\dot{\boldsymbol{F}} + \boldsymbol{f}^G) \cdot \delta_G \ddot{\boldsymbol{r}} + \sum_{i=1}^{3} \left[\frac{\mathrm{d}}{\mathrm{d}s} \left(\boldsymbol{M} \cdot \frac{\partial \boldsymbol{\omega}}{\partial \dot{q}_i} \right) - \right.$$

$$\boldsymbol{M}\cdot\frac{\partial\boldsymbol{\omega}}{\partial q_i}+\boldsymbol{F}\cdot\left(\frac{\partial\boldsymbol{\omega}}{\partial\dot{q}_i}\times\boldsymbol{e}_3\right)+\boldsymbol{m}^G\cdot\frac{\partial\boldsymbol{\omega}}{\partial\dot{q}_i}\Bigg]\delta_G\ddot{q}_i$$

$$=0 \tag{64}$$

或

$$W_G^*=(\dot{\boldsymbol{F}}+\boldsymbol{f}^G)\cdot\delta_G\ddot{\boldsymbol{r}}+\sum_{i=1}^{3}\left(\frac{\partial\Pi}{\partial\ddot{q}_i}+m_{\dot{q}_i}^G\right)\delta_G\ddot{q}_i=0 \tag{65}$$

Gauss 原理指出了截面的实际平衡状态与位置和弯扭度相同,但角加速度不同的可能平衡状态的区别.定义自由 Kirchhoff 杆的拘束函数

$$Z=\frac{1}{2}\left\{A\left[\dot{\omega}_1+\frac{1}{A}(\omega_2M_3-\omega_3M_2-F_2+m_1)\right]^2+\right.$$

$$B\left[\dot{\omega}_2+\frac{1}{B}(\omega_3M_1-\omega_1M_3+F_1+m_2)\right]^2+$$

$$\left.C\left[\dot{\omega}_3+\frac{1}{C}(\omega_1M_2-\omega_2M_1+m_3)\right]^2\right\} \tag{66}$$

式中 $m_i=\boldsymbol{m}^G\cdot\boldsymbol{e}_i$,本构关系为式(29).

Gauss **最小拘束原理**　Kirchhoff 杆的任一截面随弧坐标的实际运动和位形与弯扭度相同,但角加速度不同的可能运动比较,拘束函数在 Gauss 变分下取驻值,即有

$$\delta_G Z=0 \tag{67}$$

计算可能运动的拘束函数 Z^* 与实际运动的拘束

函数 Z 之差，有

$$\delta_G Z = Z^* - Z = \frac{1}{2}[A(\Delta\dot{\omega}_1)^2 + B(\Delta\dot{\omega}_2)^2 + C(\Delta\dot{\omega}_3)^2] > 0 \tag{68}$$

式中 $\Delta\dot{\omega}_i = \dot{\omega}_i^* - \dot{\omega}_i$. 从而证明，实际运动对应的拘束取极小值.

8. 平衡微分方程、首次积分以及微分弧段的大曲率问题

对于自由 Kirchhoff 杆，截面形心和姿态的虚位移 $\delta\boldsymbol{r}$ 和 $\delta\boldsymbol{\Phi}$ 都是独立的，从式(26)导出 Kirchhoff 方程

$$\frac{\mathrm{d}\boldsymbol{M}}{\mathrm{d}s} + \boldsymbol{e}_3 \times \boldsymbol{F} + \boldsymbol{m}^G = \boldsymbol{0}, \boldsymbol{F} + \boldsymbol{f}^G = \boldsymbol{0} \tag{69}$$

从式(32)导出 Lagrange 方程

$$\frac{\mathrm{d}}{\mathrm{d}s}\frac{\partial T}{\partial \dot{q}_i} - \frac{\partial T}{\partial q_i} + m_{\dot{q}_i}^F + m_{\dot{q}_i}^G = 0 \quad (i = 1, 2, 3) \tag{70a}$$

$$\dot{\boldsymbol{F}} + \boldsymbol{f}^G = \boldsymbol{0} \tag{70b}$$

当 $\boldsymbol{F}$ 为常矢量时，由式(36)得

$$\frac{\mathrm{d}}{\mathrm{d}s}\frac{\partial \Gamma}{\partial \dot{q}_i} - \frac{\partial \Gamma}{\partial q_i} + m_{\dot{q}_i}^G = 0 \quad (i = 1, 2, 3) \tag{71}$$

同理从式(37)和(38)可导出 Nielsen 方程和 Appell 方程. 当杆处在约束条件(18)或(20)下，诸 δq_i 不都独立，受到的约束是式(19)或(21). 从式(32)导出带乘子的 Lagrange 方程：

几何约束下

$$\frac{\mathrm{d}}{\mathrm{d}s}\frac{\partial T}{\partial \dot{q}_i}-\frac{\partial T}{\partial q_i}+m_{\dot{q}_i}^F+m_{\dot{q}_i}^G+\sum_{j=1}^{1或2}\lambda_j\frac{\partial g_j}{\partial q_i}=0\quad(i=1,2,3)$$

(72a)

$$\dot{F}_i+f_i^G+\sum_{j=1}^{1或2}\lambda_j\frac{\partial g_j}{\partial q_i}=0\quad(i=4,5,6)\qquad(72b)$$

非完整约束下

$$\frac{\mathrm{d}}{\mathrm{d}s}\frac{\partial T}{\partial \dot{q}_i}-\frac{\partial T}{\partial q_i}+m_{\dot{q}_i}^F+m_{\dot{q}_i}^G+\sum_{i=1}^{1或2}\lambda_J\frac{\partial h_j}{\partial \dot{q}_i}=0\quad(i=1,2,3)$$

(73a)

$$\dot{F}_i+f_i^G+\sum_{j=1}^{1或2}\lambda_j\frac{\partial h_j}{\partial \dot{q}_i}=0\quad(i=4,5,6)\qquad(73b)$$

也可化作 Nielsen 或 Appell 形式的方程.

当 $m_{\dot{q}_i}^G=0$ 时,式(71)与保守系统的 Lagrange 方程形式完全相同

$$\frac{\mathrm{d}}{\mathrm{d}s}\frac{\partial \Gamma}{\partial \dot{q}_i}-\frac{\partial \Gamma}{\partial q_i}=0\quad(i=1,2,3)\qquad(74)$$

其首次积分和利用首次积分使原方程降阶的问题与分析力学的积分理论完全等同.

(1)若$\partial\Gamma/\partial q_i=0$,则存在“循环积分”:$\partial\Gamma/\partial\dot{q}_i=c_i$,式中 c_i 为积分常数. 其物理意义是截面主矩关于 $\dot{q}_i$ 的分量守恒.

(2)若$\partial \Gamma / \partial s=0$,则有“能量积分”:$T-V=c$.

对于离散系统,平衡条件可用势能函数 V 表示为 $\partial V / \partial q_i=0$. 按 Kirchhoff 动力学比拟,对应于截面保持原始弯扭度,即主矩为零,平衡条件为

$$\frac{\partial F_3}{\partial q_i} \equiv 0 \quad (i=1,2,3) \tag{75}$$

稳定条件为 F_3 有极小值. 由式(33)和(75)知,$\vartheta=n\pi(n=0, \pm 1, \pm 2, \cdots)$. 当 n 为奇数时稳定,为偶数时不稳定,即带原始扭率的直杆在轴向力作用下受拉为不稳定,受压为稳定. 此处的稳定性是指关于弧坐标的 Lyapunov 稳定性.

研究 Kirchhoff 杆变形后中心线存在尖点的情形,这本质上是微分弧段内出现的大曲率问题. 设微分弧段的弧坐标分别为 s 和 $s+\Delta s$,将 $\mathrm{d}s$ 乘式(74)并在 s 和 $s+\Delta s$ 内积分

$$\int_s^{s+\Delta s}\left(\frac{\mathrm{d}}{\mathrm{d}s} \frac{\partial \Gamma}{\partial \dot{q}_i}-\frac{\partial \Gamma}{\partial q_i}+m_{\dot{q}_i}\right) \mathrm{d}s=0 \quad (i=1,2,3) \tag{76}$$

导出近似计算式

$$\Delta\left(\frac{\partial \Gamma}{\partial \dot{q}_i}\right)+\hat{m}_{\dot{q}_i}=0 \quad (i=1,2,3) \tag{77}$$

式中 $\hat{m}_{\dot{q}_i} = \int_s^{s+\Delta s} m_{\dot{q}_i} \mathrm{d}s$. 这对应于动力学中的碰撞方程. 虽然大曲率情形已越出 Kirchhoff 假定的前提,但在微分弧段内式(77)对截面弯扭度突变的近似描述是有效的.

9. 结论

(1)描述自由弹性细杆位形需要截面形心和姿态共6个广义坐标,由于存在一个非完整约束的矢量方程使自由度减为3,这个非完整约束是无须约束力就能自动实现的伪非完整约束.

(2)在单个曲面约束下,弹性细杆的广义坐标数和自由度分别减为5和2.

(3)讨论了由离散系统分析力学理论移植的超细长弹性杆平衡问题的分析力学原理,建立了 D'Alembert-Lagrange 原理、Jourdain 原理、Gauss 原理和 Gauss 最小拘束原理.

(4)根据 Kirchhoff 动力学比拟,弹性细杆的最小势能原理相似于离散系统动力学的 Hamilton 原理,建立了弹性细杆平衡问题的 Hamilton 原理和 Hamilton 正则方程.

(5)从截面平衡的变分原理导出了受约束杆的带

乘子的 Lagrange 方程.

(6)杆中心线在微小弧段内的大曲率现象可与物体的碰撞相比拟,该微小弧段杆的平衡方程与物体的碰撞方程形式相同.

北京大学数学系的郭仲衡教授早在 1979 年就证明了:非线性弹性的各变分原理(势能原理和各余能原理以及它们的推广)均可从单一的虚功原理导出,使它们在统一的框架里构成一个有机的整体.文中也给出了各原理间的内在联系及其关系图.①

1. 引言

在非线性弹性理论里,是否存在像线性理论那样应力场是唯一独立变量的余能原理,是一个近年来引起广泛兴趣的问题.讨论的焦点在于 Piola 应力张量和变形梯度的关系是否可逆,亦即相应的 Legendre 变换②③是否成立.至今被提出的非线性弹性变分理论的余能原理可概括为三大类.首先是基于 Kirchhoff 应力

---

① 摘自《应用数学和力学》,第 1 卷第 1 期,1980 年 5 月.

② COURANT R, HILBERT D. Methode der Mathematischen Physik I. 3. Auflage, bERLIN:Springer, 1968: 201-207.

③ SEWELL M J. On dual approximation principles and optimization in continuum mechanics. Philos. Trans. Roy. Soc. London, Ser., 1969(A265):319-351.

张量的 Reissner 原理[1][2][3][4][5]，它常被认为不是纯粹的余能原理，因为这里除了应力张量，位移向量也是独立变量. 其后是仅基于 Piola 应力张量 $\boldsymbol{\tau}$，一度被宣称为真正余能原理的 Levinson 原理[6]. 可是，这个原理并不总是成立的，因为，即使是各向同性的简单情形，也只有当 $\boldsymbol{\tau}^{*}\cdot\boldsymbol{\tau}$[7] 在物体的每一点有不相同的主值（难于预先知道这条件是否满足）时，Piola 应力张量和变形梯度的函数关系才是可逆的. 尽管 R. W. Ogden 给出了选择合适分支的条件，该函数的逆在整体意义下的多值性（对各向同性体至少有四个不同分支）也给实际应用带来附加困难. 总是成立的却是基于极分解的

---

① REISSNER E. On a variational theorem for finite elastic deformations. Math. Phys., 1953(32):129-135.

② TRUESDELL C, NOLL W. Non-linear field theories of mechanics, Handbuch der Physik Bd. Ⅲ/3, Berlin: Springer, 1965.

③ NEMAT-Nasser S. General variational principles in nonlinear and linear elasticity with applications. Mechanics Today vol. 1. London: Pergamon Pceas, 1972.

④ WASHIZUashizu K. Variational methods in elasticity and plasticity, ed. Ⅱ, Oxford, 1975.

⑤ WASHIZU K. Complementary Variational Principles in Elasticity and Plasticity, Lecture at the Conference on "Duality and Complementary in Mechanics of Deformable Solids", Jablonna Poland, 1977.

⑥ LEVINSON M. The complementary energy theorem in finite elasticity, Trans. ASME, Ser. E, J. Appl. Mech., 1965(87):826-828.

⑦ "＊"表示张量的共轭.

Fraeijs de Veubeke 原理[①②③④⑤⑥]. 但这里除了 Piola 应力张量,转动张量也作为独立变量出现,又没有达到人们所期待的余能原理的纯粹性. 看来,问题是否能这样提以及余能原理的其他一些问题还有待于进一步的深究.

2. 数学符号

本工作采用两点张量场法. 设物体 $\mathscr{B}$ 的参考(未变形)构型为 $\mathscr{R}$,当前(已变形)构型为 $\mathscr{r}$,分别配以两个任意的曲线坐标系 $\{X^i\}$ 和 $\{x^i\}$. 它们的基向量和度量张量相应地为

① FRAEIJS DE VEUBEKE B. A new variational principle for finite elastic displacements, Int. J. Engng. Sci., 1972(10):745-763.

② CHRISTOFFERSEN J. On Zubov's principle of stationary complementary energy and a related principle, Rep. No. 44, Danish Center for Appl. Math. and Mech., April, 1973.

③ OGDEN R W. A note on variational theorems in non-linear elastostatics, Math. Proc. Cams. Phil. Soc., 1975(77),609-615.

④ DILL E H. The complementary energy principle in nonlinear elasticity, Lett. Appl. and Engng. Sci., 1977(5): 95-106.

⑤ OGDEN R W. Inequalities associated with the inversion of elastic stress-deformation relations and their implications, Math. Proc. Camb. Phil. Soc., 1977(81):313-324.

⑥ OGDEN R W. Extremum principles in non-linear elasticity and their application to composite-I, Int. J. Solids Struct., 1978(14):265-282.

$$\boldsymbol{G}_A, \boldsymbol{G}^A, G_{AB}, G^{AB}$$

$$\boldsymbol{g}_i, \boldsymbol{g}^i, g_{ij}, g^{ij}$$

物体从 $\mathscr{B}$ 至 $\mathscr{b}$ 的变形体现为

$$\boldsymbol{x} = \boldsymbol{x}(\boldsymbol{X}) \quad 或 \quad x^i = x^i(X^A) \tag{2.1}$$

它表示物体 $\mathscr{B}$ 的质点 $\boldsymbol{X}$ 变形后占有空间位置 $\boldsymbol{x}$.

作用于点 $\boldsymbol{X}$ 的所有向量构成三维欧氏空间，不妨也记为 $\mathscr{B}$，它的任意元素均可表为 $\boldsymbol{G}_A$ 或 $\boldsymbol{G}^A$ 的线性组合，例如 $\boldsymbol{V} = V^A \boldsymbol{G}_A = V_A \boldsymbol{G}^A$. 类似地有 $\mathscr{b}$，例如 $\boldsymbol{v} = \boldsymbol{v}^i \boldsymbol{g}_i = v_i \boldsymbol{g}^i$. 又考虑四个张量积空间：$\mathscr{B} \otimes \mathscr{B}$，$\mathscr{B} \otimes \mathscr{b}$，$\mathscr{b} \otimes \mathscr{B}$，$\mathscr{b} \otimes \mathscr{b}$，它们的任意元素分别可表成 $\boldsymbol{G}_A \otimes \boldsymbol{G}_B$，$\boldsymbol{G}_A \otimes \boldsymbol{G}_i$，$\boldsymbol{g}_i \otimes \boldsymbol{G}_A$ 和 $\boldsymbol{g}_i \otimes \boldsymbol{g}_j$（也可代入以逆变基向量. 今后我们省去张量积符号而用 Gibbs 并矢记法）的线性组合，记为，例如

$$\boldsymbol{R} = R^{AB} \boldsymbol{G}_A \boldsymbol{G}_B, \boldsymbol{S} = S^{Ai} \boldsymbol{G}_A \boldsymbol{g}_i$$

$$\boldsymbol{T} = T^{iA} \boldsymbol{g}_i \boldsymbol{G}_A, \boldsymbol{U} = U^{ij} \boldsymbol{g}_i \boldsymbol{g}_j$$

通过点积运算，张量积（9 维）空间的元素变成从 3 维空间到 3 维空间的线性变换（或叫张量），例如

$$\boldsymbol{W} = \boldsymbol{R} \cdot \boldsymbol{V} = (R^{AB} \boldsymbol{G}_A \boldsymbol{G}_B) \cdot (V_D \boldsymbol{G}^D) = R^{AB} V_B \boldsymbol{G}_A = W^A \boldsymbol{G}_A$$

$$\boldsymbol{w} = \boldsymbol{T} \cdot \boldsymbol{V} = (T^{iA} \boldsymbol{g}_i \boldsymbol{G}_A) \cdot (V_B \boldsymbol{G}^B) = T^{iA} V_A \boldsymbol{g}_i = w^i \boldsymbol{g}_i$$

$\mathscr{B} \otimes \mathscr{b}$ 或 $\mathscr{b} \otimes \mathscr{B}$ 的元素，例如 $\boldsymbol{S}$ 和 $\boldsymbol{T}$，称为两点张量. 并矢的次序是本质的. “ * ”表示张量的共轭：$\boldsymbol{R}^* = R^{AB} \boldsymbol{G}_B \boldsymbol{G}_A$，$\boldsymbol{T}^* = T^{iA} \boldsymbol{G}_A \boldsymbol{g}_i$. $\mathscr{B} \otimes \mathscr{B}$ 和 $\mathscr{b} \otimes \mathscr{b}$ 的单位元素（单

位张量)是

$$\overset{\ll}{I} = G_{AB}\boldsymbol{G}^A\boldsymbol{G}^B = \delta^A_B\boldsymbol{G}_A\boldsymbol{G}^B = \cdots$$

$$\overset{\gg}{I} = g_{ij}\boldsymbol{g}^i\boldsymbol{g}^j = \delta^i_j\boldsymbol{g}_i\boldsymbol{g}^j = \cdots$$

而 $\mathscr{R}\otimes\mathscr{i}$ 和 $\mathscr{i}\otimes\mathscr{R}$ 的单位元素(称为转移张量)是

$$\overset{<>}{I} = g_A^{\cdot\, i}\boldsymbol{G}^A\boldsymbol{g}_i = g_{Ai}\boldsymbol{G}^A\boldsymbol{g}^i = \cdots$$

$$\overset{><}{I} = g_i^{\cdot\, A}\boldsymbol{g}^i\boldsymbol{G}_A = g^{iA}\boldsymbol{g}_i\boldsymbol{G}_A = \cdots$$

转移张量的几何意义是将 $\mathscr{i}$ 或 $\mathscr{R}$ 的向量不变地平移至 $\mathscr{R}$ 或 $\mathscr{i}$,其中 $g_A^{\cdot\, i} = \boldsymbol{G}_A \cdot \boldsymbol{g}^i = \boldsymbol{g}^i \cdot \boldsymbol{G}_A = g^i_{\cdot\, A}$,故可统一记为 g^i_A,g^A_i 亦然. 利用转移张量可将两点张量化成一点张量,或反之. 如果在张量的基向量并矢间用点积代替张量积,就得该张量的迹,如

$$\mathrm{tr}\,\boldsymbol{R} \overset{dj}{=\!=} R^{AB}\boldsymbol{G}_A \cdot \boldsymbol{G}_B = R^A_{\cdot\, A}$$

$$\mathrm{tr}\,\boldsymbol{S} \overset{dj}{=\!=} S^A_{\cdot\, i}\boldsymbol{G}_A \cdot \boldsymbol{g}^i = g^i_A S^A_{\cdot\, i}$$

张量的迹是标量,两张量的双点积也是标量

$$\boldsymbol{R}:\boldsymbol{S} \overset{dj}{=\!=} \mathrm{tr}(\boldsymbol{R}^* \cdot \boldsymbol{S})$$

张量场 $\boldsymbol{R}$ 和 $\boldsymbol{U}$ 的绝对微分是

$$\mathrm{d}\boldsymbol{R} = (\boldsymbol{R}\boldsymbol{\nabla}) \cdot \mathrm{d}\boldsymbol{X}$$

$$\mathrm{d}\boldsymbol{U} = (\boldsymbol{U}\mathring{\boldsymbol{\nabla}}) \cdot \mathrm{d}\boldsymbol{x}$$

其中绝对微商(梯度)

$$\boldsymbol{R}\boldsymbol{\nabla} = R^{BC}_{;A}\boldsymbol{G}_B\boldsymbol{G}_C\boldsymbol{G}^A$$

$$\boldsymbol{u}\mathring{\boldsymbol{\nabla}} = U^{jk}_{;i}\boldsymbol{g}_j\boldsymbol{g}_k\boldsymbol{g}^i$$

$(\quad)_{;A}$和$(\quad)_{;i}$分别表示在$\mathscr{R}$和$\mathscr{i}$的协变微商. 对于两点张量场,绝对微商应理解为全绝对微商

$$\boldsymbol{R}\cdot\boldsymbol{\nabla} = R^{BC}_{;A}\boldsymbol{G}_B\boldsymbol{G}_C\cdot\boldsymbol{G}^A = R^{BA}_{;A}\boldsymbol{G}_B$$

$$\boldsymbol{R}\times\boldsymbol{\nabla} = R^{BC}_{;A}\boldsymbol{G}_B\boldsymbol{G}_C\times\boldsymbol{G}^A$$

分别称为$\boldsymbol{R}$的散度和旋度. 单位张量和转移张量对于绝对微商有如常量. 在今后的全部讨论里,我们认为所有出现各量已通过转移张量预先化成在$\mathscr{R}$的一点张量,这时用$\boldsymbol{I}$表示$\overset{\ll}{\boldsymbol{I}}$. 这种做法称为 Lagrange 描述法.

3. **应变和应力**

在变形(2.1)中,物体质点$\boldsymbol{X}$的位移向量为

$$\boldsymbol{u}(\boldsymbol{X}) = \boldsymbol{x}(\boldsymbol{X}) - \boldsymbol{X} \tag{3.1}$$

变形梯度

$$\boldsymbol{F} = \boldsymbol{x}\boldsymbol{\nabla} = \boldsymbol{I} + \boldsymbol{u}\boldsymbol{\nabla} \tag{3.2}$$

的极分解为

$$\boldsymbol{F} = \boldsymbol{R}\cdot\boldsymbol{U} \tag{3.3}$$

其中$\boldsymbol{U}$是(右)伸长张量,代表纯变形,$\boldsymbol{R}$是转动张量($\boldsymbol{R}^*\cdot\boldsymbol{R} = \boldsymbol{R}\cdot\boldsymbol{R}^* = \boldsymbol{I}$). $\boldsymbol{U}$是唯一确定的正定对称张量(它的主值全大于0),满足

$$\boldsymbol{U}^2 = \boldsymbol{F}^*\cdot\boldsymbol{F} \overset{dj}{=\!=} \boldsymbol{C} \tag{3.4}$$

$\boldsymbol{C}$ 称为 Green 应变张量. 今后还用到 Almansi 应变张量

$$\boldsymbol{E}=\frac{1}{2}(\boldsymbol{C}-\boldsymbol{I})=\frac{1}{2}[\boldsymbol{u}\boldsymbol{\nabla}+\boldsymbol{\nabla}\boldsymbol{u}+(\boldsymbol{\nabla}\boldsymbol{u})\cdot(\boldsymbol{u}\boldsymbol{\nabla})] \tag{3.5}$$

其中 $\boldsymbol{\nabla}\boldsymbol{u}=(\boldsymbol{u}\boldsymbol{\nabla})^{*}$.

具有描述应力状态直接物理意义的是 Cauchy 应力张量 $\boldsymbol{t}$,它作用于构型 $\mathscr{i}$ 上的单位向量 $\boldsymbol{n}$ 给出以 $\boldsymbol{n}$ 为法向的单位面积元的接触力 $\boldsymbol{t}_n$

$$\boldsymbol{t}\cdot\boldsymbol{n}=\boldsymbol{t}_n \tag{3.6}$$

$\boldsymbol{t}$ 满足 Cauchy(力)平衡方程(为叙述简单计,设无体力)

$$\boldsymbol{t}\cdot\mathring{\boldsymbol{\nabla}}=\boldsymbol{0} \tag{3.7}$$

及力矩平衡条件

$$\boldsymbol{t}=\boldsymbol{t}^{*} \tag{3.8}$$

即 $\boldsymbol{t}$ 是对称张量. 平衡方程(3.7)是 Euler 型的. 为了得到 Lagrange 型平衡方程,人们引进了 Piola 应力张量 $\boldsymbol{\tau}$ 和 Kirchhoff 应力张量 $\boldsymbol{T}$

$$\boldsymbol{\tau}\cdot\boldsymbol{N}=\boldsymbol{F}\cdot(\boldsymbol{T}\cdot\boldsymbol{N})=\sigma_n\boldsymbol{t}_n=\boldsymbol{T}_N \tag{3.9}$$

其物理意义是:$\boldsymbol{\tau}$ 作用于参考构型 $\mathscr{R}$ 上面积元的单位法向量 $\boldsymbol{N}$ 给出相对应的以 $\boldsymbol{n}$ 为法向但面积为 σ_n 的面积元上的接触力 $\boldsymbol{T}_N$(σ_n 是面积元变形后和变形前的面积比),而 $\boldsymbol{T}$ 作用于 $\boldsymbol{N}$ 后还要经过变形梯度的作用

才给出上述接触力. 这三个应力张量的相互关系是

$$\boldsymbol{\tau}=Jt\cdot\overset{-1}{\boldsymbol{F}}{}^{*}=\boldsymbol{F}\cdot\boldsymbol{T} \tag{3.10}$$

$$\boldsymbol{T}=\overset{-1}{\boldsymbol{F}}\cdot\boldsymbol{\tau}=J\overset{-1}{\boldsymbol{F}}\cdot\boldsymbol{t}\cdot\overset{-1}{\boldsymbol{F}}{}^{*} \tag{3.11}$$

$$\boldsymbol{t}=j\boldsymbol{F}\cdot\boldsymbol{T}\cdot\boldsymbol{F}^{*}=j\boldsymbol{\tau}\cdot\boldsymbol{F}^{*} \tag{3.12}$$

其中(-1)表示逆,j 是体积元变形前后的容积比,而 $J=1/j$. Piola 张量 $\boldsymbol{\tau}$ 满足 Boussinesq(力)平衡方程

$$\boldsymbol{\tau}\cdot\boldsymbol{\nabla}=\boldsymbol{0} \tag{3.13}$$

Kirchhoff 张量 $\boldsymbol{T}$ 满足 Kirchhoff(力)平衡方程

$$(\boldsymbol{F}\cdot\boldsymbol{T})\cdot\boldsymbol{\nabla}=\boldsymbol{0} \tag{3.14}$$

从(3.12)可看到,当 $\boldsymbol{T}$ 是对称张量时,力矩平衡条件自然满足,而对 $\boldsymbol{\tau}$ 的力矩平衡条件是

$$\boldsymbol{\tau}\cdot\boldsymbol{F}^{*}=\boldsymbol{F}\cdot\boldsymbol{\tau}^{*} \tag{3.15}$$

今后认为,$\boldsymbol{t}$ 和 $\boldsymbol{T}$ 都是对称张量. 此外,下文将论及,从功的角度还引进了对余能原理起重要作用的 Jaumann 应力张量 $\boldsymbol{S}$

$$\boldsymbol{S}=\frac{1}{2}(\boldsymbol{T}\cdot\boldsymbol{U}+\boldsymbol{U}\cdot\boldsymbol{T}) \tag{3.16}$$

将

$$\boldsymbol{T}=\overset{-1}{\boldsymbol{F}}\cdot\boldsymbol{\tau}=\overset{-1}{\boldsymbol{U}}\cdot\boldsymbol{R}^{*}\cdot\boldsymbol{\tau}=\boldsymbol{\tau}^{*}\cdot\boldsymbol{R}\cdot\overset{-1}{\boldsymbol{U}}$$

代入上式,又得 Jaumann 张量的另一表达式

$$\boldsymbol{S}=\frac{1}{2}(\boldsymbol{\tau}^{*}\cdot\boldsymbol{R}+\boldsymbol{R}^{*}\cdot\boldsymbol{\tau}) \tag{3.17}$$

4. 共轭变量和 Legendre 变换

物体弹性的数学叙述为应力应变的一一对应. 超弹性材料具有贮能函数 Σ(每单位参考构型体积),它可以是任意应变度量,如 $\boldsymbol{E}$,$\boldsymbol{U}$ 等的函数,也可以通过应变度量作为变形梯度 $\boldsymbol{F}$ 的复合函数. 为使公式简短,今后常用$(\dot{\ })$代替 $\delta(\quad)$. 在任意虚位移 $\dot{\boldsymbol{u}} \equiv \delta\boldsymbol{u}$ 中,贮能函数的增量,即物体单位未变形体积吸收的虚功是

$$\dot{\Sigma} = J\boldsymbol{t} : (\dot{\boldsymbol{u}}\overset{\circ}{\boldsymbol{\nabla}}) \tag{4.1}$$

依次将式$(3.12)_{2,1}$代入,又有

$$\begin{aligned}
\dot{\Sigma}(\boldsymbol{\tau}\cdot\boldsymbol{F}^{*}) : (\dot{\boldsymbol{u}}\overset{\circ}{\boldsymbol{\nabla}}) &= \boldsymbol{\tau} : [(\dot{\boldsymbol{u}}\overset{\circ}{\boldsymbol{\nabla}})\cdot\boldsymbol{F}] \\
&= \boldsymbol{\tau} : [(\dot{\boldsymbol{u}}\overset{\circ}{\boldsymbol{\nabla}})\cdot \\
&\quad (\boldsymbol{x}\boldsymbol{\nabla})] \\
&= \boldsymbol{\tau} : (\dot{\boldsymbol{u}}\boldsymbol{\nabla}) \\
&= \boldsymbol{\tau} : [(\boldsymbol{x}+\boldsymbol{u}\boldsymbol{\nabla}-\boldsymbol{x}\boldsymbol{\nabla})] = \boldsymbol{\tau} : \dot{\boldsymbol{F}}
\end{aligned} \tag{4.2}$$

$$\begin{aligned}
\dot{\Sigma} &= (\boldsymbol{F}\cdot\boldsymbol{T}) : \dot{\boldsymbol{F}} = \boldsymbol{T} : (\boldsymbol{F}^{*}\cdot\dot{\boldsymbol{F}}) \\
&= \boldsymbol{T} : \frac{1}{2}\delta(\boldsymbol{F}^{*}\cdot\boldsymbol{F}) = \boldsymbol{T} : \dot{\boldsymbol{E}}
\end{aligned} \tag{4.3}$$

上式还可以写成$\frac{1}{2}\boldsymbol{T} : \dot{\boldsymbol{C}}$,但这并不带来新内容. 将分析力学广义坐标和广义力的概念推广至变形体力学,我

们可以把任意应变度量看作广义坐标，而从 $\dot{\Sigma}$ 的表达式得出对应的广义应力[1][2]. 取下面要用到的伸长张量 $\boldsymbol{U}$,将式(3.5)和(3.4)代入上式，得

$$\dot{\Sigma} = \boldsymbol{T}:\frac{1}{2}(\boldsymbol{U}\cdot\dot{\boldsymbol{U}}+\dot{\boldsymbol{U}}\cdot\boldsymbol{U}) = \frac{1}{2}(\boldsymbol{T}\cdot\boldsymbol{U}+\boldsymbol{U}\cdot\boldsymbol{T}):\dot{\boldsymbol{U}} = \boldsymbol{S}:\dot{\boldsymbol{U}} \tag{4.4}$$

$\boldsymbol{S}$ 就是上面所提及的 Jaumann 应力张量.

在上面虚功的四个表达式中，唯有 $\dot{\boldsymbol{E}}$ 和 $\dot{\boldsymbol{U}}$ 是应变度量的增量. 因此就本文所提及的应力张量，只有 $\boldsymbol{T}$ 和 $\boldsymbol{S}$ 才满足广义应力定义的要求. 分别将 $\boldsymbol{E}$ 和 $\boldsymbol{U}$ 作为 Σ 的变量，从(4.3),(4.4)两式我们有

$$\boldsymbol{T}(\boldsymbol{E}) = \frac{\mathrm{d}\Sigma}{\mathrm{d}\boldsymbol{E}} \tag{4.5}$$

和

$$\boldsymbol{S}(\boldsymbol{U}) = \frac{\mathrm{d}\Sigma}{\mathrm{d}\boldsymbol{U}} \tag{4.6}$$

应力应变关系的一一对应导致 $\boldsymbol{T}(\boldsymbol{E})$ 和 $\boldsymbol{S}(\boldsymbol{U})$ 是整体可逆的. 这种成对的变量我们称为共轭应力应变变量[3],并记作$(\boldsymbol{T},\boldsymbol{E})$和$(\boldsymbol{S},\boldsymbol{U})$. 对每一对共轭变量，可以

---

① HILL R. On constitutive inequalities for simple materials-I, Mech. Phys. Solids, 1968(16): 229-242.

② NOVOZHILOV V V. Theory of elasticity. London: Pergamon, (1961).

③ HILL R. On constitutive inequalities for simple materials-I, J. Mech. Phys. Solids, 1968(16): 229-242.

通过 Legendre 变换定义相应的余能

$$\boldsymbol{\Sigma}^{C}(\boldsymbol{T})=\boldsymbol{T}:\boldsymbol{E}(\boldsymbol{T})-\boldsymbol{\Sigma}[\boldsymbol{E}(\boldsymbol{T})] \tag{4.7}$$

$$\tilde{\boldsymbol{\Sigma}}^{C}(\boldsymbol{S})=\boldsymbol{S}:\boldsymbol{U}(\boldsymbol{S})-\boldsymbol{\Sigma}[\boldsymbol{U}(\boldsymbol{S})] \tag{4.8}$$

并且有

$$\boldsymbol{E}(\boldsymbol{T})=\frac{\mathrm{d}\boldsymbol{\Sigma}^{C}}{\mathrm{d}\boldsymbol{T}} \tag{4.9}$$

和

$$\boldsymbol{U}(\boldsymbol{S})=\frac{\mathrm{d}\tilde{\boldsymbol{\Sigma}}^{C}}{\mathrm{d}\boldsymbol{S}} \tag{4.10}$$

由于

$$\boldsymbol{S}:\boldsymbol{U}=(\boldsymbol{T}\cdot\boldsymbol{U}):\boldsymbol{U}=\boldsymbol{T}:\boldsymbol{U}^{2}=2\boldsymbol{T}:\boldsymbol{E}+\operatorname{tr}\boldsymbol{T}$$

我们有

$$\tilde{\boldsymbol{\Sigma}}^{C}=\boldsymbol{\Sigma}^{C}+\boldsymbol{T}:\boldsymbol{E}+\operatorname{tr}\boldsymbol{T} \tag{4.11}$$

我们约定分别称 $\boldsymbol{\Sigma}^{C}$ 和 $\tilde{\boldsymbol{\Sigma}}^{C}$ 为第一、二余能. 下面将看到,变换(4.7)引导到 Reissner 原理.

将 $\boldsymbol{\Sigma}$ 看作 $\boldsymbol{F}$ 的复合函数,式(4.2)给出

$$\boldsymbol{\tau}(\boldsymbol{F})=\frac{\mathrm{d}\boldsymbol{\Sigma}}{\mathrm{d}\boldsymbol{F}} \tag{4.12}$$

变形梯度 $\boldsymbol{F}$ 既包含应变,也包含转动,所对应的 Piola 张量 $\boldsymbol{\tau}$ 并不满足广义应力定义的要求,但由于式

(4.2)的形式,在一定程度上也可称($\boldsymbol{\tau},\dot{\boldsymbol{F}}$)为共轭变量[①]. $\dot{\boldsymbol{F}}$的转动部分对$\dot{\Sigma}$没有贡献. 从

$$\begin{aligned}\boldsymbol{\tau}:\dot{\boldsymbol{F}} &= \boldsymbol{\tau}:(\boldsymbol{R}\cdot\dot{\boldsymbol{U}}+\dot{\boldsymbol{R}}\cdot\boldsymbol{U}) \\ &= (\boldsymbol{\tau}^{*}\cdot\boldsymbol{R}):\dot{\boldsymbol{U}}+\boldsymbol{\tau}:(\dot{\boldsymbol{R}}\cdot\boldsymbol{R}^{*}\cdot\boldsymbol{F}) \\ &= \frac{1}{2}(\boldsymbol{\tau}^{*}\cdot\boldsymbol{R}+\boldsymbol{R}^{*}\cdot\boldsymbol{\tau}):\dot{\boldsymbol{U}}+ \\ &\quad (\boldsymbol{\tau}\cdot\boldsymbol{F}^{*}):(\dot{\boldsymbol{R}}\cdot\boldsymbol{R}^{*}) \\ &= \boldsymbol{S}:\dot{\boldsymbol{U}}+J\boldsymbol{t}:(\dot{\boldsymbol{R}}\cdot\boldsymbol{R}^{*})\end{aligned} \tag{4.13}$$

可以看到这一点. 式中两项分别与纯应变增量和转动增量有关. 由于$\dot{\boldsymbol{R}}\cdot\boldsymbol{R}^{*}$为反对称,$\boldsymbol{t}$为对称,后一项等于零. 和式(4.4)比较,这点也是显然的. $\boldsymbol{F}$唯一确定应变和应力状态. 但一个应力状态可以在不同的转动状态下对应同一个应变状态,在可逆的情况下,$\boldsymbol{\tau}(\boldsymbol{F})$的逆的多值性是可以理解的. 当$\boldsymbol{\tau}(\boldsymbol{F})$可逆时,根据

$$\boldsymbol{S}:\boldsymbol{U}=(\boldsymbol{\tau}^{*}\cdot\boldsymbol{R}):\boldsymbol{U}=\boldsymbol{\tau}:(\boldsymbol{R}\cdot\boldsymbol{U})=\boldsymbol{\tau}:\boldsymbol{F} \tag{4.14}$$

变换(4.8)可写为

$$\tilde{\Sigma}^{C}(\boldsymbol{\tau})=\boldsymbol{\tau}:\boldsymbol{F}(\boldsymbol{\tau})-\Sigma[\boldsymbol{F}(\boldsymbol{\tau})] \tag{4.15}$$

并且有

---

① HILL R. On constitutive inequalities for simple materials-I, J. Mech. Phys. Solids, 1968(16): 229-242.

$$\boldsymbol{F}(\boldsymbol{\tau}) = \frac{\mathrm{d}\tilde{\Sigma}^C}{\mathrm{d}\boldsymbol{\tau}} \tag{4.16}$$

应注意的是,类似于 $\Sigma(\boldsymbol{F})$,$\tilde{\Sigma}^C(\boldsymbol{\tau})$ 只能通过与转动无关的应力度量如 $\boldsymbol{\tau}^* \cdot \boldsymbol{\tau}$ 作为 $\boldsymbol{\tau}$ 的复合函数,有的作者称这为标架无差异性①. 变换(4.15)引导到 Levinson 原理.

当 $\boldsymbol{\tau}(\boldsymbol{F})$ 不可逆时,根据式(3.17)和 $\boldsymbol{S}(\boldsymbol{U})$ 的可逆性,变换(4.8)的右端依赖于同时看作独立变量的 $\boldsymbol{\tau}$ 和 $\boldsymbol{R}$. 这时第二余能可写成

$$\tilde{\Sigma}^C[\boldsymbol{S}(\boldsymbol{\tau},\boldsymbol{R})] = \boldsymbol{\tau}:[\boldsymbol{R} \cdot \boldsymbol{U}(\boldsymbol{S})] - \Sigma[\boldsymbol{U}(\boldsymbol{S})] \tag{4.17}$$

它将引导到 Fraeijs de Veubeke 原理.

5. 虚功原理

虚功原理是本文推导各变分原理的出发点.

设物体 $\mathscr{B}$ 在构型 $\mathscr{R}$ 占有区域 V,边界为 A. 在边界的 A_u 部分给定位移:$\boldsymbol{u}|_{A_u} = \mathring{\boldsymbol{u}}$;在边界的余下部分 A_t 给定每单位面积的面力;$\boldsymbol{T}_N|_{A_t} = \mathring{\boldsymbol{T}}_N$,其中 $\boldsymbol{N}$ 是构型 $\mathscr{R}$ 边界单位外法向量. 显然,面力总是作用在构型 $\mathscr{C}$ 的边

① OGDEN R W. Inequalities associated with the inversion of elastic stress-deformation relations and their implications, Math. Proc. Camb. Phil. Soc., 1977(81):313-324.

界面上. 我们只考虑,不管构型 $\mathscr{C}$ 如何,外力总取上述给定值的情形,称为死载荷.

若问题的解存在,并且是位移场 $\boldsymbol{u}$ 和 Piola 应力场 $\boldsymbol{\tau}$,就分别称之为该问题的真实位移和真实应力. “真实”二字是相对下面两个概念而言的:

(1) 可能位移场 $\overset{*}{\boldsymbol{u}}$:在 V 足够光滑;在 A_u 满足几何边条件

$$\overset{*}{\boldsymbol{u}}\,|_{A_u} = \mathring{\boldsymbol{u}} \tag{5.1}$$

(2) 可能应力场 $\underset{*}{\boldsymbol{\tau}}$:在 V 满足力平衡方程

$$\underset{*}{\boldsymbol{\tau}} \cdot \boldsymbol{\nabla}_{*} = 0 \tag{5.2}$$

在 A_t 满足力边条件

$$\underset{*}{\boldsymbol{\tau}} \cdot \boldsymbol{N}\,|_{A_t} = \mathring{\boldsymbol{T}}_{\boldsymbol{N}} \tag{5.3}$$

对任取的、互不相关的可能位移场 $\overset{*}{\boldsymbol{u}}$ 和可能应力场 $\underset{*}{\boldsymbol{\tau}}$,下列积分关系式成立

$$\int_{A_u} \mathring{\boldsymbol{u}} \cdot \boldsymbol{\tau} \cdot \boldsymbol{N} \mathrm{d}A + \int_{A_t} \overset{*}{\boldsymbol{u}} \cdot \mathring{\boldsymbol{T}}_{\boldsymbol{N}} \mathrm{d}A = \oint_{A} \overset{*}{\boldsymbol{u}} \cdot \underset{*}{\boldsymbol{\tau}} \cdot \boldsymbol{N} \mathrm{d}A$$

$$= \int_{V} (\overset{*}{\boldsymbol{u}} \cdot \underset{*}{\boldsymbol{\tau}}) \cdot \boldsymbol{\nabla} \mathrm{d}V = \int_{V} \underset{*}{\boldsymbol{\tau}} : (\overset{*}{\boldsymbol{u}} \boldsymbol{\nabla}) \mathrm{d}V \tag{5.4}$$

这里用到了条件(5.1~5.3). 关系式(5.4)首先由 Vorobyev 给出,Novozhilov 在 *Theory of Elasticity* 中作了转述.

若记 $\overset{*}{\boldsymbol{u}}=\boldsymbol{u}+\dot{\boldsymbol{u}}$,并代入式(5.4),得

$$\int_{A_u}\mathring{\boldsymbol{u}}\cdot\underset{*}{\boldsymbol{\tau}}\cdot\boldsymbol{N}\mathrm{d}A+\int_{A_t}(\boldsymbol{u}+\dot{\boldsymbol{u}})\cdot\mathring{\boldsymbol{T}}_N\mathrm{d}A=\int_V\underset{*}{\boldsymbol{\tau}}:(\boldsymbol{u}\boldsymbol{\nabla}+\dot{\boldsymbol{u}}\boldsymbol{\nabla})\mathrm{d}V \tag{5.5}$$

式(5.4)对真实位移 $\boldsymbol{u}$ 亦成立

$$\int_{A_u}\mathring{\boldsymbol{u}}\cdot\underset{*}{\boldsymbol{\tau}}\cdot\boldsymbol{N}\mathrm{d}A+\int_{A_t}\boldsymbol{u}\cdot\mathring{\boldsymbol{T}}_N\mathrm{d}A=\int_V\underset{*}{\boldsymbol{\tau}}:(\boldsymbol{u}\boldsymbol{\nabla})\mathrm{d}V \tag{5.6}$$

从式(5.5)减去式(5.6),并取真实应力,得虚位移原理

$$\int_{A_t}\dot{\boldsymbol{u}}\cdot\mathring{\boldsymbol{T}}_N\mathrm{d}A=\int_V\boldsymbol{\tau}:(\dot{\boldsymbol{u}}\boldsymbol{\nabla})\mathrm{d}V \tag{5.7}$$

又若记 $\underset{*}{\boldsymbol{\tau}}=\boldsymbol{\tau}+\dot{\boldsymbol{\tau}}$,并代入式(5.4),得

$$\int_{A_u}\mathring{\boldsymbol{u}}\cdot(\boldsymbol{\tau}+\dot{\boldsymbol{\tau}})\cdot\boldsymbol{N}\mathrm{d}A+\int_{A_t}\overset{*}{\boldsymbol{u}}\cdot\mathring{\boldsymbol{T}}_N\mathrm{d}A=\int_V(\boldsymbol{\tau}+\dot{\boldsymbol{\tau}}):(\overset{*}{\boldsymbol{u}}\boldsymbol{\nabla})\mathrm{d}V \tag{5.8}$$

式(5.4)对真实应力 $\boldsymbol{\tau}$ 亦成立

$$\int_{A_u}\mathring{\boldsymbol{u}}\cdot\boldsymbol{\tau}\cdot\boldsymbol{N}\mathrm{d}A+\int_{A_t}\overset{*}{\boldsymbol{u}}\cdot\mathring{\boldsymbol{T}}_N\mathrm{d}A=\int_V\boldsymbol{\tau}:(\overset{*}{\boldsymbol{u}}\boldsymbol{\nabla})\mathrm{d}V \tag{5.9}$$

上两式相减,并取真实位移,又得虚应力原理

$$\int_{A_u}\mathring{\boldsymbol{u}}\cdot\dot{\boldsymbol{\tau}}\cdot\boldsymbol{N}\mathrm{d}A=\int_V\dot{\boldsymbol{\tau}}:(\boldsymbol{u}\boldsymbol{\nabla})\mathrm{d}V \tag{5.10}$$

虚位移原理是推导总势能驻值原理的基础,而虚应力原理则是推导各种余能原理的基础. 我们约定泛称同源于式(5.4)的虚位移和虚应力原理为虚功原理.

6. 古典变分原理

将式(4.2)代入虚位移原理(5.7),得

$$\int_V \dot{\Sigma} \mathrm{d}V - \int_{A_t} \boldsymbol{u} \cdot \mathring{\boldsymbol{T}}_N \mathrm{d}A = 0 \tag{6.1}$$

鉴于 $\mathring{\boldsymbol{T}}_N$ 是死载荷,我们有

$$\delta\left(\int_V \Sigma \mathrm{d}V - \int_{A_t} \boldsymbol{u} \cdot \mathring{\boldsymbol{T}}_N \mathrm{d}A\right) = 0 \tag{6.2}$$

我们称定义在可能位移场函数类的泛函

$$\Pi(\boldsymbol{u}) = \int_V \Sigma(\boldsymbol{F}) \mathrm{d}V - \int_{A_t} \boldsymbol{u} \cdot \mathring{\boldsymbol{T}}_N \mathrm{d}A \tag{6.3}$$

为系统总势能. 式(6.2)的意义是,真实位移场 $\boldsymbol{u}$ 使系统总势能 Π 取驻值. 下面证明,使泛函 $\Pi(\boldsymbol{u})$ 取驻值的可能位移场 $\boldsymbol{u}$ 根据本构关系(4.12)所对应的应力场 $\boldsymbol{\tau}$ 是满足力矩平衡条件的可能应力场. 利用

$$\dot{\Sigma} = \frac{\mathrm{d}\Sigma}{\mathrm{d}\boldsymbol{F}} : \dot{\boldsymbol{F}} = \boldsymbol{\tau} : (\dot{\boldsymbol{u}}\boldsymbol{\nabla}) = (\dot{\boldsymbol{u}}\boldsymbol{\tau}) \cdot \boldsymbol{\nabla} - \dot{\boldsymbol{u}} \cdot (\boldsymbol{\tau} \cdot \boldsymbol{\nabla}) \tag{6.4}$$

得 $\Pi(\boldsymbol{u})$ 的变分

$$\dot{\Pi} = \int_V [(\dot{\boldsymbol{u}} \cdot \boldsymbol{\tau}) \cdot \boldsymbol{\nabla} - \dot{\boldsymbol{u}} \cdot (\boldsymbol{\tau} \cdot \boldsymbol{\nabla})] \mathrm{d}V - \int_{A_t} \dot{\boldsymbol{u}} \cdot \mathring{\boldsymbol{T}}_N \mathrm{d}A$$

$$= \int_{A_t} \dot{\boldsymbol{u}} \cdot (\boldsymbol{\tau} \cdot \boldsymbol{N} - \mathring{\boldsymbol{T}}_N) \, \mathrm{d}A - \int_V \dot{\boldsymbol{u}} \cdot (\boldsymbol{\tau} \cdot \boldsymbol{\nabla}) \, \mathrm{d}V \tag{6.5}$$

这里用到 $\dot{\boldsymbol{u}}|_{A_u}=0$. 由 $\dot{\Pi}=0$ 及 $\dot{\boldsymbol{u}}$ 在 V 及 A_t 的任意性，得

$$\boldsymbol{\tau} \cdot \boldsymbol{\nabla} = \boldsymbol{0} \quad (\text{在 } V \text{ 上}) \tag{6.6}$$

$$\boldsymbol{\tau} \cdot \boldsymbol{N} = \mathring{\boldsymbol{T}}_N \quad (\text{在 } A_t \text{ 上}) \tag{6.7}$$

由于 $\Sigma(\boldsymbol{F})$ 是 $\boldsymbol{F}$ 的复合函数(通过应变度量). 今取该应变度量为 $\boldsymbol{C}=\boldsymbol{F}^* \cdot \boldsymbol{F}$,则有

$$\boldsymbol{\tau} = \frac{\mathrm{d}\Sigma}{\mathrm{d}\boldsymbol{F}} = 2\boldsymbol{F} \cdot \frac{\mathrm{d}\Sigma}{\mathrm{d}\boldsymbol{C}} \tag{6.8}$$

注意到 $\frac{\mathrm{d}\Sigma}{\mathrm{d}\boldsymbol{F}}$ 的对称性,也有

$$\boldsymbol{\tau}^* = 2 \frac{\mathrm{d}\Sigma}{\mathrm{d}\boldsymbol{C}} \cdot \boldsymbol{F}^*$$

从而又得

$$\boldsymbol{\tau} \cdot \overset{*}{\boldsymbol{F}} = \boldsymbol{F} \cdot \boldsymbol{\tau}^* \quad (\text{在 } V \text{ 上}) \tag{6.9}$$

于是,该应力场 $\boldsymbol{\tau}$ 是真实应力场,因为它既对应于可能位移场 $\boldsymbol{u}$,又满足条件(6.6),(6.7),(6.9). 与泛函(6.3)相联系的变分原理称为系统总势能驻值原理.

最后,对于真实位移场和真实应力场,利用散度定理,式(6.3)可改写为

$$\Pi = \int_V \Sigma \mathrm{d}V - \oint_A \boldsymbol{u} \cdot \boldsymbol{\tau} \cdot \boldsymbol{N}\mathrm{d}A + \int_{A_u} \mathring{\boldsymbol{u}} \cdot \boldsymbol{\tau} \cdot \boldsymbol{N}\mathrm{d}A$$

$$= \int_V [\Sigma - \boldsymbol{\tau}:(\boldsymbol{u}\boldsymbol{\nabla})]\mathrm{d}A + \int_{A_u} \mathring{\boldsymbol{u}} \cdot \boldsymbol{\tau} \cdot \boldsymbol{N}\mathrm{d}A$$

考虑到

$$\boldsymbol{\tau}:(\boldsymbol{u}\boldsymbol{\nabla}) = (\boldsymbol{F} \cdot \boldsymbol{T}):(\boldsymbol{u}\boldsymbol{\nabla})$$

$$= (\boldsymbol{u}\boldsymbol{\nabla}):\boldsymbol{T} + [(\boldsymbol{\nabla}\boldsymbol{u}) \cdot (\boldsymbol{u}\boldsymbol{\nabla})]:\boldsymbol{T}$$

$$= \frac{1}{2}[\boldsymbol{u}\boldsymbol{\nabla} + \boldsymbol{\nabla}\boldsymbol{u} + (\boldsymbol{\nabla}\boldsymbol{u}) \cdot (\boldsymbol{u}\boldsymbol{\nabla})]:\boldsymbol{T} + \frac{1}{2}[(\boldsymbol{\nabla}\boldsymbol{u}) \cdot (\boldsymbol{u}\boldsymbol{\nabla})]:\boldsymbol{T}$$

$$= \boldsymbol{E}:\boldsymbol{T} + \frac{1}{2}[(\boldsymbol{\nabla}\boldsymbol{u}) \cdot (\boldsymbol{u}\boldsymbol{\nabla})]:\boldsymbol{T}$$

系统总势能在真实状态的值可用真实位移和真实应力表达为

$$\Pi'(\boldsymbol{u},\boldsymbol{T}) = \int_V \left\{\Sigma(\boldsymbol{E}) - \boldsymbol{T}:\boldsymbol{E} - \frac{1}{2}[(\boldsymbol{\nabla}\boldsymbol{u}) \cdot (\boldsymbol{u}\boldsymbol{\nabla})]:\boldsymbol{T}\right\}\mathrm{d}V + \int_{A_u} \mathring{\boldsymbol{u}} \cdot \boldsymbol{F} \cdot \boldsymbol{T} \cdot \boldsymbol{N}\mathrm{d}A \tag{6.10}$$

Π'的意义见下文.

现在我们从虚应力原理(5.10)推导系统总余能驻值原理,即 Reissner 原理. 利用式(4.7)所定义的第一余能 $\Sigma^C(\boldsymbol{T})$,式(4.9)及式(3.5),得

$$
\begin{aligned}
\dot{\boldsymbol{\tau}} : (\boldsymbol{u}\boldsymbol{\nabla}) &= \delta(\boldsymbol{F}\cdot\boldsymbol{T}) : (\boldsymbol{u}\boldsymbol{\nabla}) \\
&= (\dot{\boldsymbol{F}}\cdot\boldsymbol{T}) : (\boldsymbol{u}\boldsymbol{\nabla}) + (\boldsymbol{F}\cdot\dot{\boldsymbol{T}}) : (\boldsymbol{u}\boldsymbol{\nabla}) \\
&= [(\boldsymbol{\nabla}\boldsymbol{u})\cdot\dot{\boldsymbol{F}}] : \boldsymbol{T} + [(\boldsymbol{\nabla}\boldsymbol{u})\cdot\boldsymbol{F}] : \dot{\boldsymbol{T}} \\
&= [(\boldsymbol{\nabla}\boldsymbol{u})\cdot(\dot{\boldsymbol{u}}\boldsymbol{\nabla})] : \boldsymbol{T} + \\
&\quad [\boldsymbol{\nabla}\boldsymbol{u} + (\boldsymbol{\nabla}\boldsymbol{u})\cdot(\boldsymbol{u}\boldsymbol{\nabla})] : \dot{\boldsymbol{T}} \\
&= \frac{1}{2}\delta[(\boldsymbol{\nabla}\boldsymbol{u})\cdot(\boldsymbol{u}\boldsymbol{\nabla})] : \boldsymbol{T} + \\
&\quad \frac{1}{2}[\boldsymbol{u}\boldsymbol{\nabla} + \boldsymbol{\nabla}\boldsymbol{u} + 2(\boldsymbol{\nabla}\boldsymbol{u})\cdot(\boldsymbol{u}\boldsymbol{\nabla})] : \dot{\boldsymbol{T}} \\
&= \boldsymbol{E} : \dot{\boldsymbol{T}} + \frac{1}{2}\delta\{[(\boldsymbol{\nabla}\boldsymbol{u})\cdot(\boldsymbol{u}\boldsymbol{\nabla})] : \boldsymbol{T}\} \\
&= \delta\left\{\Sigma^C + \frac{1}{2}[(\boldsymbol{\nabla}\boldsymbol{u})\cdot(\boldsymbol{u}\boldsymbol{\nabla})] : \boldsymbol{T}\right\} \qquad (6.11)
\end{aligned}
$$

将之代入式(5.10),得

$$
\delta\left\{\int_V \left[\Sigma^C + \frac{1}{2}((\boldsymbol{\nabla}\boldsymbol{u})\cdot(\boldsymbol{u}\boldsymbol{\nabla})) : \boldsymbol{T}\right]\mathrm{d}V - \int_{A_u} \mathring{\boldsymbol{u}}\cdot\boldsymbol{F}\cdot\boldsymbol{T}\cdot\boldsymbol{N}\mathrm{d}A\right\} = 0 \qquad (6.12)
$$

这里同时出现位移场 $\boldsymbol{u}$ 和应力场 $\boldsymbol{T}$. 我们称定义在可能位移场(由于在 A_u 的面积分里已出现位移的给定边界值,可能位移的条件可削弱为不必满足将来以自然边条件出现的式(5.1))和可能应力场函数类的泛函

$$
\Pi^C(\boldsymbol{T},\boldsymbol{u}) = \int_V \left\{\Sigma^C(\boldsymbol{T}) + \frac{1}{2}[(\boldsymbol{\nabla}\boldsymbol{u})\cdot(\boldsymbol{u}\boldsymbol{\nabla})] : \boldsymbol{T}\right\}\mathrm{d}V -
$$

$$\int_{A_u} \mathring{\boldsymbol{u}}\cdot\boldsymbol{F}\cdot\boldsymbol{T}\cdot\boldsymbol{N}\mathrm{d}A \tag{6.13}$$

为系统总余能. 式(6.12)的含义是:真实应力场 $\boldsymbol{T}$ 和真实位移场 $\boldsymbol{u}$ 使系统总余能取驻值. 下面证明,使泛函 $\Pi^C(\boldsymbol{T},\boldsymbol{u})$ 取驻值的可能应力场 $\boldsymbol{T}$ 和可能位移场 $\boldsymbol{u}$ 是互相对应的,并且 $\boldsymbol{u}$ 满足边条件(5.1),因而是问题的解. 变量 $\boldsymbol{T}$ 在 V 内受力平衡方程的约束,证明时需先解除约束. 我们将应力边条件也引进泛函,使 $\boldsymbol{T}$ 成为完全的自由变量. 引进 Lagrange 乘子 $\boldsymbol{\eta}$ 和 $\boldsymbol{\xi}$(分别是定义在 V 和 A_t 的向量场),得无约束条件的泛函

$$\overset{*}{\Pi}{}^C = \int_V \left\{\Sigma^C(\boldsymbol{T}) + \frac{1}{2}[(\nabla\boldsymbol{u})\cdot(\boldsymbol{u}\nabla)]:\boldsymbol{T} + \boldsymbol{\eta}\cdot[(\boldsymbol{F}\cdot\boldsymbol{T})\cdot\nabla]\right\}\mathrm{d}V - \int_{A_u}\mathring{\boldsymbol{u}}\cdot\boldsymbol{F}\cdot\boldsymbol{T}\cdot\boldsymbol{N}\mathrm{d}A + \int_{A_t}\boldsymbol{\xi}\cdot(\mathring{\boldsymbol{T}}_N - \boldsymbol{F}\cdot\boldsymbol{T}\cdot\boldsymbol{N})\mathrm{d}A \tag{6.14}$$

其中 $\boldsymbol{T},\boldsymbol{u},\boldsymbol{\eta}$ 和 $\boldsymbol{\xi}$ 是独立的自由变量. 首先暂且令 $\boldsymbol{T},\boldsymbol{\eta}$ 和 $\boldsymbol{\xi}$ 保持不变地求 $\overset{*}{\Pi}{}^C$ 的变分

$$\delta\overset{*}{\Pi}{}^C = \int_V \{[(\nabla\boldsymbol{u})\cdot(\dot{\boldsymbol{u}}\nabla)]:\boldsymbol{T} + \boldsymbol{\eta}\cdot[(\dot{\boldsymbol{F}}\cdot\boldsymbol{T})\cdot\nabla]\}\mathrm{d}V - \int_{A_u}\mathring{\boldsymbol{u}}\cdot\dot{\boldsymbol{F}}\cdot\boldsymbol{T}\cdot\boldsymbol{N}\mathrm{d}A -$$

$$\int_{A_t} \boldsymbol{\xi} \cdot \dot{\boldsymbol{F}} \cdot \boldsymbol{T} \cdot \boldsymbol{N} \mathrm{d}A$$

$$= \int_V [(\boldsymbol{u} - \boldsymbol{\eta})\boldsymbol{\nabla}] : \dot{\boldsymbol{\tau}} \mathrm{d}V +$$

$$\int_{A_u} (\boldsymbol{\eta} - \mathring{\boldsymbol{u}}) \cdot \dot{\boldsymbol{\tau}} \cdot \boldsymbol{N} \mathrm{d}A +$$

$$\int_{A_t} (\boldsymbol{\eta} - \boldsymbol{\xi}) \cdot \dot{\boldsymbol{\tau}} \cdot \boldsymbol{N} \mathrm{d}A \tag{6.15}$$

这里 $\dot{\boldsymbol{\tau}} = \dot{\boldsymbol{F}} \cdot \boldsymbol{T}$ 是仅由 $\dot{\boldsymbol{u}}$ 而引起的 Piola 张量场 $\boldsymbol{\tau}$ 的变化,它完全是自由的. 由此,$\delta \overset{*}{\Pi}{}^C = 0$ 就给出 Lagrange 乘子 $\boldsymbol{\eta}$ 和 $\boldsymbol{\xi}$ 的意义

$$(\boldsymbol{u} - \boldsymbol{\eta})\boldsymbol{\nabla} = \boldsymbol{0} \quad (\text{在 } V \text{ 上}) \tag{6.16}$$

$$\boldsymbol{\eta} = \mathring{\boldsymbol{u}} \quad (\text{在 } A_u \text{ 上}) \tag{6.17}$$

$$\boldsymbol{\xi} = \boldsymbol{\eta} \quad (\text{在 } A_t \text{ 上}) \tag{6.18}$$

从式(6.16)可知,向量场 $\boldsymbol{\eta}$ 和位移场 $\boldsymbol{u}$ 至多可差一个常向量场. 根据式(6.17)及连续性进一步可知,在整个区域 V 内 $\boldsymbol{\eta} = \boldsymbol{u}$. 而由式(6.18),在 A_t 上又有 $\boldsymbol{\xi} = \boldsymbol{u}$. 将 $\boldsymbol{\eta}$ 和 $\boldsymbol{\xi}$ 的意义代回式(6.14)就得

$$\overset{*}{\Pi}{}^C(\boldsymbol{T}, \boldsymbol{u}) = \int_V \left\{ \Sigma^C(\boldsymbol{T}) + \frac{1}{2} [(\boldsymbol{\nabla}\boldsymbol{u}) \cdot (\boldsymbol{u}\boldsymbol{\nabla})] : \boldsymbol{T} + \boldsymbol{u} \cdot [(\boldsymbol{F} \cdot \boldsymbol{T}) \cdot \boldsymbol{\nabla}] \right\} \mathrm{d}V -$$

$$\int_{A_u} \mathring{\boldsymbol{u}} \cdot \boldsymbol{F} \cdot \boldsymbol{T} \cdot \boldsymbol{N} \mathrm{d}A - \int_{A_t} \boldsymbol{u} \cdot (\boldsymbol{F} \cdot \boldsymbol{T} \cdot \boldsymbol{N} - \mathring{\boldsymbol{T}}_N) \mathrm{d}A \tag{6.19}$$

这是一个具有两个自由变量 $\boldsymbol{T}$ 和 $\boldsymbol{u}$ 的无约束泛函. 现在用它来进行上面所提及的证明. 先将 $\overset{*}{\Pi}{}^C(\boldsymbol{T},\boldsymbol{u})$ 改写为

$$\overset{*}{\Pi}{}^C(\boldsymbol{T},\boldsymbol{u})=\int_V\left\{\Sigma^C(\boldsymbol{T})+\frac{1}{2}[(\boldsymbol{\nabla u})\cdot(\boldsymbol{u\nabla})]:\boldsymbol{T}-(\boldsymbol{u\nabla}):(\boldsymbol{F}\cdot\boldsymbol{T})\right\}\mathrm{d}V+\int_{A_u}(\boldsymbol{u}-\mathring{\boldsymbol{u}})\cdot\boldsymbol{F}\cdot\boldsymbol{T}\cdot\boldsymbol{N}\mathrm{d}A+\int_{A_t}\boldsymbol{u}\cdot\mathring{\boldsymbol{T}}\mathrm{d}A \quad (6.20)$$

它的变分是

$$\delta\overset{*}{\Pi}{}^C=\int_V\left\{\frac{\mathrm{d}\Sigma^C}{\mathrm{d}\boldsymbol{T}}:\dot{\boldsymbol{T}}+\frac{1}{2}[(\boldsymbol{\nabla u})\cdot(\boldsymbol{u\nabla})]:\dot{\boldsymbol{T}}+[(\boldsymbol{\nabla u})\cdot(\dot{\boldsymbol{u}}\boldsymbol{\nabla})]:\boldsymbol{T}-(\dot{\boldsymbol{u}}\boldsymbol{\nabla}):(\boldsymbol{F}\cdot\boldsymbol{T})-(\boldsymbol{u\nabla}):(\dot{\boldsymbol{F}}\cdot\boldsymbol{T})-(\boldsymbol{u\nabla}):(\boldsymbol{F}\cdot\dot{\boldsymbol{T}})\right\}\mathrm{d}V+\int_{A_u}\dot{\boldsymbol{u}}\cdot\boldsymbol{F}\cdot\boldsymbol{T}\cdot\boldsymbol{N}\mathrm{d}A+\int_{A_u}(\boldsymbol{u}-\mathring{\boldsymbol{u}})\cdot\delta(\boldsymbol{F}\cdot\boldsymbol{T})\cdot\boldsymbol{N}\mathrm{d}A+\int_{A_t}\dot{\boldsymbol{u}}\cdot\mathring{\boldsymbol{T}}_N\mathrm{d}A$$

$$=\int_V\left\{\left[\frac{\mathrm{d}\Sigma^C}{\mathrm{d}\boldsymbol{T}}-\frac{1}{2}(\boldsymbol{u\nabla}+\boldsymbol{\nabla u}+(\boldsymbol{\nabla u})\cdot(\boldsymbol{u\nabla}))\right]:\dot{\boldsymbol{T}}+[(\boldsymbol{F}\cdot\boldsymbol{T})\cdot\boldsymbol{\nabla}]\cdot\mathring{\boldsymbol{u}}\right\}\mathrm{d}V+\int_{A_u}(\boldsymbol{u}-\mathring{\boldsymbol{u}})\cdot\delta(\boldsymbol{F}\cdot\boldsymbol{T})\cdot\boldsymbol{N}\mathrm{d}A-\int_{A_t}\boldsymbol{u}\cdot(\boldsymbol{F}\cdot\boldsymbol{T}\cdot\boldsymbol{N}-\mathring{\boldsymbol{T}}_N)\mathrm{d}A \quad (6.21)$$

于是得

$$(\boldsymbol{F}\cdot\boldsymbol{T})\cdot\boldsymbol{\nabla}=\boldsymbol{0}\quad(\text{在 }V\text{ 上})\qquad(6.22)$$

$$\boldsymbol{F}\cdot\boldsymbol{T}\cdot\boldsymbol{N}=\mathring{\boldsymbol{T}}_N\quad(\text{在 }A_t\text{ 上})\qquad(6.23)$$

$$\boldsymbol{u}=\mathring{\boldsymbol{u}} \quad (\text{在 } A_u \text{ 上}) \tag{6.24}$$

$$\frac{\mathrm{d}\Sigma^C}{\mathrm{d}\boldsymbol{T}}=\frac{1}{2}[\boldsymbol{u}\boldsymbol{\nabla}+\boldsymbol{\nabla}\boldsymbol{u}+(\boldsymbol{\nabla}\boldsymbol{u})\cdot(\boldsymbol{u}\boldsymbol{\nabla})] \quad (\text{在 } V \text{ 上}) \tag{6.25}$$

式(6.25)说明,使 $\overset{*}{\Pi}{}^C(\boldsymbol{T},\boldsymbol{u})$(亦即使 $\Pi^C(\boldsymbol{T},\boldsymbol{u})$)取驻值的 $\boldsymbol{T}$ 通过本构关系(4.9)与使 $\overset{*}{\Pi}{}^C$ 取驻值的 $\boldsymbol{u}$ 相对应. 这就是所需证明的.

利用式(4.7),比较(6.10)和(6.13)两式,我们有 $\Pi^C \underset{-}{r} \Pi'$. 即对于真实位移和真实应力,系统总势能和系统总余能间存在互补关系

$$\Pi+\Pi^C \underset{-}{r} 0 \tag{6.26}$$

这里"$\underset{-}{r}$"表示仅在真实状态才成立的等式. 应注意的是,$\Pi(\boldsymbol{u})$ 和 $\Pi^C(\boldsymbol{T},\boldsymbol{u})$ 是两个不同的泛函,式(6.26)只说明它们在驻点的值相同而已.

这样,我们从虚功原理出发,统一地推导并充分论证了非线性弹性的两个古典变分原理,还指出了它们间的相互关系.

7. 广义变分原理

两个古典变分原理都是有条件的变分原理,在应用上有时未必方便. 后来发展了广义变分的思想. 这个思想起初是通过猜测,后来是通过 Lagrange 乘子实现

的. 在这方面我国学者做过相当的工作①②. Lagrange 乘子将附加条件合并到原始泛函而得到无约束泛函. 其实,在证明 Reissner 原理时,我们已经这样做了. 可以说,广义变分的思想就是:证明带约束的变分问题时,在弄清用以解除约束的 Lagrange 乘子的物理意义后所得的无约束泛函,作为原始问题的泛函而加以应用.

现从总势能泛函推导广义变分泛函. 总势能驻值原理的独立变量是 $\boldsymbol{u}$,但 Π 所包含的贮能函数 Σ 实际上是通过应变度量,如式(3.5)的 $\boldsymbol{E}$,依赖于 $\boldsymbol{u}$ 的. 我们引入待定的 Lagrange 乘子 $\boldsymbol{\sigma}$(定义在 V 的对称张量场)和 $\boldsymbol{\xi}$(定义在 A_u 的向量场),将条件(3.5)和约束位移的边条件(5.1)合并到总势能(6.3),得

$$\overset{*}{\Pi} = \int_V \Sigma(\boldsymbol{E})\mathrm{d}V + \int_V \left\{\frac{1}{2}[\boldsymbol{u}\boldsymbol{\nabla} + \boldsymbol{\nabla}\boldsymbol{u} + (\boldsymbol{\nabla}\boldsymbol{u})\cdot(\boldsymbol{u}\boldsymbol{\nabla})] - \boldsymbol{E}\right\}:\boldsymbol{\sigma}\mathrm{d}V - \int_{A_t} \boldsymbol{u}\cdot\mathring{\boldsymbol{T}}_N\mathrm{d}A + \int_{A_u} (\mathring{\boldsymbol{u}} - \boldsymbol{u})\cdot\boldsymbol{\xi}\mathrm{d}A \tag{7.1}$$

这里的独立变量是 $\boldsymbol{E}, \boldsymbol{u}, \boldsymbol{\sigma}$ 和 $\boldsymbol{\xi}$. 暂且令 $\boldsymbol{\sigma}$ 和 $\boldsymbol{\xi}$ 保持不变地求 $\overset{*}{\Pi}$ 的变分. 利用式(4.3),得

$$\delta\overset{*}{\Pi} = \int_V \{(\boldsymbol{T} - \boldsymbol{\sigma}):\dot{\boldsymbol{E}} + (\dot{\boldsymbol{u}}\boldsymbol{\nabla}):\boldsymbol{\sigma} +$$

---

① 胡海昌. 弹塑性理论中的一些变分原理. 中国科学,1955(4):33-54.

② 钱伟长. 弹性理论中广义变分原理的研究及其在有限元计算中的应用. 清华大学科学报告,1978.

$$(\dot{\boldsymbol{u}}\boldsymbol{\nabla}):[(\boldsymbol{u}\boldsymbol{\nabla})\cdot\boldsymbol{\sigma}]\}\mathrm{d}V-$$

$$\int_{A_t}\dot{\boldsymbol{u}}\cdot\mathring{\boldsymbol{T}}_N\mathrm{d}A-\int_{A_u}\dot{\boldsymbol{u}}\cdot\boldsymbol{\xi}\mathrm{d}A \tag{7.2}$$

利用散度定理，又得

$$\int_V\{(\dot{\boldsymbol{u}}\boldsymbol{\nabla}):\boldsymbol{\sigma}+(\dot{\boldsymbol{u}}\boldsymbol{\nabla}):[(\boldsymbol{u}\boldsymbol{\nabla})\cdot\boldsymbol{\sigma}]\}\mathrm{d}V$$

$$=\int_V(\dot{\boldsymbol{u}}\boldsymbol{\nabla}):(\boldsymbol{F}\cdot\boldsymbol{\sigma})\mathrm{d}V$$

$$=\oint_A\dot{\boldsymbol{u}}\cdot\boldsymbol{F}\cdot\boldsymbol{\sigma}\cdot\boldsymbol{N}\mathrm{d}A-\int_V\dot{\boldsymbol{u}}\cdot[(\boldsymbol{F}\cdot\boldsymbol{\sigma})\cdot\boldsymbol{\nabla}]\mathrm{d}V$$

将之代入式(7.2)，有

$$\delta\overset{*}{\Pi}=\int_V\{(\boldsymbol{T}-\boldsymbol{\sigma}):\dot{\boldsymbol{E}}-[(\boldsymbol{F}\cdot\boldsymbol{\sigma})\cdot\boldsymbol{\nabla}]\cdot\dot{\boldsymbol{u}}\}\mathrm{d}V+$$

$$\int_{A_t}(\boldsymbol{F}\cdot\boldsymbol{\sigma}\cdot\boldsymbol{N}-\mathring{\boldsymbol{T}}_N)\cdot\dot{\boldsymbol{u}}\mathrm{d}A+$$

$$\int_{A_u}(\boldsymbol{F}\cdot\boldsymbol{\sigma}\cdot\boldsymbol{N}-\boldsymbol{\xi})\cdot\dot{\boldsymbol{u}}\mathrm{d}A \tag{7.3}$$

由 $\boldsymbol{\delta}\overset{*}{\Pi}=0$，从体积分的第一项及最后一个面积分，得 Lagrange 乘子的物理意义

$$\boldsymbol{\sigma}=\boldsymbol{T}\quad(\text{在 }V\text{ 上}) \tag{7.4}$$

$$\boldsymbol{\xi}=\boldsymbol{F}\cdot\boldsymbol{T}\cdot\boldsymbol{N}\quad(\text{在 }A_u\text{ 上}) \tag{7.5}$$

代回式(7.1)，就得

$$\overset{*}{\Pi}(\boldsymbol{u},\boldsymbol{T})=\int_V\left\{\Sigma(\boldsymbol{E})-\left[\boldsymbol{E}-\frac{1}{2}(\boldsymbol{u}\boldsymbol{\nabla}+\boldsymbol{\nabla}\boldsymbol{u}+(\boldsymbol{\nabla}\boldsymbol{u})\cdot(\boldsymbol{u}\boldsymbol{\nabla}))\right]:\boldsymbol{T}\right\}\mathrm{d}V-$$

$$\int_{A_t} \boldsymbol{u}\cdot\mathring{\boldsymbol{T}}_N \mathrm{d}A - \int_{A_u} (\boldsymbol{u}-\mathring{\boldsymbol{u}})\cdot\boldsymbol{F}\cdot\boldsymbol{T}\cdot\boldsymbol{N}\mathrm{d}A \quad (7.6)$$

尽管 $\boldsymbol{E}$ 也在泛函里出现，但它是通过本构关系(4.5)依赖于 $\boldsymbol{T}$ 的. 可以称具有两个自由变量 $\boldsymbol{u}$ 和 $\boldsymbol{T}$ 的无约束泛函 $\overset{*}{\Pi}(\boldsymbol{u},\boldsymbol{T})$ 为系统广义总势能. $\boldsymbol{E}$ 不是独立变量这一点也可以从式(7.6)的变分看出，由于式(4.3)而 $\dot{\boldsymbol{E}}$ 不出现. $\overset{*}{\Pi}(\boldsymbol{u},\boldsymbol{T})$ 的变分是

$$\begin{aligned}\delta\overset{*}{\Pi} = &\int_V \left\{\left[\frac{1}{2}(\boldsymbol{u}\boldsymbol{\nabla}+\boldsymbol{\nabla}\boldsymbol{u}+(\boldsymbol{\nabla}\boldsymbol{u})\cdot(\boldsymbol{u}\boldsymbol{\nabla}))-\boldsymbol{E}\right]:\dot{\boldsymbol{T}}-[(\boldsymbol{F}\cdot\boldsymbol{T})\cdot\boldsymbol{\nabla}]\cdot\dot{\boldsymbol{u}}\right\}\mathrm{d}V+\\ &\int_{A_t}(\boldsymbol{F}\cdot\boldsymbol{T}\cdot\boldsymbol{N}-\mathring{\boldsymbol{T}}_N)\cdot\dot{\boldsymbol{u}}\mathrm{d}A+\\ &\int_{A_u}(\mathring{\boldsymbol{u}}-\boldsymbol{u})\cdot\delta(\boldsymbol{F}\cdot\boldsymbol{T})\cdot\boldsymbol{N}\mathrm{d}A \end{aligned}\quad (7.7)$$

由此可知，使泛函 $\overset{*}{\Pi}(\boldsymbol{u},\boldsymbol{T})$ 取驻值的 $\boldsymbol{u}$，$\boldsymbol{T}$ 以及 $\boldsymbol{E}$（通过 $\boldsymbol{T}=\dfrac{\mathrm{d}\Sigma}{\mathrm{d}\boldsymbol{E}}$ 对应于 $\boldsymbol{T}$）满足条件

$$\boldsymbol{u}=\mathring{\boldsymbol{u}} \quad (\text{在 } A_u \text{ 上}) \quad (7.8)$$

$$\boldsymbol{F}\cdot\boldsymbol{T}\cdot\boldsymbol{N}=\mathring{\boldsymbol{T}}_N \quad (\text{在 } A_t \text{ 上}) \quad (7.9)$$

$$(\boldsymbol{F}\cdot\boldsymbol{T})\cdot\boldsymbol{\nabla}=\boldsymbol{0} \quad (\text{在 } V \text{ 上}) \quad (7.10)$$

$$\boldsymbol{E}=\frac{1}{2}[\boldsymbol{u}\boldsymbol{\nabla}+\boldsymbol{\nabla}\boldsymbol{u}+(\boldsymbol{\nabla}\boldsymbol{u})\cdot(\boldsymbol{u}\boldsymbol{\nabla})] \quad (\text{在 } V \text{ 上}) \quad (7.11)$$

式(7.11)表明，$\boldsymbol{T}$ 通过本构关系(4.5)和 $\boldsymbol{u}$ 相对应；式

(7.8) ~ (7.10)则表明,这两相对应的 $\boldsymbol{u}$ 和 $\boldsymbol{T}$ 同时是可能位移场和可能应力场,因而是问题的解. 泛函为 $\overset{*}{\Pi}(\boldsymbol{u},\boldsymbol{T})$ 的无约束条件的变分原理称为广义总势能驻值原理.

利用下述结果

$$\begin{aligned}
&\frac{1}{2}[\boldsymbol{u}\boldsymbol{\nabla}+\boldsymbol{\nabla}\boldsymbol{u}+(\boldsymbol{\nabla}\boldsymbol{u})\cdot(\boldsymbol{u}\boldsymbol{\nabla})]:\boldsymbol{T}\\
&=(\boldsymbol{u}\boldsymbol{\nabla}):(\boldsymbol{F}\cdot\boldsymbol{T})-\frac{1}{2}[(\boldsymbol{\nabla}\boldsymbol{u})\cdot(\boldsymbol{u}\boldsymbol{\nabla})]:\boldsymbol{T}\\
&=(\boldsymbol{u}\cdot\boldsymbol{F}\cdot\boldsymbol{T})\cdot\boldsymbol{\nabla}-\boldsymbol{u}\cdot[(\boldsymbol{F}\cdot\boldsymbol{T})\cdot\boldsymbol{\nabla}]-\\
&\quad\frac{1}{2}[(\boldsymbol{\nabla}\boldsymbol{u})\cdot(\boldsymbol{u}\boldsymbol{\nabla})]:\boldsymbol{T}
\end{aligned}\tag{7.12}$$

散度定理和式(4.7),并改变符号,广义总势能(7.6)又可化成另一种等价形式

$$\begin{aligned}
\overset{*}{\Pi}{}^{C}(\boldsymbol{u},\boldsymbol{T})=&\int_V\left\{\Sigma^C(\boldsymbol{T})+\frac{1}{2}[(\boldsymbol{\nabla}\boldsymbol{u})\cdot(\boldsymbol{u}\boldsymbol{\nabla})]:\boldsymbol{T}+\boldsymbol{u}\cdot[(\boldsymbol{F}\cdot\boldsymbol{T})\cdot\boldsymbol{\nabla}]\right\}\mathrm{d}V-\\
&\int_{A_t}\boldsymbol{u}\cdot(\boldsymbol{F}\cdot\boldsymbol{T}\cdot\boldsymbol{N}-\mathring{\boldsymbol{T}}_N)\mathrm{d}A-\int_{A_u}\mathring{\boldsymbol{u}}\cdot\boldsymbol{F}\cdot\boldsymbol{T}\cdot\boldsymbol{N}\mathrm{d}A
\end{aligned}\tag{7.13}$$

这正是证明 Reissner 原理时出现的无约束泛函(6.19). 假如类似地称泛函(6.19)为系统广义总余能,并相应地有广义总余能驻值原理,则广义总势能和广义总余能间也存在互补关系

$$\overset{*}{\Pi} + \overset{*}{\Pi}{}^{C} = 0 \tag{7.14}$$

式(7.12)对任意无关的 $\boldsymbol{u}$ 和 $\boldsymbol{T}$ 成立,从而互补关系(7.14)也对任意 $\boldsymbol{u}$ 和 $\boldsymbol{T}$ 成立(不同于式(6.26)),故广义总势能原理和广义总余能原理实质上是同一泛函(差一符号)的驻值原理,因此可统称为广义变分原理.

8. Levinson **原理**

在 $\boldsymbol{\tau}(\boldsymbol{F})$ 可逆的条件下,采用变换(4.15)定义的第二余能,得

$$\dot{\tilde{\Sigma}}{}^{C} = \boldsymbol{F} : \dot{\boldsymbol{\tau}} = \dot{\boldsymbol{\tau}} : (\boldsymbol{u}\boldsymbol{\nabla}) + \mathrm{tr}\,\dot{\boldsymbol{\tau}} \tag{8.1}$$

代入虚应力原理(5.10),我们有

$$\delta\left\{ \int_V [\tilde{\Sigma}^{C}(\boldsymbol{\tau}) - \mathrm{tr}\,\boldsymbol{\tau}]\mathrm{d}V - \int_{A_u} \mathring{\boldsymbol{u}} \cdot \boldsymbol{\tau} \cdot \boldsymbol{N}\mathrm{d}A \right\} = 0 \tag{8.2}$$

对于定义在可能应力场函数类的泛函

$$\Pi_1^{C}(\boldsymbol{\tau}) = \int_V [\tilde{\Sigma}^{C}(\boldsymbol{\tau}) - \mathrm{tr}\,\boldsymbol{\tau}]\mathrm{d}V - \int_{A_u} \mathring{\boldsymbol{u}} \cdot \boldsymbol{\tau} \cdot \boldsymbol{N}\mathrm{d}A \tag{8.3}$$

式(8.2)的意义是,真实应力场 $\boldsymbol{\tau}$ 使 $\Pi_1^C(\boldsymbol{\tau})$ 取驻值. 下面证明,使 Π_1^C 取驻值的可能应力场 $\boldsymbol{\tau}$ 满足力矩平衡条件,且有满足边条件的位移场 $\boldsymbol{u}$ 与之相对应. 由于 $\tilde{\Sigma}^{C}(\boldsymbol{\tau})$ 是 $\boldsymbol{\tau}$ 的标架无差异的复合函数(通过 $\boldsymbol{K} = \boldsymbol{\tau}^{*} \cdot$

$\boldsymbol{\tau}$)，故

$$\boldsymbol{F}=\frac{\mathrm{d}\tilde{\Sigma}^{C}}{\mathrm{d}\boldsymbol{\tau}}=2\boldsymbol{\tau}\cdot\frac{\mathrm{d}\tilde{\Sigma}^{C}}{\mathrm{d}\boldsymbol{K}} \tag{8.4}$$

考虑到$\frac{\mathrm{d}\tilde{\Sigma}^{C}}{\mathrm{d}\boldsymbol{K}}$的对称性，即得力矩平衡条件

$$\boldsymbol{\tau}\cdot\boldsymbol{F}^{*}=\boldsymbol{F}\cdot\boldsymbol{\tau}^{*}\quad（在 V 上） \tag{8.5}$$

只有当式(8.4)的 $\boldsymbol{F}$ 满足条件

$$\boldsymbol{F}\times\boldsymbol{\nabla}=\boldsymbol{0}\quad（在 V 上） \tag{8.6}$$

才有可能从

$$\boldsymbol{u}(\boldsymbol{X})=\int_{x_0}^{y}(\boldsymbol{F}-\boldsymbol{I})\cdot\mathrm{d}\boldsymbol{X}+\mathring{\boldsymbol{u}}(\boldsymbol{X}_0) \tag{8.7}$$

求得位移场. 为了证明，我们这次不用 Lagrange 乘子解除约束而引进自然满足力平衡方程(3.13)的应力张量函数 $\boldsymbol{\Phi}$

$$\boldsymbol{\tau}=\boldsymbol{\Phi}\times\boldsymbol{\nabla} \tag{8.8}$$

以代替受约束的变量 $\boldsymbol{\tau}$. 这样，Π_1^C 的变分是

$$\dot{\Pi}_1^C=\int_V(\boldsymbol{F}-\boldsymbol{I}):(\dot{\boldsymbol{\Phi}}\times\boldsymbol{\nabla})\mathrm{d}V-\int_{A_u}\mathring{\boldsymbol{u}}\cdot\dot{\boldsymbol{\tau}}\cdot\boldsymbol{N}\mathrm{d}A \tag{8.9}$$

利用恒等式

$$(\boldsymbol{F}-\boldsymbol{I}):(\dot{\boldsymbol{\Phi}}\times\boldsymbol{\nabla})=(\boldsymbol{F}\times\boldsymbol{\nabla}):\dot{\boldsymbol{\Phi}}+\mathrm{tr}\{[(\boldsymbol{F}^{*}-\boldsymbol{I})\cdot\dot{\boldsymbol{\Phi}}]\times\boldsymbol{\nabla}\} \tag{8.10}$$

及散度定理

$$\int_V \mathrm{tr}[(\boldsymbol{F}^* \cdot \dot{\boldsymbol{\Phi}}) \times \boldsymbol{\nabla}]\mathrm{d}V = \oint_A \boldsymbol{F} : (\dot{\boldsymbol{\Phi}} \times \boldsymbol{N})\mathrm{d}A \tag{8.11}$$

式(8.9)成为

$$\dot{\Pi}_1^C = \int_V (\boldsymbol{F} \times \boldsymbol{\nabla}) : \dot{\boldsymbol{\Phi}}\mathrm{d}V + \oint_A (\boldsymbol{F} - \boldsymbol{I}) : (\dot{\boldsymbol{\Phi}} \times \boldsymbol{N})\mathrm{d}A - \int_{A_u} \mathring{\boldsymbol{u}} \cdot \dot{\boldsymbol{\tau}} \cdot \boldsymbol{N}\mathrm{d}A \tag{8.12}$$

从体积分项得可积条件(8.6). 因此，可积分式(8.7)而得向量场 $\boldsymbol{u}(\boldsymbol{X})$. 当这个向量场满足几何边条件时，就成为对应的位移场. 将 $\boldsymbol{u}(\boldsymbol{X})$ 代入上式的第一个面积分，它的被积函数化为

$$(\boldsymbol{F} - \boldsymbol{I}) : (\dot{\boldsymbol{\Phi}} \times \boldsymbol{N}) = (\boldsymbol{u}\boldsymbol{\nabla}) : (\dot{\boldsymbol{\Phi}} \times \boldsymbol{N})$$
$$= \boldsymbol{u} \cdot (\dot{\boldsymbol{\Phi}} \times \boldsymbol{\nabla}) \cdot \boldsymbol{N} - [(\boldsymbol{u} \cdot \dot{\boldsymbol{\Phi}}) \times \boldsymbol{\nabla}] \cdot \boldsymbol{N} \tag{8.13}$$

根据 Stokes 定理，任何向量场的旋度在闭曲面上的通量等于零，于是式(8.12)的两个面积分变为

$$\oint_A \boldsymbol{u} \cdot (\dot{\boldsymbol{\Phi}} \times \boldsymbol{\nabla}) \cdot \boldsymbol{N}\mathrm{d}A - \int_{A_u} \mathring{\boldsymbol{u}} \cdot \dot{\boldsymbol{\tau}} \cdot \boldsymbol{N}\mathrm{d}A = \int_{A_u} (\boldsymbol{u} - \mathring{\boldsymbol{u}}) \cdot \dot{\boldsymbol{\tau}} \cdot \boldsymbol{N}\mathrm{d}A \tag{8.14}$$

于是得

$$\boldsymbol{u}|_{A_u} = \mathring{\boldsymbol{u}} \tag{8.15}$$

这里用到 $\dot{\boldsymbol{\tau}}|_{A_t} = 0$. 于是，Levinson 原理得到全部证明(恒等式(8.10)，(8.13)及散度定理(8.11)都不难用

分量形式证明).

同样可以用 Lagrange 乘子将可能应力场的约束条件合并到泛函(8.3),略去中间过程,最终得无约束泛函

$$\overset{*}{\Pi}_1^C(\boldsymbol{\tau},\boldsymbol{u}) = \int_V [\tilde{\Sigma}^C(\boldsymbol{\tau}) - \mathrm{tr}\,\boldsymbol{\tau} + \boldsymbol{u}\cdot(\boldsymbol{\tau}\cdot\boldsymbol{\nabla})]\mathrm{d}V - \int_{A_u} \mathring{\boldsymbol{u}}\cdot\boldsymbol{\tau}\cdot\boldsymbol{N}\mathrm{d}A - \int_{A_t} \boldsymbol{u}\cdot(\boldsymbol{\tau}\cdot\boldsymbol{N} - \mathring{\boldsymbol{T}}_N)\mathrm{d}A \tag{8.16}$$

它的变分是

$$\delta\overset{*}{\Pi}_1^C(\boldsymbol{\tau},\boldsymbol{u}) = \int_V [(\boldsymbol{F} - \boldsymbol{I} - \boldsymbol{u}\boldsymbol{\nabla}):\dot{\boldsymbol{\tau}} + (\boldsymbol{\tau}\cdot\boldsymbol{\nabla})\cdot\dot{\boldsymbol{u}}]\mathrm{d}V + \int_{A_u} (\boldsymbol{u} - \mathring{\boldsymbol{u}})\cdot\dot{\boldsymbol{\tau}}\cdot\boldsymbol{N}\mathrm{d}A - \int_{A_t} \dot{\boldsymbol{u}}\cdot(\boldsymbol{\tau}\cdot\boldsymbol{N} - \mathring{\boldsymbol{T}}_N)\mathrm{d}A \tag{8.17}$$

$\delta\overset{*}{\Pi}_1^C = 0$ 给出

$$\begin{cases} \boldsymbol{\tau}\cdot\boldsymbol{\nabla} = \boldsymbol{0} & (\text{在 } V \text{ 上}) \\ \boldsymbol{u}\boldsymbol{\nabla} = \boldsymbol{F} - \boldsymbol{I} & (\text{在 } V \text{ 上}) \\ \boldsymbol{\tau}\cdot\boldsymbol{N} = \mathring{\boldsymbol{T}}_N & (\text{在 } A_t \text{ 上}) \\ \boldsymbol{u} = \mathring{\boldsymbol{u}} & (\text{在 } A_u \text{ 上}) \end{cases} \tag{8.18}$$

(8.18)各式加上力矩平衡条件(8.5)及可积条件(8.6)说明,$\boldsymbol{\tau}$ 和 $\boldsymbol{u}$ 就是真实应力场和真实位移场. 可以称与泛函(8.16)相联系的变分原理为广义 Levinson

原理.

对于真实状态,下列等式成立

$$\tilde{\Sigma}^{C}-\operatorname{tr}\boldsymbol{\tau}=\Sigma^{C}+\frac{1}{2}[(\nabla\boldsymbol{u})\cdot(\boldsymbol{u}\nabla)]:\boldsymbol{T} \quad (8.19)$$

故亦存在两个互补关系

$$\Pi+\Pi_1^C\overset{r}{=}0,\quad \overset{*}{\Pi}+\overset{*}{\Pi}{}_1^C\overset{r}{=}0 \quad (8.20)$$

9. Fraeijs de Veubeke **原理**

当 $\boldsymbol{\tau}(\boldsymbol{F})$ 不可逆时, Levinson 原理失效. 采用式(4.17)的第二余能

$$\tilde{\Sigma}^{C}[\boldsymbol{S}(\boldsymbol{\tau},\boldsymbol{R})]=\boldsymbol{\tau}:[\boldsymbol{R}\cdot\boldsymbol{U}(\boldsymbol{S})]-\Sigma[\boldsymbol{U}(\boldsymbol{S})] \quad (9.1)$$

虚应力原理就引导到基于极分解的 Fraeijs de Veubeke 原理. 将仍然成立的式(8.1)代入虚应力原理(5.10), 我们就有

$$\delta\left\{\int_V[\tilde{\Sigma}^{C}(\boldsymbol{S})-\operatorname{tr}\boldsymbol{\tau}]\mathrm{d}V-\int_{A_u}\mathring{\boldsymbol{u}}\cdot\boldsymbol{\tau}\cdot\boldsymbol{N}\mathrm{d}A\right\}=0 \quad (9.2)$$

对于以 Piola 应力张量 $\boldsymbol{\tau}$ 和转动张量 $\boldsymbol{R}$ 作为独立变量的泛函

$$\Pi_2^C(\boldsymbol{\tau},\boldsymbol{R})=\int_V\{\tilde{\Sigma}^{C}[\boldsymbol{S}(\boldsymbol{\tau},\boldsymbol{R})]-\operatorname{tr}\boldsymbol{\tau}\}\mathrm{d}V-\int_{A_u}\mathring{\boldsymbol{u}}\cdot\boldsymbol{\tau}\cdot\boldsymbol{N}\mathrm{d}A \quad (9.3)$$

式(9.2)的意义是:真实应力场 $\boldsymbol{\tau}$ 和真实转动场 $\boldsymbol{R}$ 使

泛函 Π_1^C 取驻值. 逆命题也是对的:使 Π_2^C 取驻值的可能应力场 $\boldsymbol{\tau}$ 和转动场 $\boldsymbol{R}$ 满足力矩平衡条件. 且有唯一的,满足几何边条件的位移场 $\boldsymbol{u}$ 与之对应,因而是问题的解. 注意,这里力矩平衡条件已不能像 Levinson 原理那样由 $\tilde{\Sigma}^C$ 函数本身的形式所保证. 所谓对应的位移场,就是指从下述步骤求出的向量场 $\boldsymbol{u}$:从 $\boldsymbol{\tau}$ 和 $\boldsymbol{R}$ 出发,根据式(3.17)及 $\boldsymbol{S}(\boldsymbol{U})$ 的可逆性算出 $\boldsymbol{R}\cdot\boldsymbol{U}$,然后根据(3.2),(3.3)积分得

$$\boldsymbol{u}(\boldsymbol{X}) = \int_{x_0}^{y} (\boldsymbol{R}\cdot\boldsymbol{U}-\boldsymbol{I})\cdot \mathrm{d}\boldsymbol{X} + \mathring{\boldsymbol{u}}(\boldsymbol{X}_0) \qquad (9.4)$$

可积条件是

$$(\boldsymbol{R}\cdot\boldsymbol{U})\times\boldsymbol{\nabla}=\boldsymbol{0} \quad (\text{在 } V \text{ 上}) \qquad (9.5)$$

利用式(4.10)和(3.17),考虑到伸长张量 $\boldsymbol{U}$ 的对称性,并引入应力张量函数 $\boldsymbol{\Phi}$ 如式(8.8),则泛函 Π_2^C 变分的体积分被积函数是

$$\begin{aligned}
\frac{\mathrm{d}\tilde{\Sigma}^C}{\mathrm{d}\boldsymbol{S}}:\dot{\boldsymbol{S}}-\mathrm{tr}\,\dot{\boldsymbol{\tau}} &= \boldsymbol{U}:(\dot{\boldsymbol{\tau}}^{*}\cdot\boldsymbol{R}+\boldsymbol{\tau}^{*}\cdot\dot{\boldsymbol{R}})-\mathrm{tr}\,\dot{\boldsymbol{\tau}}\\
&= \dot{\boldsymbol{\tau}}:(\boldsymbol{R}\cdot\boldsymbol{U}-\boldsymbol{I})+\boldsymbol{\tau}:(\boldsymbol{R}\cdot\boldsymbol{U})\\
&= (\boldsymbol{R}\cdot\boldsymbol{U}-\boldsymbol{I}):(\dot{\boldsymbol{\Phi}}\times\boldsymbol{\nabla})+\\
&\quad \frac{1}{2}(\boldsymbol{\tau}\cdot\boldsymbol{U}\cdot\boldsymbol{R}^{*}-\boldsymbol{R}\cdot\boldsymbol{U}\cdot\boldsymbol{\tau}^{*}):\\
&\quad (\dot{\boldsymbol{R}}\cdot\boldsymbol{R}^{*})
\end{aligned} \qquad (9.6)$$

利用恒等式(8.10)及散度定理(8.11),得 Π_2^C 的变分

$$\dot{\Pi}_2^C = \int_V \{[(\boldsymbol{R}\cdot\boldsymbol{U})\times\boldsymbol{\nabla}]:\dot{\boldsymbol{\Phi}} + \frac{1}{2}(\boldsymbol{\tau}\cdot\boldsymbol{U}\cdot\boldsymbol{R}^* - \boldsymbol{R}\cdot\boldsymbol{U}\cdot\boldsymbol{\tau}^*):(\dot{\boldsymbol{R}}\cdot\boldsymbol{R}^*)\}\mathrm{d}V + \oint_A (\boldsymbol{R}\cdot\boldsymbol{U}-\boldsymbol{I}):(\dot{\boldsymbol{\Phi}}\times\boldsymbol{N})\mathrm{d}A - \int_{A_u}\mathring{\boldsymbol{u}}\cdot\dot{\boldsymbol{\tau}}\cdot\boldsymbol{N}\mathrm{d}A \tag{9.7}$$

从体积分项得可积条件(9.5)及力矩平衡条件

$$\boldsymbol{\tau}\cdot\boldsymbol{U}\cdot\boldsymbol{R}^* = \boldsymbol{R}\cdot\boldsymbol{U}\cdot\boldsymbol{\tau}^* \quad (\text{在 } V \text{ 上}) \tag{9.8}$$

只要用 $\boldsymbol{R}\cdot\boldsymbol{U}$ 代替 $\boldsymbol{F}$,就可用完全类似于证 Levinson 原理的步骤求得满足几何边条件的位移场 $\boldsymbol{u}(\boldsymbol{X})$. 从而 Fraeijs de Veubeke 原理得到全部证明.

类似地,相应的无约束泛函是

$$\overset{*}{\Pi}{}_2^C(\boldsymbol{\tau},\boldsymbol{R},\boldsymbol{u}) = \int_V \{\tilde{\Sigma}^C[\boldsymbol{s}(\boldsymbol{\tau},\boldsymbol{R})] - \mathrm{tr}\,\boldsymbol{\tau} + \boldsymbol{u}\cdot(\boldsymbol{\tau}\cdot\boldsymbol{\nabla})\}\mathrm{d}V - \int_{A_u}\mathring{\boldsymbol{u}}\cdot\boldsymbol{\tau}\cdot\boldsymbol{N}\mathrm{d}A - \int_{A_t}\boldsymbol{u}\cdot(\boldsymbol{\tau}\cdot\boldsymbol{N} - \mathring{\boldsymbol{T}}_N)\mathrm{d}A \tag{9.9}$$

它的变分是

$$\delta\overset{*}{\Pi}{}_2^C = \int_V [(\boldsymbol{R}\cdot\boldsymbol{U} - \boldsymbol{I} - \boldsymbol{u}\boldsymbol{\nabla}):\dot{\boldsymbol{\tau}} + \frac{1}{2}(\boldsymbol{\tau}\cdot\boldsymbol{U}\cdot\boldsymbol{R}^* - \boldsymbol{R}\cdot\boldsymbol{U}\cdot\boldsymbol{\tau}^*):(\dot{\boldsymbol{R}}\cdot\boldsymbol{R}^*) + (\boldsymbol{\tau}\cdot\boldsymbol{\nabla})\cdot\dot{\boldsymbol{u}}]\mathrm{d}V + \int_{A_u}(\boldsymbol{u}-\mathring{\boldsymbol{u}})\cdot\dot{\boldsymbol{\tau}}\cdot\boldsymbol{N}\mathrm{d}A - \int_{A_t}\dot{\boldsymbol{u}}\cdot(\boldsymbol{\tau}\cdot\boldsymbol{N} - \mathring{\boldsymbol{T}}_N)\mathrm{d}A \tag{9.10}$$

$\delta \overset{*}{\Pi}{}_2^C = 0$ 给出

$$\begin{cases} \boldsymbol{\tau} \cdot \boldsymbol{\nabla} = \boldsymbol{0} \quad (\text{在 } V \text{ 上}) \\ \boldsymbol{\tau} \cdot \boldsymbol{U} \cdot \boldsymbol{R}^* = \boldsymbol{R} \cdot \boldsymbol{U} \cdot \boldsymbol{\tau}^* \quad (\text{在 } V \text{ 上}) \\ \boldsymbol{u} \boldsymbol{\nabla} = \boldsymbol{R} \cdot \boldsymbol{U} - \boldsymbol{I} \quad (\text{在 } V \text{ 上}) \\ \boldsymbol{u} \cdot \boldsymbol{N} = \mathring{\boldsymbol{T}}_N \quad (\text{在 } A_t \text{ 上}) \\ \boldsymbol{u} = \mathring{\boldsymbol{u}} \quad (\text{在 } A_u \text{ 上}) \end{cases} \tag{9.11}$$

它们加上本构关系

$$\boldsymbol{U} = \boldsymbol{U}(\boldsymbol{S}) = \boldsymbol{U}\left[\frac{1}{2}(\boldsymbol{\tau}^* \cdot \boldsymbol{R} + \boldsymbol{R}^* \cdot \boldsymbol{\tau})\right] \tag{9.12}$$

及转动张量的正交性

$$\boldsymbol{R} \cdot \boldsymbol{R}^* = \boldsymbol{I} \tag{9.13}$$

构成了问题的全部关系式. 可以称与泛函 $\overset{*}{\Pi}{}_2^C$ 相联系的变分原理为广义 Fraeijs de Veubeke 原理.

类似地也可证明两个关系

$$\Pi + \Pi_2^C \overset{r}{=} 0, \quad \overset{*}{\Pi} + \overset{*}{\Pi}{}_2^C = 0$$

10. 关系图

这里给出统一从虚功原理推导出来的各变分原理的关系图(图 1). 它使我们更能看清各原理间的纵横内在联系:纵向是势能原理、余能原理、Levinson 原理(L-原理)和 Fraeijs de Veubeke 原理(FV-原理)四条线,古典变分原理构成一闭回路,而 L-原理和 FV-原理则各单独成一开分支;各原理间又存在着横向的互补关系.

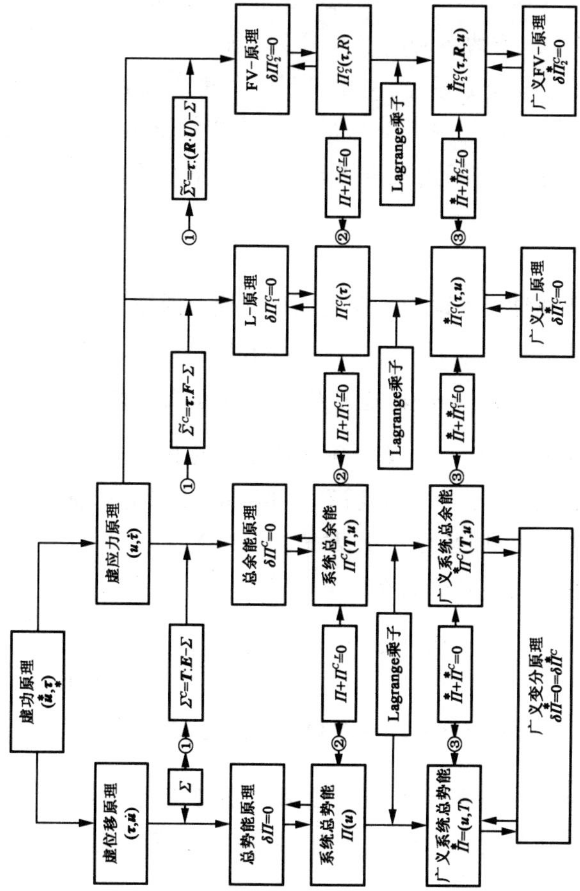
虚功原理 $(\overset{*}{u},\overset{*}{\tau})$
虚位移原理 $(\tau,\dot{u})$
虚应力原理 $(u,\dot{\tau})$
Σ
$\Sigma^C=T{:}E-\Sigma$
$\widetilde{\Sigma}^C=\tau{:}F-\Sigma$
$\widetilde{\Sigma}^C=\tau{:}(R\cdot U)-\Sigma$
总势能原理 $\delta\Pi=0$
总余能原理 $\delta\Pi^C=0$
L-原理 $\delta\Pi_1^C=0$
FV-原理 $\delta\Pi_2^C=0$
系统总势能 $\Pi(u)$
系统总余能 $\Pi^C(T,u)$
$\Pi_1^C(\tau)$
$\Pi_2^C(\tau,R)$
$\Pi+\Pi^C=0$
$\Pi+\Pi_1^C=0$
$\Pi+\Pi_2^C=0$
Lagrange乘子
广义系统总势能 $\overset{*}{\Pi}=(u,T)$
广义系统总余能 $\overset{*}{\Pi}{}^C(T,u)$
$\overset{*}{\Pi}{}_1^C(\tau,u)$
$\overset{*}{\Pi}{}_2^C(\tau,R,u)$
$\overset{*}{\Pi}+\overset{*}{\Pi}{}^C=0$
$\overset{*}{\Pi}+\overset{*}{\Pi}{}_1^C=0$
$\overset{*}{\Pi}+\overset{*}{\Pi}{}_2^C=0$
广义变分原理 $\delta\overset{*}{\Pi}=0=\delta\overset{*}{\Pi}{}^C$
广义L-原理 $\delta\overset{*}{\Pi}{}_1^C=0$
广义FV-原理 $\delta\overset{*}{\Pi}{}_2^C=0$

图1

西南交通大学和辽宁大学的截天民教授1982年发表论文《论非线性弹性理论的各种变分原理》以推导非线性弹性理论的各种可能的主要变分原理,文中采用郭仲衡[①②]所用的记法和符号,必要时另作说明.并且只给出相应的变分泛函形式.线性弹性理论是其特殊情形,只在需要时提一下,不另写出相应的变分泛函形式[③].

1. 非线性弹性理论的完全的广义变分原理

设弹性体的区域为 V,边界为 A;在边界的 A_u 部分上给定位移:$\boldsymbol{u}|_{A_u}=\mathring{\boldsymbol{u}}$;在边界的另一部分上给定每单位面积的面力:$\boldsymbol{T}_N|_{A_t}=\mathring{\boldsymbol{T}}_N$;每单位质量的体力为 f;ρ_0 为质量密度;$\boldsymbol{F}$ 为变形梯度;$\boldsymbol{T}$ 为 Kirchhoff 应力张量:$\boldsymbol{\tau}=\boldsymbol{F}\cdot\boldsymbol{T}$ 为 Piola 应力张量;$\boldsymbol{E}$ 为应变张量.

现在定义弹性体的全能量泛函 Φ 如下

$$\Phi=\int_V(\boldsymbol{F}\cdot\boldsymbol{T}):(\boldsymbol{u}\boldsymbol{\nabla})\,\mathrm{d}V-\int_V\boldsymbol{u}\cdot f\rho_0\mathrm{d}V-\int_{A_u}\mathring{\boldsymbol{u}}\cdot(\boldsymbol{F}\cdot\boldsymbol{T})\cdot\boldsymbol{N}\mathrm{d}A-\int_{A_t}\boldsymbol{u}\cdot\mathring{\boldsymbol{T}}_N\mathrm{d}A \tag{1.1}$$

设应力张量 $\boldsymbol{T}$ 和应变张量 $\boldsymbol{E}$ 的关系是可逆的并引进应变能密度 $\Sigma^p(\boldsymbol{E})$ 和余应变能密度 $\Sigma(\boldsymbol{T})$,使得

---

① 胡海昌.弹性力学的变分原理及其应用.北京:科学出版社,1981.

② 郭仲衡.非线性弹性理论.北京:科学出版社,1980.

③ 摘自《应用数学和力学》,1982年,第3卷第5期.

$$\Sigma^{p}(\boldsymbol{E})+\Sigma(\boldsymbol{T})=\boldsymbol{T}:\boldsymbol{E} \tag{1.2}$$

已知①

$$(\boldsymbol{F}\cdot\boldsymbol{T}):(\boldsymbol{u}\boldsymbol{\nabla})=\boldsymbol{E}:\boldsymbol{T}+\frac{1}{2}[(\boldsymbol{\nabla}\boldsymbol{u})\cdot(\boldsymbol{u}\boldsymbol{\nabla})]:\boldsymbol{T} \tag{1.3}$$

故由(1.2)可把上式写成下列形式

$$(\boldsymbol{F}\cdot\boldsymbol{T}):(\boldsymbol{u}\boldsymbol{\nabla})=\Sigma^{p}(\boldsymbol{E})+\Sigma(\boldsymbol{T})+\frac{1}{2}[(\boldsymbol{\nabla}\boldsymbol{u})\cdot(\boldsymbol{u}\boldsymbol{\nabla})]:\boldsymbol{T} \tag{1.4}$$

或写成下列推广的形式

$$(\boldsymbol{F}\cdot\boldsymbol{T}):(\boldsymbol{u}\boldsymbol{\nabla})=\boldsymbol{E}:\boldsymbol{T}-\Sigma^{p}(\boldsymbol{T})+\boldsymbol{T}:\boldsymbol{E}-\Sigma^{p}(\boldsymbol{E})+\frac{1}{2}[(\boldsymbol{\nabla}\boldsymbol{u})\cdot(\boldsymbol{u}\boldsymbol{\nabla})]:\boldsymbol{T} \tag{1.5}$$

把(1.5)代入(1.1),则可把全能量泛函 $\boldsymbol{\Phi}$ 改写成下列形式

$$\boldsymbol{\Phi}_0=\left\{\iint_V[\boldsymbol{E}:\boldsymbol{T}-\Sigma^{c}(\boldsymbol{T})]\mathrm{d}V-\int_V\boldsymbol{u}\cdot\boldsymbol{f}\rho_0\mathrm{d}V-\int_{A_t}\boldsymbol{u}\cdot\mathring{\boldsymbol{T}}_N\mathrm{d}A\right\}+\left\{\iint_V[\boldsymbol{T}:\boldsymbol{E}-\Sigma^{p}(\boldsymbol{E})]\mathrm{d}V+\int_V\frac{1}{2}[(\boldsymbol{\nabla}\boldsymbol{u})\cdot(\boldsymbol{u}\boldsymbol{\nabla})]:\boldsymbol{T}\mathrm{d}V-\int_{A_u}\mathring{\boldsymbol{u}}\cdot\boldsymbol{F}\cdot\boldsymbol{T}\cdot\boldsymbol{N}\mathrm{d}A\right\}=\boldsymbol{\Phi}_0^{p}+\boldsymbol{\Phi}_0^{c} \tag{1.6}$$

这便是本文的出发点. 以后将会看到这样写法的目的.

---

① 郭仲衡. 非线性弹性理论变分原理的统一理论. 应用数学和力学,1980,1(1):5-23.

这里我们有意识地把全能量泛函 Φ_0 用花括号分成 Φ_0^p 和 Φ_0^c 两部分,即

$$\Phi_0^p = \int_V [\boldsymbol{E}:\boldsymbol{T} - \Sigma^c(\boldsymbol{T})]\,\mathrm{d}V - \int_V \boldsymbol{u}\cdot \boldsymbol{f}\rho_0\,\mathrm{d}V - \int_{A_t} \boldsymbol{u}\cdot \mathring{\boldsymbol{T}}_N\,\mathrm{d}A \tag{1.7a}$$

$$\Phi_0^c = \int_V [\boldsymbol{T}:\boldsymbol{E} - \Sigma^p(\boldsymbol{E})]\,\mathrm{d}V + \int_V \frac{1}{2}[(\nabla\boldsymbol{u})\cdot(\boldsymbol{u}\nabla)]:\boldsymbol{T}\,\mathrm{d}V - \int_{A_u} \mathring{\boldsymbol{u}}\cdot\boldsymbol{F}\ \boldsymbol{T}\cdot\boldsymbol{N}\,\mathrm{d}A \tag{1.7b}$$

它们分别表示弹性理论古典变分原理的总位能泛函和总余能泛函的推广形式.

下面列出非线性弹性理论的基本关系式和边界条件.

平衡条件:

平衡方程

$$(\boldsymbol{F}\cdot\boldsymbol{T})\cdot\nabla + \boldsymbol{f}\rho_0 = 0 \quad (在\ V\ 内) \tag{1.8}$$

力边界条件

$$\boldsymbol{F}\cdot\boldsymbol{T}\cdot\boldsymbol{N} = \mathring{\boldsymbol{T}}_N \quad (在\ A_t\ 上) \tag{1.9}$$

连续条件:

应变位移关系

$$\boldsymbol{E} = \frac{1}{2}[\nabla\boldsymbol{u} + \boldsymbol{u}\nabla + (\nabla\boldsymbol{u})\cdot(\boldsymbol{u}\nabla)] \quad (在\ V\ 内) \tag{1.10}$$

位移边界条件

$$\boldsymbol{u}=\mathring{\boldsymbol{u}} \quad (\text{在 } A_u \text{ 上}) \tag{1.11}$$

本构关系:

应变能形式

$$\boldsymbol{T}=\frac{\mathrm{d}\Sigma^p(\boldsymbol{E})}{\mathrm{d}\boldsymbol{E}} \quad (\text{在 } V \text{ 内}) \tag{1.12a}$$

余应变能形式

$$\boldsymbol{E}=\frac{\mathrm{d}\Sigma^c(\boldsymbol{T})}{\mathrm{d}\boldsymbol{T}} \quad (\text{在 } V \text{ 内}) \tag{1.12b}$$

现从弹性体的全能量泛函 Φ_0 推导广义全能量泛函 $\overset{*}{\Phi}_0$. 此时广义全能量泛函的独立变量是 $\boldsymbol{u},\boldsymbol{E},\boldsymbol{T}$,它们共有 15 个分量. 它们必须满足上述基本关系式和边界条件(1.8)~(1.12). 这是一个在 15 个约束条件下的条件极值问题. 为此可以采用钱伟长[①]于 1964 年提出的 Lagrange 乘子法. 引入待定的 Lagrange 乘子:$\boldsymbol{\sigma}$(定义在 V 内的对称张量场),$\boldsymbol{\xi}$(定义在 A_u 上的向量场),$\boldsymbol{\eta}$(定义在 V 内的向量场)和 $\boldsymbol{\zeta}$(定义在 A_t 上的向量场),它们也共有 15 个分量.

把平衡条件(1.8)和(1.9)及连续条件(1.10)和(1.11)合并到弹性体的全能量泛函 Φ_0 以后,则得广义全能量泛函 $\overset{*}{\Phi}_0$ 如下:

---

① 钱伟长. 变分法及有限元(上册). 北京:科学出版社,1980.

$$\overset{*}{\Phi}_0=\Big\{\int_V[\boldsymbol{E}:\boldsymbol{T}-\Sigma^c(\boldsymbol{T})]\mathrm{d}V-\int_V\boldsymbol{u}\cdot\boldsymbol{f}\rho_0\mathrm{d}V-\int_{A_t}\boldsymbol{u}\cdot\mathring{\boldsymbol{T}}_N\mathrm{d}A+$$

$$\int_V[\frac{1}{2}(\nabla\boldsymbol{u}+\boldsymbol{u}\nabla+(\nabla\boldsymbol{u})\cdot(\boldsymbol{u}\nabla))-\boldsymbol{E}]:\boldsymbol{\sigma}\,\mathrm{d}V+\int_{A_u}(\mathring{\boldsymbol{u}}-\boldsymbol{u})\cdot\boldsymbol{\xi}\mathrm{d}A\Big\}+$$

$$\Big\{\int_V[\boldsymbol{T}:\boldsymbol{E}-\Sigma^p(\boldsymbol{E})]\mathrm{d}V+\int_V\frac{1}{2}[(\nabla\boldsymbol{u})\cdot(\boldsymbol{u}\nabla)]:\boldsymbol{T}\mathrm{d}V-$$

$$\int_{A_u}\mathring{\boldsymbol{u}}\cdot\boldsymbol{F}\cdot\boldsymbol{T}\cdot\boldsymbol{N}\mathrm{d}A+\int_V\boldsymbol{\eta}\cdot[(\boldsymbol{F}\cdot\boldsymbol{T})\cdot\nabla+\boldsymbol{f}\rho_0]\mathrm{d}V+$$

$$\int_{A_t}\boldsymbol{\zeta}\cdot(\mathring{\boldsymbol{T}}_N-\boldsymbol{F}\cdot\boldsymbol{T}\cdot\boldsymbol{N})\mathrm{d}A\Big\}$$

$$=\overset{*}{\Phi}{}_0^p+\overset{*}{\Phi}{}_0^c \tag{1.13}$$

问题的真正解可由(1.13)定义的广义全能量泛函的驻值条件给出．此时广义全能量泛函 $\overset{*}{\Phi}_0$ 中的独立变量是 $\boldsymbol{u},\boldsymbol{E},\boldsymbol{T},\boldsymbol{\sigma},\boldsymbol{\xi},\boldsymbol{\eta},\boldsymbol{\zeta}$．对这些变量进行变分，则有

$$\delta\overset{*}{\Phi}_0=\delta\overset{*}{\Phi}{}_0^p+\delta\overset{*}{\Phi}{}_0^c \tag{1.14}$$

其中

$$\delta\overset{*}{\Phi}{}_0^p=\int_V\Big[\boldsymbol{E}-\frac{\mathrm{d}\Sigma^c(\boldsymbol{T})}{\mathrm{d}\boldsymbol{T}}\Big]:\delta\boldsymbol{T}\mathrm{d}V+\int_V(\boldsymbol{T}-\boldsymbol{\sigma}):\delta\boldsymbol{E}\mathrm{d}V+$$

$$\int_V\Big[\frac{1}{2}(\nabla\boldsymbol{u}+\boldsymbol{u}\nabla+(\nabla\boldsymbol{u})\cdot(\boldsymbol{u}\nabla))-\boldsymbol{E}\Big]:\delta\boldsymbol{\sigma}+$$

$$\int_V[(\delta\boldsymbol{u}\nabla:\boldsymbol{\sigma}+(\delta\boldsymbol{u}\nabla):(\boldsymbol{u}\nabla)\cdot\boldsymbol{\sigma}]\mathrm{d}V-\int_V\delta\boldsymbol{u}\cdot\boldsymbol{f}\rho_0\mathrm{d}V-$$

$$\int_{A_t}\delta\boldsymbol{u}\cdot\boldsymbol{T}_N\mathrm{d}A-\int_{A_u}\delta\boldsymbol{u}\cdot\boldsymbol{\xi}\mathrm{d}A+\int_{A_u}(\mathring{\boldsymbol{u}}-\boldsymbol{u})\cdot\delta\boldsymbol{\xi}\mathrm{d}A \tag{1.15a}$$

$$\delta\overset{*}{\Phi}{}_0^c = \int_V \left[\boldsymbol{T} - \frac{\mathrm{d}\Sigma^p(\boldsymbol{E})}{\mathrm{d}\boldsymbol{E}} \right] : \delta\boldsymbol{E}\mathrm{d}V + \int_V \left[\boldsymbol{E} + \frac{1}{2}(\boldsymbol{\nabla u}) \cdot (\boldsymbol{u\nabla}) \right] :$$

$$\delta\boldsymbol{T}\mathrm{d}V + \int_V [(\delta\boldsymbol{u\nabla}) : (\boldsymbol{u\nabla})] \cdot \boldsymbol{T}\mathrm{d}V + \int_V \delta\boldsymbol{\eta} \cdot [(\boldsymbol{F} \cdot \boldsymbol{T}) \cdot$$

$$\boldsymbol{\nabla} + \boldsymbol{f}\rho_0]\mathrm{d}V - \int_V \delta(\boldsymbol{F} \cdot \boldsymbol{T}) : (\boldsymbol{\eta\nabla})\mathrm{d}V - \int_{A_u} \mathring{\boldsymbol{u}} \cdot \delta(\boldsymbol{F} \cdot \boldsymbol{T}) \cdot$$

$$\boldsymbol{N}\mathrm{d}A + \int_{A_t} \boldsymbol{\eta} \cdot \delta(\boldsymbol{F} \cdot \boldsymbol{T}) \cdot \boldsymbol{N}\mathrm{d}A + \int_{A_u} \boldsymbol{\eta} \cdot \delta(\boldsymbol{F} \cdot \boldsymbol{T}) \cdot \boldsymbol{N}\mathrm{d}A +$$

$$\int_{A_t} \delta\boldsymbol{\zeta} \cdot (\mathring{\boldsymbol{T}}_N - \boldsymbol{F} \cdot \boldsymbol{T} \cdot \boldsymbol{N})\mathrm{d}A - \int_{A_t} \boldsymbol{\zeta} \cdot \delta(\boldsymbol{F} \cdot \boldsymbol{T}) \cdot \boldsymbol{N}\mathrm{d}A \tag{1.15b}$$

但是

$$\int_V [(\delta\boldsymbol{u\nabla}) : \boldsymbol{\sigma} + (\delta\boldsymbol{u\nabla}) : ((\boldsymbol{u\nabla}) \cdot \boldsymbol{\sigma})]\mathrm{d}V$$

$$= \int_A \delta\boldsymbol{u} \cdot \boldsymbol{F} \cdot \sigma \cdot \boldsymbol{N}\mathrm{d}A - \int_V \delta\boldsymbol{u} \cdot [(\boldsymbol{F} \cdot \boldsymbol{\sigma}) \cdot \boldsymbol{\nabla}]\mathrm{d}V \text{①} \tag{1.16}$$

而

$$-\int_V \delta(\boldsymbol{F} \cdot \boldsymbol{T}) : (\boldsymbol{\eta\nabla})\mathrm{d}V = -\int_V (\delta\boldsymbol{F} \cdot \boldsymbol{T}) : (\boldsymbol{\eta\nabla})\mathrm{d}V$$

$$-\int_V (\boldsymbol{F} \cdot \delta\boldsymbol{T}) : (\boldsymbol{\eta\nabla})\mathrm{d}V$$

$$= -\int_V (\boldsymbol{\nabla\eta} \cdot \delta\boldsymbol{F}) : \boldsymbol{T}\mathrm{d}V$$

---

① 郭仲衡. 非线性弹性理论变分原理的统一理论. 应用数学和力学, 1980, 1(1):5-23.

$$
\begin{aligned}
&-\int_V (\nabla\boldsymbol{\eta}\cdot\boldsymbol{F}):\delta\boldsymbol{F}\mathrm{d}V \\
&=-\int_V [\nabla\boldsymbol{\eta}\cdot\delta(\boldsymbol{I}+\boldsymbol{u}\nabla)]:\boldsymbol{T}\mathrm{d}V \\
&\quad-\int_V [\nabla\boldsymbol{\eta}\cdot(\boldsymbol{I}+\boldsymbol{u}\nabla)]:\delta\boldsymbol{T}\mathrm{d}V \\
&=-\int_V [(\delta\boldsymbol{u}\nabla):(\boldsymbol{\eta}\nabla)]\cdot\boldsymbol{T}\mathrm{d}V \\
&\quad-\int_V [\nabla\boldsymbol{\eta}+(\nabla\boldsymbol{\eta})\cdot(\boldsymbol{u}\nabla)]:\delta\boldsymbol{T}\mathrm{d}V
\end{aligned}
\tag{1.17}
$$

把(1. 16)和(1. 17)分别代入(1. 15a)和(1. 15b)内并把结果稍加整理,则可把(1. 14)写成下列形式

$$
\begin{aligned}
\delta\overset{*}{\Phi}_0 &=\delta\overset{*}{\Phi}{}_0^p+\delta\overset{*}{\Phi}{}_0^c \\
&=\Bigg\{\int_V\left[\boldsymbol{E}-\frac{\mathrm{d}\Sigma^c(\boldsymbol{T})}{\mathrm{d}\boldsymbol{T}}\right]:\delta\boldsymbol{T}\mathrm{d}V+\int_V(\boldsymbol{T}-\boldsymbol{\sigma}):\delta\boldsymbol{E}\mathrm{d}V+ \\
&\quad\int_V\left[\frac{1}{2}(\nabla\boldsymbol{u}+\boldsymbol{u}\nabla+(\nabla\boldsymbol{u})\cdot(\boldsymbol{u}\nabla))-\boldsymbol{E}\right]:\delta\boldsymbol{\sigma}\mathrm{d}V- \\
&\quad\int_V\delta\boldsymbol{u}\cdot[(\boldsymbol{F}\cdot\boldsymbol{\sigma})\cdot\nabla+f\rho_0]\mathrm{d}V+\int_{A_t}\delta\boldsymbol{u}\cdot(\boldsymbol{F}\cdot\boldsymbol{\sigma}\cdot\boldsymbol{N}-\mathring{\boldsymbol{T}}_N)\mathrm{d}A+ \\
&\quad\int_{A_u}\delta\boldsymbol{u}\cdot(\boldsymbol{F}\cdot\boldsymbol{\sigma}\cdot\boldsymbol{N}-\boldsymbol{\xi})\mathrm{d}A+\int_{A_u}(\mathring{\boldsymbol{u}}-\boldsymbol{u})\cdot\delta\boldsymbol{\xi}\mathrm{d}A\Bigg\}+ \\
&\quad\Bigg\{\int_V\left[\boldsymbol{T}-\frac{\mathrm{d}\Sigma^p(\boldsymbol{E})}{\mathrm{d}\boldsymbol{E}}\right]:\delta\boldsymbol{E}\mathrm{d}V+\int_V[(\delta\boldsymbol{u}\nabla):(\boldsymbol{u}-\eta)\nabla]\cdot\boldsymbol{T}\mathrm{d}V+ \\
&\quad\int_V\left[\boldsymbol{E}-(\nabla\boldsymbol{\eta}+(\nabla\boldsymbol{\eta})\cdot(\boldsymbol{u}\nabla))-\frac{1}{2}(\nabla\boldsymbol{u})\cdot(\boldsymbol{u}\nabla))\right]:\delta\boldsymbol{T}\mathrm{d}V+
\end{aligned}
$$

$$\int_V \delta\boldsymbol{\eta}\cdot[(\boldsymbol{F}\cdot\boldsymbol{T})\cdot\nabla+\boldsymbol{f}\rho_0]\mathrm{d}V+\int_{A_u}(\boldsymbol{\eta}-\mathring{\boldsymbol{u}})\cdot\delta(\boldsymbol{F}\cdot\boldsymbol{T})\cdot\boldsymbol{N}\mathrm{d}A+$$

$$\int_{A_t}\delta\boldsymbol{\zeta}\cdot(\mathring{\boldsymbol{T}}_N-\boldsymbol{F}\cdot\boldsymbol{T}\cdot\boldsymbol{N})\mathrm{d}A+\int_{A_t}(\boldsymbol{\eta}-\boldsymbol{\zeta})\cdot\delta(\boldsymbol{F}\cdot\boldsymbol{T})\cdot\boldsymbol{N}\mathrm{d}A\Big\}$$

(1.18)

考虑到(1.18)中的 $\delta\boldsymbol{u},\delta\boldsymbol{E},\delta\boldsymbol{T},\delta\boldsymbol{\sigma},\delta\boldsymbol{\xi},\delta\boldsymbol{\eta},\delta\boldsymbol{\zeta}$ 都是作为独立变量的变分，故得下列两组对应的条件：

表1

位能型		余能型	
$\boldsymbol{E}=\dfrac{\mathrm{d}\Sigma^c(\boldsymbol{T})}{\mathrm{d}\boldsymbol{T}}$	(在 V 内)	$\boldsymbol{T}=\dfrac{\mathrm{d}\Sigma^p(\boldsymbol{E})}{\mathrm{d}\boldsymbol{E}}$	(在 V 内)
$\boldsymbol{\sigma}=\boldsymbol{T}$	(在 V 内)	$(\boldsymbol{u}-\boldsymbol{\eta})\nabla=0$	(在 V 内)
$\boldsymbol{E}=\dfrac{1}{2}[\nabla\boldsymbol{u}+\boldsymbol{u}\nabla+(\nabla\boldsymbol{u})\cdot(\boldsymbol{u}\nabla)]$	(在 V 内)	$(\boldsymbol{F}\cdot\boldsymbol{T})\cdot\nabla+\boldsymbol{f}\rho_0=0$	(在 V 内)
$(\boldsymbol{F}\cdot\boldsymbol{\sigma})\cdot\nabla+\boldsymbol{f}\rho_0=0$	(在 V 内)	$\boldsymbol{E}=\dfrac{1}{2}[\nabla\boldsymbol{\eta}+\boldsymbol{\eta}\nabla+(\nabla\boldsymbol{\eta})\cdot(\boldsymbol{u}\nabla)-(\nabla\boldsymbol{u})\cdot(\boldsymbol{u}\nabla)]$	(在 V 内)
$\boldsymbol{u}=\mathring{\boldsymbol{u}}$	(在 A_u 上)	$\boldsymbol{F}\cdot\boldsymbol{T}\cdot\boldsymbol{N}=\mathring{\boldsymbol{T}}_N$	(在 A_t 上)
$\boldsymbol{F}\cdot\boldsymbol{\sigma}\cdot\boldsymbol{N}=\mathring{\boldsymbol{T}}_N$	(在 A_t 上)	$\boldsymbol{\eta}=\mathring{\boldsymbol{u}}$	(在 A_u 上)
$\boldsymbol{\xi}=\boldsymbol{F}\cdot\boldsymbol{\sigma}\cdot\boldsymbol{N}$	(在 A_u 上)	$\boldsymbol{\zeta}=\boldsymbol{\eta}$	(在 A_t 上)
(1.19a)		(1.19b)	

由此可以确定所有 Lagrange 乘子的物理意义，即 $\boldsymbol{\sigma}=\boldsymbol{T},\boldsymbol{\xi}=\boldsymbol{F}\cdot\boldsymbol{T}\cdot\boldsymbol{N},\boldsymbol{\eta}=\boldsymbol{u},\boldsymbol{\zeta}=\boldsymbol{u}$. 于是两组条件(1.19a)和

(1.19b)表示完全对应的条件，这些条件正是非线性弹性理论的全部基本关系式和边界条件(1.8)~(1.12). 把已经确定的 Lagrange 乘子代回广义全能量泛函的变分 $\delta\overset{*}{\Phi}_0$，则有

$$
\begin{aligned}
\delta\overset{*}{\Phi}_0 = & \Big\{\int_V \left[\boldsymbol{E}-\frac{\mathrm{d}\Sigma^c(\boldsymbol{T})}{\mathrm{d}\boldsymbol{T}}\right]:\delta\boldsymbol{T}\mathrm{d}V+ \\
& \int_V\left[\frac{1}{2}(\boldsymbol{\nabla u}+\boldsymbol{u\nabla}+(\boldsymbol{\nabla}u)\cdot(\boldsymbol{u\nabla}))-\boldsymbol{E}\right]:\delta\boldsymbol{T}\mathrm{d}V- \\
& \int_V\delta\boldsymbol{u}\cdot[(\boldsymbol{F}\cdot\boldsymbol{T})\cdot\boldsymbol{\nabla}+\boldsymbol{f}\rho_0]\mathrm{d}V+\int_{A_u}(\mathring{\boldsymbol{u}}-\boldsymbol{u})\cdot \\
& \delta(\boldsymbol{F}\cdot\boldsymbol{T})\cdot\boldsymbol{N}\mathrm{d}A+\int_{A_t}\delta\boldsymbol{u}\cdot(\boldsymbol{F}\cdot\boldsymbol{T}\cdot\boldsymbol{N}-\mathring{\boldsymbol{T}}_N)\mathrm{d}A\Big\}+ \\
& \Big\{\int_V\left[\boldsymbol{T}-\frac{\mathrm{d}\Sigma^p(\boldsymbol{E})}{\mathrm{d}\boldsymbol{E}}\right]:\delta\boldsymbol{E}\mathrm{d}V+ \\
& \int_V\left[\boldsymbol{E}-\frac{1}{2}(\boldsymbol{\nabla u}+\boldsymbol{u\nabla}+(\boldsymbol{\nabla u})\cdot(\boldsymbol{u\nabla}))\right]:\ \delta\boldsymbol{T}\mathrm{d}V+ \\
& \int_V\delta\boldsymbol{u}[(\boldsymbol{F}\cdot\boldsymbol{T})\cdot\boldsymbol{N}+\boldsymbol{f}\rho_0]\mathrm{d}V+\int_{A_t}\delta\boldsymbol{u}\cdot(\mathring{\boldsymbol{T}}_N-\boldsymbol{F}\cdot \\
& \boldsymbol{T}\cdot\boldsymbol{N})\mathrm{d}A+\int_{A_u}(\boldsymbol{u}-\mathring{\boldsymbol{u}})\cdot\delta(\boldsymbol{F}\cdot\boldsymbol{T})\cdot\boldsymbol{N}\mathrm{d}A\Big\} \\
= & \delta\overset{*}{\Phi}{}_0^p+\delta\overset{*}{\Phi}{}_0^c \qquad (1.20)
\end{aligned}
$$

实际上，由于

$$\boldsymbol{T}:\delta\boldsymbol{E}+\boldsymbol{E}:\delta\boldsymbol{T}=\frac{\mathrm{d}\Sigma^{p}(\boldsymbol{E})}{\mathrm{d}\boldsymbol{E}}:\delta\boldsymbol{E}+\frac{\mathrm{d}\Sigma^{c}(\boldsymbol{T})}{\mathrm{d}\boldsymbol{T}}:\delta\boldsymbol{T} \tag{1.21}$$

确实

$$\delta\,\overset{*}{\Phi}{}_0^p+\delta\,\overset{*}{\Phi}{}_0^c=\delta\,\overset{*}{\Phi}_0=0 \tag{1.22}$$

显然，它包括

$$\delta\,\overset{*}{\Phi}{}_0^p=0 \tag{1.23a}$$

$$\delta\,\overset{*}{\Phi}{}_0^c=0 \tag{1.23b}$$

分别是非线性弹性理论的完全的广义位能和广义余能驻值变分原理.

把已经确定的 Lagrange 乘子代回广义全能量泛函(1.13)，则有

$$\begin{aligned}\overset{*}{\Phi}{}_0^p=&\int_V[\boldsymbol{E}:\boldsymbol{T}-\Sigma^c(\boldsymbol{T})]\mathrm{d}V-\int_V\boldsymbol{u}\cdot\boldsymbol{f}\rho_0\mathrm{d}V-\int_{A_t}\boldsymbol{u}\cdot\mathring{\boldsymbol{T}}_N\mathrm{d}A+\\&\int_V\Big[\frac{1}{2}(\nabla\boldsymbol{u}+\boldsymbol{u}\nabla+(\nabla\boldsymbol{u})\cdot(\boldsymbol{u}\nabla))-\boldsymbol{E}\Big]:\boldsymbol{T}\mathrm{d}V+\\&\int_{A_t}(\mathring{\boldsymbol{u}}-\boldsymbol{u})\cdot\boldsymbol{F}\cdot\boldsymbol{T}\cdot\boldsymbol{N}\mathrm{d}A\end{aligned} \tag{1.24a}$$

$$\begin{aligned}\overset{*}{\Phi}{}_0^c=&\int_V[\boldsymbol{T}:\boldsymbol{E}-\Sigma^p(\boldsymbol{E})]\mathrm{d}V+\int_V\frac{1}{2}[(\nabla\boldsymbol{u})\cdot(\boldsymbol{u}\nabla)]:\boldsymbol{T}\mathrm{d}V-\\&\int_{A_u}\mathring{\boldsymbol{u}}\cdot\boldsymbol{F}\cdot\boldsymbol{T}\cdot\boldsymbol{N}\mathrm{d}A+\int_V\boldsymbol{u}\cdot[(\boldsymbol{F}\cdot\boldsymbol{T})\cdot\nabla+\boldsymbol{f}\rho_0]\mathrm{d}V+\\&\int_{A_t}\boldsymbol{u}\cdot(\mathring{\boldsymbol{T}}_N-\boldsymbol{F}\cdot\boldsymbol{T}\cdot\boldsymbol{N})\mathrm{d}A\end{aligned} \tag{1.24b}$$

这是从形式上所能推出的非线性弹性理论的完全的广义位能和广义余能变分原理的泛函. 实际上,(1.24a)归化为下列形式

$$\overset{*}{\Phi}{}_{0R}^{p} = -\int_V \Sigma^c(\boldsymbol{T})\,\mathrm{d}V - \int_V \boldsymbol{u}\cdot \boldsymbol{f}\rho_0\,\mathrm{d}V - \int_{A_t} \boldsymbol{u}\cdot \mathring{\boldsymbol{T}}_N\,\mathrm{d}A + \int_V \left[\frac{1}{2}[\boldsymbol{\nabla u} + \boldsymbol{\nabla u} + (\boldsymbol{\nabla u})\cdot(\boldsymbol{u\nabla})]\right] : \boldsymbol{T}\,\mathrm{d}V + \int_{A_u} (\mathring{\boldsymbol{u}} - \boldsymbol{u})\cdot \boldsymbol{F}\cdot \boldsymbol{T}\cdot \boldsymbol{N}\,\mathrm{d}A \tag{1.25}$$

这便是鹫津久一郎[①]给出的广义变分原理的泛函;在线性弹性理论中上式便是 Reissner 原来提出的形式.

由此可见,从形式上似乎是可以从(1.24a)找到《弹性力学的变分原理及其应用》(古海昌著,1981)中表6.1的第11类变分原理,但实际上并没有能够达到原来的目的. 同时也可以说明,到目前为止,(1.24b)是弹性理论中真正完全的广义变分原理的泛函,亦即所有其他的变分泛函都不能像(1.24b)那样能够完整地反映物理问题的全部客观规律.

线性弹性理论的变分泛函形式(1.24b)早已被胡海昌和鹫津久一郎先后提出,后来大家常把这类变分

① WASHIZU K. Variational methods in elasticity and plasticity. London: Pergamon Press,1968.

原理称为胡海昌－鹫津久一郎广义变分原理[①];钱伟长[②]于1978年首先给出了非线性弹性理论的这类完全的广义变分泛函,即(1.24b).

2. 非线性弹性理论的各种不完全的互补的广义位能和广义余能变分泛函

(1)不完全的广义位能和广义余能变分泛函之一(事先分别满足互补的本构关系的情况).

在(1.13)中取

$$\boldsymbol{E}:\boldsymbol{T}-\Sigma^{c}(\boldsymbol{T})=\Sigma^{p}(\boldsymbol{E}) \tag{2.1a}$$

和

$$\boldsymbol{E}:\boldsymbol{T}-\Sigma^{p}(\boldsymbol{E})=\Sigma(\boldsymbol{T}) \tag{2.1b}$$

则得下列一对互补[③]的变分泛函

$$\overset{*}{\Phi}{}_{1}^{p}=\int_{V}\Sigma^{p}(\boldsymbol{E})\mathrm{d}V-\int_{V}\boldsymbol{u}\cdot\boldsymbol{f}\rho_{0}\mathrm{d}V-\int_{A_t}\boldsymbol{u}\cdot\mathring{\boldsymbol{T}}_{N}\mathrm{d}A+\int_{V}\left[\frac{1}{2}(\nabla\boldsymbol{u}+\boldsymbol{u}\nabla+(\nabla\boldsymbol{u})\cdot(\boldsymbol{u}\nabla))-\boldsymbol{E}\right]:\boldsymbol{\sigma}\mathrm{d}V-\int_{A_u}(\mathring{\boldsymbol{u}}-\boldsymbol{u})\cdot\boldsymbol{\xi}\mathrm{d}A \tag{2.2a}$$

① 胡海昌.弹性力学的变分原理及其应用.北京:科学出版社1981.

② 钱伟长.变分法及有限元(上册).北京:科学出版社,1980.

③ 本文中“互补的泛函”一词意指这一对泛函在互补条件下反映相同的物理问题的客观规律.

$$\overset{*}{\Phi}{}_1^c = \int_V \Sigma(\boldsymbol{T})\,\mathrm{d}V + \int_V \frac{1}{2}[(\nabla \boldsymbol{u}) \cdot (\boldsymbol{u}\nabla)] : \boldsymbol{T}\,\mathrm{d}V - \int_{A_u} \mathring{\boldsymbol{u}} \cdot \boldsymbol{F} \cdot \boldsymbol{T} \cdot \boldsymbol{N}\,\mathrm{d}A + \int_V \boldsymbol{\eta} \cdot [(\boldsymbol{F} \cdot \boldsymbol{T}) \cdot \nabla + \boldsymbol{f}\rho_0]\,\mathrm{d}V + \int_{A_t} \boldsymbol{\zeta} \cdot (\mathring{\boldsymbol{T}}_N - \boldsymbol{F} \cdot \boldsymbol{T} \cdot \boldsymbol{N})\,\mathrm{d}A \tag{2.2b}$$

对独立变量进行变分，则得

$$\delta\overset{*}{\Phi}{}_0^p = \int_V \left[\frac{\mathrm{d}\Sigma^c(\boldsymbol{E})}{\mathrm{d}\boldsymbol{E}} - \boldsymbol{\sigma}\right] : \delta\boldsymbol{E}\,\mathrm{d}V + \int_V \left[\frac{1}{2}\nabla\boldsymbol{u} + \boldsymbol{u}\nabla + (\nabla\boldsymbol{u}) \cdot (\boldsymbol{u}\nabla)) - \boldsymbol{E}\right] : \delta\boldsymbol{\sigma}\,\mathrm{d}V - \int_V [(\boldsymbol{F} \cdot \boldsymbol{\sigma}) \cdot \nabla + \boldsymbol{f}\rho_0] \cdot \delta\boldsymbol{u}\,\mathrm{d}V + \int_{A_u} (\mathring{\boldsymbol{u}} - \boldsymbol{u}) \cdot \delta\boldsymbol{\xi}\,\mathrm{d}A + \int_{A_t} \delta\boldsymbol{u} \cdot (\boldsymbol{F} \cdot \boldsymbol{\sigma} \cdot \boldsymbol{N} - \mathring{\boldsymbol{T}}_N)\,\mathrm{d}A + \int_{A_u} \delta\boldsymbol{u} \cdot (\boldsymbol{F} \cdot \boldsymbol{\sigma} \cdot \boldsymbol{N} - \boldsymbol{\xi})\,\mathrm{d}A \tag{2.3a}$$

$$\delta\overset{*}{\Phi}{}_1^c = \int_V \left[\frac{\mathrm{d}\Sigma(\boldsymbol{T})}{\mathrm{d}\boldsymbol{T}} - (\nabla\boldsymbol{\eta} + (\nabla\boldsymbol{\eta}) \cdot (\boldsymbol{u}\nabla) - \frac{1}{2}(\nabla\boldsymbol{u}) \cdot (\boldsymbol{u}\nabla))\right] : \delta\boldsymbol{T}\,\mathrm{d}V + \int_V \delta\boldsymbol{\eta} \cdot [(\boldsymbol{F} \cdot \boldsymbol{T}) \cdot \nabla + \boldsymbol{f}\rho_0]\,\mathrm{d}V + \int_{A_t} \delta\boldsymbol{\zeta} \cdot (\mathring{\boldsymbol{T}}_N - \boldsymbol{F} \cdot \boldsymbol{T} \cdot \boldsymbol{N})\,\mathrm{d}A + \int_{A_u} (\boldsymbol{\eta} - \mathring{\boldsymbol{u}}) \cdot \delta(\boldsymbol{F} \cdot \boldsymbol{T}) \cdot \boldsymbol{N}\,\mathrm{d}A + \int_A (\boldsymbol{\eta} - \boldsymbol{\zeta}) \cdot \delta(\boldsymbol{F} \cdot \boldsymbol{T}) \cdot \boldsymbol{N}\,\mathrm{d}A \tag{2.3b}$$

于是驻值条件给出表 2 中两组对应的条件：

表2

位能形式	余能形式
$\frac{\mathrm{d}\Sigma^{p}(\boldsymbol{E})}{\mathrm{d}\boldsymbol{E}}=0$ （在 V 内）	$\frac{\mathrm{d}\Sigma^{c}(\boldsymbol{T})}{\mathrm{d}\boldsymbol{T}}=\boldsymbol{\varepsilon}$ （在 V 内）
$(\boldsymbol{F}\cdot\boldsymbol{\sigma})\boldsymbol{\nabla}+\boldsymbol{f}\rho_0=0$ （在 V 内）	$\boldsymbol{\varepsilon}=\frac{1}{2}[(\boldsymbol{\nabla\eta}+\boldsymbol{\eta\nabla}+2(\boldsymbol{\nabla\eta})\cdot(\boldsymbol{u\nabla})-(\boldsymbol{\nabla u})\cdot(\boldsymbol{u\nabla}))]$ （在 V 内）
$\boldsymbol{E}=\frac{1}{2}[\boldsymbol{\nabla u}+\boldsymbol{u\nabla}+(\boldsymbol{\nabla u})\cdot(\boldsymbol{u\nabla})]$ （在 V 内）	$(\boldsymbol{F}\cdot\boldsymbol{T})\cdot\boldsymbol{\nabla}+\boldsymbol{f}\rho_0=0$ （在 V 内）
$\boldsymbol{u}=\mathring{\boldsymbol{u}}$ （在 A_u 上）	$\boldsymbol{F}\cdot\boldsymbol{T}\cdot\boldsymbol{N}=\mathring{\boldsymbol{T}}_N$ （在 A_t 上）
$\boldsymbol{F}\cdot\boldsymbol{\sigma}\cdot\boldsymbol{N}=\mathring{\boldsymbol{T}}_N$ （在 A_t 上）	$\boldsymbol{\eta}=\mathring{\boldsymbol{u}}$ （在 A_u 上）
$\boldsymbol{\xi}=\boldsymbol{F}\cdot\boldsymbol{\sigma}\cdot\boldsymbol{N}$ （在 A_u 上）	$\boldsymbol{\zeta}=\boldsymbol{\eta}$ （在 A_t 上）
(2.4a)	(2.4b)

由(2.4a)和(2.4b)可见，这里必须利用事先假定的互补的本构关系，即$\frac{\mathrm{d}\Sigma^{p}(\boldsymbol{E})}{\mathrm{d}\boldsymbol{E}}=\boldsymbol{T}$和$\frac{\mathrm{d}\Sigma(\boldsymbol{T})}{\mathrm{d}\boldsymbol{T}}=\boldsymbol{E}$，才能确定所有的 Lagrange 乘子的物理意义，于是得 $\boldsymbol{\varepsilon}=\boldsymbol{E}$，$\boldsymbol{\sigma}=\boldsymbol{T}$，$\boldsymbol{\xi}=\boldsymbol{F}\cdot\boldsymbol{T}\cdot\boldsymbol{N}$，$\boldsymbol{\eta}=\boldsymbol{u}$，$\boldsymbol{\zeta}=\boldsymbol{u}$.

把已经确定的 Lagrange 乘子代回(2.2a)和(2.2b)，则得下列一对互补的广义位能和广义余能变分泛函

$$\overset{*}{\Phi}{}_1^{p}=\int_V\Sigma^{p}(\boldsymbol{E})\mathrm{d}V-\int_V\boldsymbol{u}\cdot\boldsymbol{f}\rho_0\mathrm{d}V-\int_{A_t}\boldsymbol{u}\cdot\mathring{\boldsymbol{T}}_N\mathrm{d}A+$$

$$\int_V \left[\frac{1}{2}(\nabla \boldsymbol{u} + \boldsymbol{u}\nabla + (\nabla \boldsymbol{u}) \cdot (\boldsymbol{u}\nabla)) - \boldsymbol{E}\right] : \boldsymbol{T}\mathrm{d}V +$$

$$\int_{A_u} (\mathring{\boldsymbol{u}} - \boldsymbol{u}) \cdot \boldsymbol{F} \cdot \boldsymbol{T} \cdot \boldsymbol{N}\mathrm{d}A \tag{2.5a}$$

$$\overset{*}{\Phi}{}_1^c = \int_V \Sigma^c(\boldsymbol{T})\mathrm{d}V + \int_V \frac{1}{2}[(\nabla \boldsymbol{u}) \cdot (\boldsymbol{u}\nabla)] : \boldsymbol{T}\mathrm{d}V -$$

$$\int_{A_u} \mathring{\boldsymbol{u}} \cdot \boldsymbol{F} \cdot \boldsymbol{T} \cdot \boldsymbol{N}\mathrm{d}A + \int_V \boldsymbol{u} \cdot [(\boldsymbol{F} \cdot \boldsymbol{T}) \cdot \nabla + \boldsymbol{f}\rho_0]\mathrm{d}V +$$

$$\int_{A_t} \boldsymbol{u} \cdot (\mathring{\boldsymbol{T}}_N - \boldsymbol{F} \cdot \boldsymbol{T} \cdot \boldsymbol{N})\mathrm{d}A \tag{2.5b}$$

此时易证下列互补关系成立

$$\overset{*}{\Phi}{}_1^p + \overset{*}{\Phi}{}_1^c = 0 \tag{2.6}$$

事实上,在(2.5a)中的独立变量是 $\boldsymbol{u}$ 和 $\boldsymbol{T}$. 尽管 $\boldsymbol{E}$ 也在泛函中出现,但它不是独立的,[①]故本文把这种情况称为拟完全的,以示有别于前述完全的情况以及下面就要提到的其他各种不完全的情况.

(2)不完全的广义位能和广义余能变分泛函之二(事先分别满足应变位移关系和平衡方程的情况).

此时可直接由(1.24a)和(1.24b)写出下列一对互补的变分泛函

$$\overset{*}{\Phi}{}_2^p = \int_V [\boldsymbol{E} : \boldsymbol{T} - \Sigma^c(\boldsymbol{T})]\mathrm{d}V - \int_V \boldsymbol{u} \cdot \boldsymbol{f}\rho_0\mathrm{d}V -$$

---

① 郭仲衡. 非线性弹性理论. 北京:科学出版社,1980.

$$\int_{A_t} \boldsymbol{u} \cdot \mathring{\boldsymbol{T}}_N \mathrm{d}A + \int_{A_u} (\mathring{\boldsymbol{u}} - \boldsymbol{u}) \cdot \boldsymbol{F} \cdot \boldsymbol{T} \cdot \boldsymbol{N} \mathrm{d}A \tag{2.7a}$$

$$\overset{*}{\Phi}{}_2^c = \int_V [\boldsymbol{T}:\boldsymbol{E} - \Sigma^p(\boldsymbol{E})] \mathrm{d}V + \int_V \frac{1}{2}[(\nabla \boldsymbol{u}) \cdot (\boldsymbol{u}\nabla)]:\boldsymbol{T}\mathrm{d}V - \int_{A_u} \mathring{\boldsymbol{u}} \cdot \boldsymbol{F} \cdot \boldsymbol{T} \cdot \boldsymbol{N} \mathrm{d}A + \int_{A_t} \boldsymbol{u} \cdot (\mathring{\boldsymbol{T}}_N - \boldsymbol{F} \cdot \boldsymbol{T} \cdot \boldsymbol{N}) \mathrm{d}A \tag{2.7b}$$

易证下列互补关系成立

$$\overset{*}{\Phi}{}_2^p + \overset{*}{\Phi}{}_2^c = 0 \tag{2.8}$$

本文以此为例说明这对变分泛函(2.7a)和(2.7b)的互补性,其他情况不再赘述.

对(2.7a)和(2.7b)进行变分并考虑到①

$$\int_V \boldsymbol{T}:\delta\boldsymbol{E}\mathrm{d}V = \int_V (\boldsymbol{F} \cdot \boldsymbol{T} \cdot \delta\boldsymbol{u}) \cdot \nabla \mathrm{d}V - \int_V \delta\boldsymbol{u} \cdot (\boldsymbol{F} \cdot \boldsymbol{T}) \cdot \nabla \mathrm{d}V \tag{2.9}$$

和

$$\int_V [\nabla \boldsymbol{u} \cdot (\delta \boldsymbol{u} \nabla)]:\boldsymbol{T}\mathrm{d}V = \int_V \delta(\boldsymbol{F} \cdot \boldsymbol{T}):\boldsymbol{u}\nabla \mathrm{d}V - \int_V [\nabla \boldsymbol{u} + (\nabla \boldsymbol{u}) \cdot (\boldsymbol{u}\nabla)]:\delta\boldsymbol{T}\mathrm{d}V \tag{2.10}$$

---

① 郭仲衡. 非线性弹性理论. 北京:科学出版社,1980.

则可写出下列两式

$$\delta\overset{*}{\Phi}{}_2^p=\int_V\left[\boldsymbol{E}-\frac{\mathrm{d}\Sigma^c(\boldsymbol{T})}{\mathrm{d}\boldsymbol{T}}\right]:\delta\boldsymbol{T}\mathrm{d}V-\int_V[(\boldsymbol{F}\cdot\boldsymbol{T})\cdot\boldsymbol{\nabla}+\boldsymbol{f}\rho_0]\cdot\delta\boldsymbol{u}\mathrm{d}V+\int_{A_u}(\mathring{\boldsymbol{u}}-\boldsymbol{u})\cdot\delta(\boldsymbol{F}\cdot\boldsymbol{T})\cdot\boldsymbol{N}\mathrm{d}A+\int_{A_t}(\boldsymbol{F}\cdot\boldsymbol{T}\cdot\boldsymbol{N}-\mathring{\boldsymbol{T}}_N)\cdot\delta\boldsymbol{u}\mathrm{d}A=0 \tag{2.11a}$$

$$\delta\overset{*}{\Phi}{}_2^c=\int_V\left[\boldsymbol{T}-\frac{\mathrm{d}\Sigma^p(\boldsymbol{E})}{\mathrm{d}\boldsymbol{E}}\right]:\delta\boldsymbol{E}\mathrm{d}V-\int_V\left[\boldsymbol{E}-\frac{1}{2}(\boldsymbol{\nabla}\boldsymbol{u}+\boldsymbol{u}\boldsymbol{\nabla}+(\boldsymbol{\nabla}\boldsymbol{u})\cdot(\boldsymbol{u}\boldsymbol{\nabla}))\right]:\delta\boldsymbol{T}\mathrm{d}V+\int_{A_t}\delta\boldsymbol{u}\cdot(\mathring{\boldsymbol{T}}_N-\boldsymbol{F}\cdot\boldsymbol{T}\cdot\boldsymbol{N})\mathrm{d}A+\int_{A_u}(\boldsymbol{u}-\mathring{\boldsymbol{u}})\cdot\delta(\boldsymbol{F}\cdot\boldsymbol{T})\cdot\boldsymbol{N}\mathrm{d}A=0 \tag{2.11b}$$

由此可见，在事先满足应变位移关系的条件下，$\overset{*}{\Phi}{}_2^p$ 所反映的物理客观规律是余应变能形式的本构方程、平衡方程和两个边界条件；而在事先满足平衡方程的条件下，$\overset{*}{\Phi}{}_2^c$ 所反映的物理客观规律是应变能形式的本构方程、应变位移关系和两个边界条件.

(3) 不完全的广义位能和广义余能变分泛函之三（事先分别满足位移边界条件和力边界条件的情况）.

此时可直接由(1.24a)和(1.24b)写出下列一对互补的变分泛函

$$\overset{*}{\Phi}{}_3^p=\int_V[\boldsymbol{E}:\boldsymbol{T}-\Sigma^c(\boldsymbol{T})]\mathrm{d}V-\int_V\boldsymbol{u}\cdot\boldsymbol{f}\rho_0\mathrm{d}V-\int_{A_t}\boldsymbol{u}\cdot\mathring{\boldsymbol{T}}_N\mathrm{d}A+$$

$$\int_V \left[\frac{1}{2}(\boldsymbol{\nabla u}+\boldsymbol{u\nabla}+(\boldsymbol{\nabla u})\cdot(\boldsymbol{u\nabla}))-\boldsymbol{E}\right]:\boldsymbol{T}\mathrm{d}V$$

$$(2.12\text{a})$$

$$\overset{*}{\Phi}{}_3^c=\int_V[\boldsymbol{T}:\boldsymbol{E}-\Sigma^p(\boldsymbol{E})]\mathrm{d}V+\int_V\frac{1}{2}[(\boldsymbol{\nabla u})\cdot(\boldsymbol{u\nabla})]:\boldsymbol{T}\mathrm{d}V-\int_{A_u}\mathring{\boldsymbol{u}}\cdot\boldsymbol{F}\cdot\boldsymbol{T}\cdot\boldsymbol{N}\mathrm{d}A+\int_V\boldsymbol{u}\cdot[(\boldsymbol{F}\cdot\boldsymbol{T})\cdot\boldsymbol{\nabla}+\boldsymbol{f}\rho_0]\mathrm{d}V \quad (2.12\text{b})$$

在这种情况下,实际上,(2.12a)归化为下列形式

$$\overset{*}{\Phi}{}_{3R}^p=-\int_V\Sigma^c(\boldsymbol{T})\mathrm{d}V-\int_V\boldsymbol{u}\cdot\boldsymbol{f}\rho_0\mathrm{d}V-\int_{A_t}\boldsymbol{u}\cdot\mathring{\boldsymbol{T}}_N\mathrm{d}A+\int_V\frac{1}{2}[\boldsymbol{\nabla u}+\boldsymbol{u\nabla}+(\boldsymbol{\nabla u})\cdot(\boldsymbol{u\nabla})]:\boldsymbol{T}\mathrm{d}V$$

$$(2.13)$$

于是,本文原来企图用(2.12a)给出《弹性力学的变分原理及其应用》,(胡海昌,1981)中表6.1的第6类变分原理的目的也没有达到.

(4)不完全的广义位能和广义余能变分泛函之四(事先分别满足应变位移关系和位移边界条件及平衡方程和力边界条件的情况).

此时可直接由(1.24a)和(1.24b)写出下列一对互补的变分泛函

$$\overset{*}{\Phi}{}_4^p = \int_V [\boldsymbol{E} : \boldsymbol{T} - \Sigma^c(\boldsymbol{T})]\mathrm{d}V - \int_V \boldsymbol{u} \cdot \boldsymbol{f}\rho_0 \mathrm{d}V - \int_{A_t} \boldsymbol{u} \cdot \mathring{\boldsymbol{T}}_N \mathrm{d}A \tag{1.7a}$$

$$\overset{*}{\Phi}{}_4^c = \int_V [\boldsymbol{T} : \boldsymbol{E} - \Sigma^p(\boldsymbol{E})]\mathrm{d}V - \int_V \frac{1}{2}[(\boldsymbol{\nabla u}) \cdot (\boldsymbol{u\nabla})] : \boldsymbol{T}\mathrm{d}V - \int_{A_u} \mathring{\boldsymbol{u}} \cdot \boldsymbol{F} \cdot \boldsymbol{T} \cdot \boldsymbol{N}\mathrm{d}A \tag{1.7b}$$

易证下列互补关系成立

$$\overset{*}{\Phi}{}_4^p + \overset{*}{\Phi}{}_4^c = 0 \tag{2.14}$$

(5)不完全的广义位能和广义余能变分泛函之五(事先分别满足余应变能形式的本构关系和位移边界条件及应变能形式的本构关系和力边界条件的情况).

此时可直接由(2.5a)和(2.5b)写出下列一对互补的变分泛函

$$\overset{*}{\Phi}{}_5^p = \int_V \Sigma^p(\boldsymbol{E})\mathrm{d}V - \int_V \boldsymbol{u} \cdot \boldsymbol{f}\rho_0 \mathrm{d}V - \int_{A_t} \boldsymbol{u} \cdot \mathring{\boldsymbol{T}}_N \mathrm{d}A + \int_V \left[\frac{1}{2}(\boldsymbol{\nabla u} + \boldsymbol{u\nabla} + (\boldsymbol{\nabla u}) \cdot (\boldsymbol{u\nabla})) - \boldsymbol{E}\right] : \boldsymbol{T}\mathrm{d}V \tag{2.15a}$$

$$\overset{*}{\Phi}{}_5^c = \int_V \Sigma^c(\boldsymbol{T})\mathrm{d}V + \int_V \frac{1}{2}[(\boldsymbol{\nabla u}) \cdot (\boldsymbol{u\nabla})] : \boldsymbol{T}\mathrm{d}V -$$

$$\int_{A_u} \mathring{\boldsymbol{u}} \cdot \boldsymbol{F} \cdot \boldsymbol{T} \cdot \boldsymbol{N} \mathrm{d}A + \int_V \boldsymbol{u} \cdot [(\boldsymbol{F} \cdot \boldsymbol{T}) \cdot \boldsymbol{\nabla} + \boldsymbol{f}\rho_0] \mathrm{d}V$$

(2.15b)

易证下列互补关系成立

$$\overset{*}{\Phi}{}_5^p + \overset{*}{\Phi}{}_5^c = 0 \qquad (2.16)$$

6. 不完全的广义位能和广义余能变分泛函之六(事先分别满足余应变能形式的本构关系和应变位移关系及应变能形式的本构关系和平衡方程的情况).

此时可直接由(2.5a)和(2.5b)写出下列一对互补的变分泛涵

$$\overset{*}{\Phi}{}_6^p = \int_V \Sigma^p(\boldsymbol{E}) \mathrm{d}V - \int_V \boldsymbol{u} \cdot \boldsymbol{f}\rho_0 \mathrm{d}V - \int_{A_t} \boldsymbol{u} \cdot \mathring{\boldsymbol{T}}_N \mathrm{d}A + \int_{A_u} (\mathring{\boldsymbol{u}} - \boldsymbol{u}) \cdot \boldsymbol{F} \cdot \boldsymbol{T} \cdot \boldsymbol{N} \mathrm{d}A \qquad (2.17a)$$

$$\overset{*}{\Phi}{}_6^c = \int_V \Sigma^c(\boldsymbol{T}) \mathrm{d}V + \int_V \frac{1}{2} [(\boldsymbol{\nabla}\boldsymbol{u}) \cdot (\boldsymbol{u}\boldsymbol{\nabla})] : \boldsymbol{T} \mathrm{d}V - \int_{A_u} \mathring{\boldsymbol{u}} \cdot \boldsymbol{F} \cdot \boldsymbol{T} \cdot \boldsymbol{N} \mathrm{d}A + \int_{A_t} \boldsymbol{u} \cdot (\mathring{\boldsymbol{T}}_N - \boldsymbol{F} \cdot \boldsymbol{T} \cdot \boldsymbol{N}) \mathrm{d}A$$

(2.17b)

易证下列互补关系成立

$$\overset{*}{\Phi}{}_6^p + \overset{*}{\Phi}{}_6^c = 0 \qquad (2.18)$$

(7)不完全的位能和余能变分泛函之七(最小位能和最小余能变分泛函).

此时可直接由(2.5a)和(2.5b)写出下列一对互补的变分泛函

$$\boldsymbol{\Phi}_7^p = \int_V \Sigma^p(\boldsymbol{E})\,\mathrm{d}V - \int_V \boldsymbol{u}\cdot\boldsymbol{f}\rho_0\,\mathrm{d}V - \int_{A_t} \boldsymbol{u}\cdot\mathring{\boldsymbol{T}}_N\,\mathrm{d}A \tag{2.19a}$$

$$\Phi_7^c = \int_V \Sigma^c(\boldsymbol{T})\,\mathrm{d}V + \int_V \frac{1}{2}[(\nabla\boldsymbol{u})\cdot(\boldsymbol{u}\nabla)]:\boldsymbol{T}\,\mathrm{d}V - \int_{A_u} \mathring{\boldsymbol{u}}\cdot\boldsymbol{F}\cdot\boldsymbol{T}\cdot\boldsymbol{N}\,\mathrm{d}A \tag{2.19b}$$

易证下列互补关系成立

$$\boldsymbol{\Phi}_7^p + \boldsymbol{\Phi}_7^c = 0 \tag{2.20}$$

本书的4.1节介绍了弹性杆的变分公式,清华大学基础部力学教研组的钱伟长院士曾以《弹性理论中广义变分原理的研究及其在有限元计算中的应用》为题说明怎样系统地建立各种广义变分原理,怎样合理地使用各种广义变分原理来改进有限元计算的成效.为了易于说明问题,他指出只局限于弹性理论的各种广义变分原理,但其推广并不困难.

他指出,广义变分原理的泛函,可以系统地采用Lagrange乘子法,把一般有条件的变分原理化为无条件的变分原理来唯一地决定.Lagrange乘子所代表的物理量,可以通过变分求极值或驻值的过程求得,从而消除了在建立广义变分原理的泛函时,人们经常陷入像猜谜一样的困境.

他也指出：我们同样可以用 Lagrange 乘子法把一般有多个条件的变分原理，化为条件个数较少的变分原理. 我们称变分条件减少了的变分原理为各级不完全的广义变分原理. 凡是把全部变分条件都消除了的变分原理，称为完全的广义变分原理，或简称广义变分原理；实际上是完全无条件的变分原理.

他建立了弹性小位移变形理论中的各级不完全的广义位能原理和各级不完全的广义余能原理，包括从最小位能原理和最小余能原理分别导出的最完全的广义变分原理；并且证明了这两个弹性力学广义变分原理的泛函是等同的. 在这些广义变分原理中，包括了 Hellinger-Reissner(1950)①②，胡海昌－鹫津久一郎(1955)③④的广义变分原理.

他也建立了弹性大位移变形理论中的位能原理和余能原理，并建立了有关位能余能的各级不完全的广义变分原理，包括以大位移变形的最小位能和最小余能原理分别导出的弹性力学广义变分原理，并且也证明了在大位移变形情况下，这两个弹性力学的广义变分原理也是等同的.

他除了列举广义变分原理在有限元法上的众所周知的应用

---

① 胡海昌. 弹性力学的变分原理及其应用. 北京：科学出版社，1981.

② 鹫津久一郎. 弹性学の变分原理概论. ユンヒコ——タによる构造工学讲座Ⅱ－3－A. 日本钢构造协会编. 培风馆，昭和 47 年，1972.

③ 钱伟长. 变分法及有限元(上册). 北京：科学出版社，1980.

④ 金宝桢，杨式德，朱宝华. 结构力学(下册)第 2 版. 北京：人民教育出版社，1961.

外,还补充了三个比较重要的应用范围.

1. 小位移变形弹性理论的最小位能原理和最小余能原理

本文只限于考虑弹性静力学问题,当然不难推广至弹性动力学和有关的塑性力学问题. 设在卡氏坐标系 $x_i(i=1,2,3)$ 中,弹性体内各点位移为 $\boldsymbol{u}_i(i=1,2,3)$,应变和应力分别为 $\boldsymbol{e}_{ij},\boldsymbol{\sigma}_{ij}(i,j=1,2,3)$,应变与位移服从下列小位移变形关系

$$\boldsymbol{e}_{ij}=\frac{1}{2}(\boldsymbol{u}_{i,j}+\boldsymbol{u}_{j,i}) \tag{1}$$

$\boldsymbol{u}_{i,j}$代表 $\boldsymbol{u}_i$ 对 x_j 的偏导数,对于小应变的线性弹性而言,最一般的应力应变关系为

$$\boldsymbol{\sigma}_{ij}=a_{ijkl}\boldsymbol{e}_{kl} \tag{2}$$

$$\boldsymbol{e}_{ij}=b_{ijkl}\boldsymbol{\sigma}_{kl} \tag{3}$$

其中 a_{ijkl}称为弹性常数,b_{ijkl}称为劲度系数,它们有下列对称性质

$$\begin{cases}a_{ijkl}=a_{jikl}=a_{ijlk}=a_{klij}\\ b_{ijkl}=b_{jikl}=b_{ijlk}=b_{klij}\end{cases} \tag{4}$$

在弹性体的体积内,弹性体的应力 $\boldsymbol{\sigma}_{ij}$和体积力 $\boldsymbol{F}_i(i=1,2,3)$之间,满足静力平衡方程

$$\boldsymbol{\sigma}_{ij,j}+\boldsymbol{F}_i=\mathbf{0} \qquad (\text{在 }\tau\text{ 内},i=1,2,3) \tag{5}$$

在弹性体的边界面上,满足相应的边界条件. 边界表面

S 一般可以分为两部分：在 S_p 上，外力 $\bar{\boldsymbol{p}}_i$ 已知；在 S_u 上，位移 $\bar{\boldsymbol{u}}_i$ 已知.

$$S = S_p + S_u \tag{6}$$

而

$$\boldsymbol{\sigma}_{ij} n_j = \bar{\boldsymbol{p}}_i \qquad (\text{在 } S_p \text{ 上}) \tag{7a}$$

$$\boldsymbol{u}_i = \bar{\boldsymbol{u}}_i \qquad (\text{在 } S_u \text{ 上}) \tag{7b}$$

其中 n_j 为 S_p 上的外法线矢量的方向余弦.

从上述诸关系中看到，弹性体的平衡问题是一个边界值问题，共有 15 个未知数，即 6 个应力分量 $\boldsymbol{\sigma}_{ij}$，6 个应变分量 $\boldsymbol{e}_{ij}$，3 个位移分量 $\boldsymbol{u}_i$，它们通过(1)，(2)，(5)中 15 个方程求解，它们必须满足边界条件(7).

设 $A(\boldsymbol{e}_{ij})$代表弹性体的应变能密度，它是应变分量的函数，而且根据应变能密度的定义，我们有

$$\frac{\partial A}{\partial \boldsymbol{e}_{ij}} = \boldsymbol{\sigma}_{ij} \qquad (i,j = 1,2,3) \tag{8}$$

在小应变的情况下，将式(2)中的 $\boldsymbol{\sigma}_{ij}$代入上式，积分求得线性理论的弹性应变能密度

$$A(\boldsymbol{e}_{ij}) = \frac{1}{2} a_{ijkl} \boldsymbol{e}_{ij} \boldsymbol{e}_{kl} \tag{9}$$

同样，在小应变条件下，我们有余能密度 $B(\boldsymbol{\sigma}_{ij})$，它和应变能密度 $A(\boldsymbol{e}_{ij})$的关系为

$$A(\boldsymbol{e}_{ij}) + B(\boldsymbol{\sigma}_{ij}) = \boldsymbol{e}_{kl} \boldsymbol{\sigma}_{kl} \tag{10}$$

我们很易证明

$$\frac{\partial B}{\partial \boldsymbol{\sigma}_{ij}} = \boldsymbol{e}_{ij} \tag{11}$$

或者通过积分

$$B(\boldsymbol{\sigma}_{ij}) = \frac{1}{2} b_{ijkl} \boldsymbol{\sigma}_{ij} \boldsymbol{\sigma}_{kl} \tag{12}$$

于是,我们有众所周知的小位移变形线性弹性体的最小位能原理和最小余能原理.

变分原理Ⅰ(小位移变形线性弹性理论的最小位能原理) 在满足小位移应变关系(1)式和边界位移已知的条件(7b)的所有容许的应变 $\boldsymbol{e}_{ij}$ 和位移 $\boldsymbol{u}_i$ 中,实际的 $\boldsymbol{e}_{ij}$ 和 $\boldsymbol{u}_i$ 必使弹性体的总位能

$$\Pi_{\mathrm{I}} = \iiint_{\tau} [A(\boldsymbol{e}_{ij}) - \boldsymbol{F}_i \boldsymbol{u}_i] \mathrm{d}\tau - \iint_{S_n} \bar{\boldsymbol{p}}_i \boldsymbol{u}_i \mathrm{d}s \tag{13}$$

为最小. 亦即,使式(13)的泛函 Π_{I} 为最小的 $\boldsymbol{u}_i$,$\boldsymbol{e}_{ij}$ 必满足平衡方程(5)和边界外力已知条件(7a);在证明中,我们认为式(13)中的 $A(\boldsymbol{e}_{ij})$ 是 $1/2a_{ijkl}\boldsymbol{e}_{ij}\boldsymbol{e}_{kl}$ 的简写,并利用了应力应变关系式(2),或认为式(8)成立. 证明是众所周知的.

变分原理Ⅱ(小位移变形线性弹性理论的最小余能原理) 在满足小位移变形的平衡方程(5)和边界外力已知的条件(7a)的所有容许的应力 $\boldsymbol{\sigma}_{ij}$ 中,实际的应力 $\boldsymbol{\sigma}_{ij}$ 必使弹性总余能

$$\Pi_{\mathrm{II}} = \iiint_{\tau} B(\boldsymbol{\sigma}_{ij}) \mathrm{d}\tau - \iint_{S_u} \bar{\boldsymbol{u}}_i \boldsymbol{\sigma}_{ij} n_j \mathrm{d}s \tag{14}$$

为最小. 亦即,使式(14)的泛函 Π_{II} 为最小的 $\boldsymbol{\sigma}_{ij}$ 必满足边界位移已知的条件(7b);在证明中,我们认为式(14)中的 $B(\boldsymbol{\sigma}_{ij})$ 为 $1/2b_{ijkl}\boldsymbol{\sigma}_{ij}\boldsymbol{\sigma}_{kl}$ 的简写,并利用了应力应变关系式(3)和应变位移关系式(1),或利用(11)和(1). 证明是众所周知的.

不论变分原理Ⅰ或Ⅱ,都是有条件的变分原理,最小位能原理的条件是式(1)和式(7b). 最小余能原理的条件是式(5)和式(7a). 当然在前者,我们还利用了应力应变关系式(2)或式(8);在后者,我们也利用了应力应变关系式(3)或式(11),和应变位移关系式(1). 其实这些被利用的关系,也可以当作某种形式的条件.

2. 小位移弹性理论的完全的广义变分原理和各级不完全的广义变分原理

现在让我们采用 Lagrange 乘子法把有两组条件(1),(7b)的最小位能原理(变分原理Ⅰ)化为无条件的广义变分原理.

设 $\boldsymbol{\lambda}_{ij}$ 和 $\boldsymbol{\mu}_i$ 为待定的 Lagrange 乘子,于是根据式(13)导出的无条件的广义变分原理的泛函为

$$\Pi_{\mathrm{I}}^{*}=\iiint_{\tau}[A(\boldsymbol{e}_{ij})-\boldsymbol{F}_i\boldsymbol{u}_i]\mathrm{d}\tau-\iint_{S_p}\bar{\boldsymbol{p}}_i\boldsymbol{u}_i\mathrm{d}s+$$

$$\iiint_{\tau}\left[\boldsymbol{e}_{ij}-\frac{1}{2}(\boldsymbol{u}_{i,j}+\boldsymbol{u}_{j,i})\right]\boldsymbol{\lambda}_{ij}\mathrm{d}\tau+\iint_{S_u}(\boldsymbol{u}_i-\overline{\boldsymbol{u}}_i)\boldsymbol{\mu}_i\mathrm{d}s \tag{15}$$

把 $\boldsymbol{e}_{ij},\boldsymbol{u}_i,\boldsymbol{\lambda}_{ij},\boldsymbol{\mu}_i$ 都当作独立变量进行变分，当 Π_{I}^{*} 达到驻值时，有 $\delta\Pi_{\mathrm{I}}^{*}=0$，即

$$\delta\Pi_{\mathrm{I}}^{*}=\iiint_{\tau}\left[\left(\frac{\partial A}{\partial \boldsymbol{e}_{ij}}+\boldsymbol{\lambda}_{ij}\right)\delta\boldsymbol{e}_{ij}+\left(\boldsymbol{e}_{ij}-\frac{1}{2}\boldsymbol{u}_{i,j}-\frac{1}{2}\boldsymbol{u}_{j,i}\right)\delta\boldsymbol{\lambda}_{ij}-\boldsymbol{\lambda}_{ij}\delta\boldsymbol{u}_{i,j}-\boldsymbol{F}_i\delta\boldsymbol{u}_i\right]\mathrm{d}\tau+$$
$$\iint_{S_u}[\boldsymbol{\mu}_i\delta\boldsymbol{u}_i+(\boldsymbol{u}_i-\overline{\boldsymbol{u}}_i)\delta\boldsymbol{\mu}_i]\mathrm{d}s-\iint_{S_p}\overline{\boldsymbol{p}}_i\delta\boldsymbol{u}_i\mathrm{d}s=0 \tag{16}$$

其中，利用式(8)，我们有 $\dfrac{\partial A}{\partial \boldsymbol{e}_{ij}}=\boldsymbol{\sigma}_{ij}$，再利用 Green 公式

$$\iiint_{\tau}\boldsymbol{\lambda}_{ij}\delta\boldsymbol{u}_{i,j}\mathrm{d}\tau=\iint_{S_p+S_u}\boldsymbol{\lambda}_{ij}n_j\delta\boldsymbol{u}_i\mathrm{d}s-\iiint_{\tau}\boldsymbol{\lambda}_{ij,j}\delta\boldsymbol{u}_i\mathrm{d}\tau \tag{17}$$

其中 n_j 为表面 S_p+S_u 的外法线单位矢量的方向余弦. 把式(8)和式(17)代入式(16)，得

$$\delta\Pi_{\mathrm{I}}^{*}=\iiint_{\tau}\left[(\boldsymbol{\sigma}_{ij}+\boldsymbol{\lambda}_{ij})\delta\boldsymbol{e}_{ij}+\left(\boldsymbol{e}_{ij}-\frac{1}{2}\boldsymbol{u}_{i,j}-\frac{1}{2}\boldsymbol{u}_{j,i}\right)\delta\boldsymbol{\lambda}_{ij}+\right.$$
$$\left.(\boldsymbol{\lambda}_{ij,j}-\boldsymbol{F}_i)\delta\boldsymbol{u}_i\right]\mathrm{d}\tau+\iint_{S_u}[(\boldsymbol{\mu}_i-\boldsymbol{\lambda}_{ij}n_j)\delta\boldsymbol{u}_i+$$
$$(\boldsymbol{u}_i-\overline{\boldsymbol{u}}_i)\delta\boldsymbol{\mu}_i]\mathrm{d}s-\iint_{S_p}(\boldsymbol{\lambda}_{ij}n_j+\overline{\boldsymbol{p}}_i)\delta\boldsymbol{u}_i\mathrm{d}s=0 \tag{17a}$$

由于 τ 内的 $\delta \boldsymbol{e}_{ij}, \delta \boldsymbol{\lambda}_{ij}, \delta \boldsymbol{u}_j$ 和 S_u 上的 $\delta \boldsymbol{u}_i, \delta \boldsymbol{\mu}_i$，以及 S_p 上的 $\delta \boldsymbol{u}_i$ 都是独立的，所以 $\delta \Pi_{\mathrm{I}}^{*}$ 的驻值条件导出了

$$\boldsymbol{\lambda}_{ij} = -\boldsymbol{\sigma}_{ij}, \boldsymbol{e}_{ij} = \frac{1}{2}(\boldsymbol{u}_{i,j} + \boldsymbol{u}_{j,i}), \boldsymbol{\lambda}_{ij,j} - \boldsymbol{F}_i = \mathbf{0} \quad (\text{在 } \tau \text{ 中}) \tag{18a,b,c}$$

$$\boldsymbol{\mu}_i = \boldsymbol{\lambda}_{ij} n_j, \boldsymbol{u}_i - \bar{\boldsymbol{u}}_i = \mathbf{0} \quad (\text{在 } S_u \text{ 上}) \tag{18d,c}$$

$$\boldsymbol{\lambda}_{ij} n_j + \bar{p}_i = \mathbf{0} \quad (\text{在 } S_p \text{ 上}) \tag{18f}$$

(18b,c)为原来的变分条件，(18a)，(18d)给出了待定的 Lagrange 乘子所代表的物理量，即

$$\boldsymbol{\lambda}_{ij} = -\boldsymbol{\sigma}_{ij} \quad (\text{在 } \tau \text{ 中}), \boldsymbol{\mu}_i = \boldsymbol{\lambda}_{ij} n_j = -\boldsymbol{\sigma}_{ij} n_j \quad (\text{在 } S_u \text{ 上}) \tag{19}$$

把式(19)代入式(18c)，即得平衡方程，把式(19)代入式(18f)，即得外力已给的边界条件. 把式(19)代入式(15)，即得无条件的弹性理论广义变分原理的泛函.

变分原理Ⅲ(完全的小位移线性弹性理论的广义变分原理——从最小位能原理导出的)　满足(1)，(2)，(5)，(7)诸关系的 $\boldsymbol{u}_i, \boldsymbol{e}_{ij}, \boldsymbol{\sigma}_{ij}$ 的解，必使下述泛函 Π_{I}^{*} 为驻值

$$\Pi_{\mathrm{I}}^{*} = \iiint_{\tau} \left[A(\boldsymbol{e}_{ij}) - \left(\boldsymbol{e}_{ij} - \frac{1}{2}\boldsymbol{u}_{i,j} - \frac{1}{2}\boldsymbol{u}_{j,i} \right) \boldsymbol{\sigma}_{ij} - \boldsymbol{F}_i \boldsymbol{u}_i \right] \mathrm{d}\tau - \iint_{S_p} \bar{\boldsymbol{p}}_i \boldsymbol{u}_i \mathrm{d}s - \iint_{S_u} (\boldsymbol{u}_i - \bar{\boldsymbol{u}}_i) \boldsymbol{\sigma}_{ij} n_j \mathrm{d}s \tag{20}$$

这里必须指出，首先，在使用这个变分原理时，$A(\boldsymbol{e}_{ij})$

认为是$\frac{1}{2}a_{ijkl}\boldsymbol{e}_{ij}\boldsymbol{e}_{kl}$的简写. 其次,所有 $\boldsymbol{e}_{ij}$,$\boldsymbol{\sigma}_{ij}$,$\boldsymbol{u}_i$ 诸量都是独立的变分量. 再次,这只是一个驻值原理,不是一个“最小”原理. 最后,我们应该指出,这个广义变分原理是胡海昌(1955)首先提出的,但胡海昌只是证明了这个广义变分原理的驻值解相当于小位移变形问题的解,并没有说明泛函是怎样求得的. 钱伟长曾在1964年提出用Lagrange乘子法寻找广义变分泛函的方法,并指出这个方法有广泛的实用价值. 当时人们并不理解Lagrange乘子的作用,在建立广义变分原理的泛函时,只能用假设—试验—修正—再试验的方法进行工作. 在1964年《力学学报》曾展开过一场关于极限设计广义变分原理中某一乘子究竟代表什么的争论,就是这种原因所引起的. 这场争论只有在正确利用了Lagrange乘子法(1974)以后,才得结束. 在国际上,人们利用Lagrange乘子法建立广义变分的泛函,从国内接触到的文献看,大概在1968—1969年,首先是鹫津久一郎(1968),其次是卞学璜(1969)曾明确提出了这个观点,但并没有充分利用这个观点. 例如,在鹫津久一郎的著作中,在建立泛函(19)时,Lagrange乘子 $\boldsymbol{\lambda}_{ij}$就直接使用了 $-\boldsymbol{\sigma}_{ij}$;在这一点上,他和胡海昌的工作并无什么不同,也没有什么改进. 在本文以后的讨论中,都是用待定的Lagrange乘子,通过变分,而认识其所代表的物理量的. 当然,为了节省篇幅有时并没有一一推导.

变分原理Ⅳ(完全的小位移线性弹性理论的广义变分原理——从最小余能原理导出的)　满足(1),(2),(5),(7)诸方程的 $\boldsymbol{\sigma}_{ij}, \boldsymbol{e}_{ij}, \boldsymbol{u}_i$ 的解,必使下述泛函 $\Pi^*_{\mathrm{II a}}$ 或 $\Pi^*_{\mathrm{II b}}$ 为驻值

$$\Pi^*_{\mathrm{II a}} = \iiint_\tau [B(\boldsymbol{\sigma}_{ij}) + (\boldsymbol{\sigma}_{ij,j} + \boldsymbol{F}_i)\boldsymbol{u}_i]\mathrm{d}\tau - \iint_{S_p} (\boldsymbol{\sigma}_{ij} n_j - \bar{\boldsymbol{p}}_i)\boldsymbol{u}_i \mathrm{d}s - \iint_{S_u} \boldsymbol{\sigma}_{ij} n_j \bar{\boldsymbol{u}}_i \mathrm{d}s \tag{21}$$

或在利用了式(10)以后

$$\Pi^*_{\mathrm{II b}} = \iiint_\tau [\boldsymbol{e}_{ij}\boldsymbol{\sigma}_{ij} - A(\boldsymbol{e}_{ij}) + (\boldsymbol{\sigma}_{ij,j} + \boldsymbol{F}_i)\boldsymbol{u}_i]\mathrm{d}\tau - \iint_{S_p} (\boldsymbol{\sigma}_{ij} n_j - \bar{\boldsymbol{p}}_i)\boldsymbol{u}_i \mathrm{d}s - \iint_{S_u} \boldsymbol{\sigma}_{ij} n_j \bar{\boldsymbol{u}}_i \mathrm{d}s \tag{22}$$

其中 $\boldsymbol{u}_i, \boldsymbol{e}_{ij}, \boldsymbol{\sigma}_{ij}$ 都是独立变量,$A(\boldsymbol{e}_{ij})$ 代表 $\frac{1}{2}a_{ijkl}\boldsymbol{e}_{ij}\boldsymbol{e}_{kl}$ 的简写,$B(\boldsymbol{\sigma}_{ij})$ 代表 $\frac{1}{2}b_{ijkl}\boldsymbol{\sigma}_{ij}\boldsymbol{\sigma}_{kl}$ 的简写.

证明式(21)或式(22)时,我们可以从最小余能原理(14)开始. 它有两组条件,即平衡方程(5)和边界外力已知的条件(7a). 用 Lagrange 乘子法,引进待定的 Lagrange 乘子 $\boldsymbol{\lambda}_i, \boldsymbol{\mu}_i$,广义变分的泛函为

$$\Pi^*_{\text{II b}} = \iiint_\tau [\boldsymbol{e}_{ij}\boldsymbol{\sigma}_{ij} - A(\boldsymbol{e}_{ij}) + (\boldsymbol{\sigma}_{ij,j} + \boldsymbol{F}_i)\boldsymbol{\lambda}_i]\mathrm{d}\tau + \iint_{S_p} (\boldsymbol{\sigma}_{ij}n_j - \bar{\boldsymbol{p}}_i)\boldsymbol{\mu}_i\mathrm{d}s - \iint_{S_u} \boldsymbol{\sigma}_{ij}n_j\bar{\boldsymbol{u}}_i\mathrm{d}s \tag{23}$$

其中也已用了式(10),把 $B(\boldsymbol{\sigma}_{ij})$ 写成 $\boldsymbol{e}_{ij}\boldsymbol{\sigma}_{ij} - A(\boldsymbol{e}_{ij})$,把 $\boldsymbol{\sigma}_{ij}, \boldsymbol{e}_{ij}, \boldsymbol{\lambda}_i, \boldsymbol{\mu}_i$ 作为独立变量进行变分,得

$$\delta\Pi^*_{\text{IIb}} = \iiint_\tau [(\boldsymbol{\sigma}_{ij} - a_{ijkl}\boldsymbol{e}_{kl})\delta\boldsymbol{e}_{ij} + (\boldsymbol{e}_{ij} - \boldsymbol{\lambda}_{i,j})\delta\boldsymbol{\sigma}_{ij} + (\boldsymbol{\sigma}_{ij,j} + \boldsymbol{F}_i)\delta\boldsymbol{\lambda}_i]\mathrm{d}\tau + \iint_{S_p} [(\boldsymbol{\sigma}_{ij}n_j - \bar{\boldsymbol{p}}_i)\delta\boldsymbol{\mu}_i + (\boldsymbol{\mu}_i + \boldsymbol{\lambda}_i)n_j\delta\boldsymbol{\sigma}_{ij}]\mathrm{d}s + \iiint_{S_u} (\boldsymbol{\lambda}_i - \boldsymbol{u}_i)n_j\delta\boldsymbol{\sigma}_{ij}\mathrm{d}s \tag{24}$$

在利用了式(1)后

$$\iiint_\tau (\boldsymbol{\sigma}_{ij} - \boldsymbol{\lambda}_{i,j})\delta\boldsymbol{\sigma}_{ij}\mathrm{d}\tau = \iiint_\tau (\boldsymbol{u}_i - \boldsymbol{\lambda}_i)_{,j}\delta\boldsymbol{\sigma}_{ij}\mathrm{d}\tau \tag{25}$$

于是,从 $\delta\Pi^*_{\text{II b}} = 0$ 的驻值条件证明

$$\boldsymbol{\sigma}_{ij} = a_{ijkl}\boldsymbol{e}_{kl}, \boldsymbol{\sigma}_{ij,j} + \boldsymbol{F}_i = \boldsymbol{0}, \boldsymbol{u}_i = \boldsymbol{\lambda}_i \quad (\text{在 } \tau \text{ 内}) \tag{26a,b,c}$$

$$\boldsymbol{\sigma}_{ij}n_j = \bar{\boldsymbol{p}}_i, \boldsymbol{\mu}_i = -\boldsymbol{\lambda}_i \quad (\text{在 } S_p \text{ 上}) \tag{26d,e}$$

$$\boldsymbol{\lambda}_i - \bar{\boldsymbol{u}}_i = \boldsymbol{0} \quad (\text{在 } S_u \text{ 上}) \tag{26f}$$

其中(26a),(26b)分别代表应力应变关系式(2)和平衡方程(5),(26c)给出了 $\boldsymbol{\lambda}_i$ 所代表的物理量 $\boldsymbol{u}_i$,(26e)给出了 $\boldsymbol{\mu}_i$ 所代表的物理量 $-\boldsymbol{u}_i$. 最后,(26d),(26f)分别代表边界条件(7a),(7b),把式(23)的 $\boldsymbol{\lambda}_i$,

$\boldsymbol{\mu}_i$ 用 $\boldsymbol{u}_i$ 及 $-\boldsymbol{u}_i$ 代替，即得 Π^*_{IIb}的式(22)，即证明了变分原理Ⅳ.

现在让我们证明变分原理Ⅲ和Ⅳ是等价的. 因为

$$\Pi^*_{\mathrm{I}} + \Pi^*_{\mathrm{IIb}} = \iiint_{\tau} \left[\frac{1}{2}(\boldsymbol{u}_{i,j} + \boldsymbol{u}_{j,i})\boldsymbol{\sigma}_{ij} + \boldsymbol{\sigma}_{ij,j}\boldsymbol{u}_i\right]\mathrm{d}\tau - \iint_{S_p} \boldsymbol{\sigma}_{ij}\boldsymbol{u}_i n_j \mathrm{d}s - \iint_{S_u} \boldsymbol{u}_i\boldsymbol{\sigma}_{ij} n_j \mathrm{d}s \tag{27}$$

但是

$$\begin{aligned}
&\iiint_{\tau} \left[\frac{1}{2}(\boldsymbol{u}_{i,j} + \boldsymbol{u}_{j,i})\boldsymbol{\sigma}_{ij} + \boldsymbol{\sigma}_{ij,j}\boldsymbol{u}_i\right]\mathrm{d}\tau \\
&= \iiint_{\tau} (\boldsymbol{u}_{i,j}\boldsymbol{\sigma}_{ij} + \boldsymbol{\sigma}_{ij,j}\boldsymbol{u}_i)\mathrm{d}\tau \\
&= \iiint_{\tau} (\boldsymbol{u}_i\boldsymbol{\sigma}_{ij})_{,j}\mathrm{d}\tau = \iint_{S_p+S_u} \boldsymbol{\sigma}_{ij} n_j \boldsymbol{u}_i \mathrm{d}s
\end{aligned} \tag{28}$$

这就证明了

$$\Pi^*_{\mathrm{I}} = -\Pi^*_{\mathrm{IIb}} \tag{29}$$

这就证明了原理Ⅲ和原理Ⅳ的等同性. 胡海昌(1955)[①]称原理Ⅲ为广义位能原理，称原理Ⅳ为广义余能原理，显然这是容易引起误会的. 钱伟长(1964)曾证明了这个等同性原理. 鹫津久一郎(1968)也发觉了这个问题，但认为这种等同性原理只限于小位移问题，这是因为没有能在大位移问题中找到相应的余能

① 钱伟长. 变分法及有限元(上册). 北京：科学出版社，1980.

原理所致.

上面证明了原理Ⅲ和原理Ⅳ只在形式上不同,在实质上是相等的. 它们所解决的物理问题相同,所用的变分泛函相等(只差一个正负号,这对于驻值条件是没有关系的),所以,我们在以后,将称原理Ⅲ为小位移线性弹性理论的广义变分原理的位能形式,原理Ⅳ则为小位移线性弹性理论的广义变分原理的余能形式. 它们都是完全的广义变分原理,可以统称为小位移线性弹性理论的完全的广义变分原理.

用相同的方法可以证明下列诸小位移线性弹性原理的不完全的广义位能原理:

变分原理 ⅠA　在满足边界位移已知的条件(7b)的所有容许的 $\boldsymbol{u}_i,\boldsymbol{e}_{ij},\boldsymbol{\sigma}_{ij}$ 中,实际的 $\boldsymbol{u}_i,\boldsymbol{e}_{ij},\boldsymbol{\sigma}_{ij}$ 必使下列广义泛函为驻值

$$\Pi_{\mathrm{IA}}=\iiint_{\tau}\left\{A(\boldsymbol{e}_{ij})-\left[\boldsymbol{e}_{ij}-\frac{1}{2}(\boldsymbol{u}_{i,j}+\boldsymbol{u}_{j,i})\right]\boldsymbol{\sigma}_{ij}-\boldsymbol{F}_i\boldsymbol{u}_i\right\}\mathrm{d}\tau-\iint_{S_u}\bar{\boldsymbol{p}}_i\boldsymbol{u}_i\mathrm{d}s \tag{30}$$

在这里,原来最小位能原理有两组条件,现在只有一组条件,关于应变位移关系的条件,可以采用 Lagrange 乘子 $\boldsymbol{\lambda}_{ij}$,使它归纳入广义泛函,最后通过变分,决定它是 $-\boldsymbol{\sigma}_{ij}$.

变分原理 ⅠB　在满足应变位移关系式(1)的所有

容许的 $\boldsymbol{u}_i,\boldsymbol{e}_{ij},\boldsymbol{\sigma}_{ij}$ 中，实际的 $\boldsymbol{u}_i,\boldsymbol{e}_{ij},\boldsymbol{\sigma}_{ij}$ 必使下述泛函为驻值

$$\Pi_{\mathrm{IB}} = \iiint_{\tau} [A(\boldsymbol{e}_{ij}) - \boldsymbol{F}_i\boldsymbol{u}_i]\mathrm{d}\tau - \iint_{S_p} \overline{\boldsymbol{p}}_i\boldsymbol{u}_i\mathrm{d}s - \iint_{S_u} (\boldsymbol{u}_i - \overline{\boldsymbol{u}}_i)\boldsymbol{\sigma}_{ij}n_j\mathrm{d}s \tag{31}$$

这个广义变分原理实质上就是 Jones(1964)等在有限元法中使用的广义变分原理.

变分原理 IC　在满足一个边界位移已知的条件 $\boldsymbol{u}_1 - \overline{\boldsymbol{u}}_1 = \boldsymbol{0}$ 的所有容许的位移 $\boldsymbol{u}_i$，应变 $\boldsymbol{e}_{ij}$ 和应力 $\boldsymbol{\sigma}_{ij}$ 中，实际的 $\boldsymbol{u}_i,\boldsymbol{e}_{ij},\boldsymbol{\sigma}_{ij}$ 必使下列广义泛函为驻值

$$\Pi_{\mathrm{IC}} = \iiint_{\tau} \left\{A(\boldsymbol{e}_{ij}) - \left[\boldsymbol{e}_{ij} - \frac{1}{2}(\boldsymbol{u}_{i,j} + \boldsymbol{u}_{j,i})\right]\boldsymbol{\sigma}_{ij} - F_i\boldsymbol{u}_i\right\}\mathrm{d}\tau - \iint_{S_p} \overline{\boldsymbol{p}}_i\boldsymbol{u}_i\mathrm{d}s - \iint_{S_u} [(\boldsymbol{u}_2 - \overline{\boldsymbol{u}}_2)\boldsymbol{\sigma}_{2j}n_j + (\boldsymbol{u}_3 - \overline{\boldsymbol{u}}_3)\boldsymbol{\sigma}_{3j}n_j]\mathrm{d}s \tag{32}$$

也还有满足任意一个或二个边界位移已知的条件的不完全广义位能原理.

变分原理 ID　在满足一个应变位移关系 $\boldsymbol{e}_{11} - \boldsymbol{u}_{1,1} = \boldsymbol{0}$ 的所有容许的 $\boldsymbol{u}_i,\boldsymbol{e}_{ij},\boldsymbol{\sigma}_{ij}$ 中，实际的 $\boldsymbol{u}_i,\boldsymbol{e}_{ij},\boldsymbol{\sigma}_{ij}$ 必使下列广义泛函为驻值

$$\Pi_{\mathrm{ID}} = \iiint_{\tau} \left\{A(\boldsymbol{e}_{ij}) - \left[\boldsymbol{e}_{ij} - \frac{1}{2}(\boldsymbol{u}_{i,j} + \boldsymbol{u}_{j,i})\right]\boldsymbol{\sigma}_{ij} + (\boldsymbol{e}_{11} - \boldsymbol{u}_{1,1})\boldsymbol{\sigma}_{11} - \boldsymbol{F}_i\boldsymbol{u}_i\right\}\mathrm{d}\tau -$$

$$\iint_{S_p} \bar{\boldsymbol{p}}_i \boldsymbol{u}_i \mathrm{d}s - \iint_{S_u} (\boldsymbol{u}_i - \bar{\boldsymbol{u}}_i) \boldsymbol{\sigma}_{ij} n_j \mathrm{d}s \tag{33}$$

也可以有满足任意个(五个以下)应变位移关系的变分,它们都是不完全的广义位能原理;也可以有满足一部分位移应变关系和一部分边界位移已知条件的不完全的广义位能原理.

下面是一些小位移非线性弹性理论的不完全的广义余能原理:

变分原理ⅡA 在满足边界外力已知的条件(7a)的所有容许的 $\boldsymbol{u}_i, \boldsymbol{e}_{ij}, \boldsymbol{\sigma}_{ij}$ 中,实际的 $\boldsymbol{u}_i, \boldsymbol{e}_{ij}, \boldsymbol{\sigma}_{ij}$ 必使下述广义泛函为驻值

$$\Pi_{\mathrm{II\,A}} = \iiint_{\tau} [\boldsymbol{e}_{ij} \boldsymbol{\sigma}_{ij} - A(\boldsymbol{e}_{ij}) + (\boldsymbol{\sigma}_{ij,j} + \boldsymbol{F}_i) \boldsymbol{u}_i] \mathrm{d}\tau - \iint_{S_u} \boldsymbol{\sigma}_{ij} n_j \bar{\boldsymbol{u}}_i \mathrm{d}s \tag{34}$$

变分原理ⅡB 在满足平衡方程(5)的所有容许的 $\boldsymbol{u}_i, \boldsymbol{e}_{ij}, \boldsymbol{\sigma}_{ij}$ 中,实际的 $\boldsymbol{u}_i, \boldsymbol{e}_{ij}, \boldsymbol{\sigma}_{ij}$ 必使下述广义泛函为驻值

$$\Pi_{\mathrm{II\,B}} = \iiint_{\tau} [\boldsymbol{e}_{ij} \boldsymbol{\sigma}_{ij} - A(\boldsymbol{e}_{ij})] \mathrm{d}\tau - \iint_{S_p} (\boldsymbol{\sigma}_{ij} n_j - \bar{\boldsymbol{p}}_i) \boldsymbol{u}_i \mathrm{d}s - \iint_{S_u} \boldsymbol{\sigma}_{ij} n_j \bar{\boldsymbol{u}}_i \mathrm{d}s \tag{35}$$

变分原理ⅡC 在满足一个边界外力已知的条件

$\overline{\boldsymbol{p}}_1=\boldsymbol{\sigma}_{1j}n_j$ 的所有容许的 $\boldsymbol{u}_i,\boldsymbol{e}_{ij},\boldsymbol{\sigma}_{ij}$ 中，实际的 $\boldsymbol{u}_i,\boldsymbol{e}_{ij},\boldsymbol{\sigma}_{ij}$ 必使下列广义泛函为驻值

$$\Pi_{\text{IIC}}=\iiint_{\tau}[\boldsymbol{e}_{ij}\boldsymbol{\sigma}_{ij}-A(\boldsymbol{e}_{ij})+(\boldsymbol{\sigma}_{ij,j}+\boldsymbol{F}_i)\boldsymbol{u}_i]\mathrm{d}\tau-\iint_{S_u}\boldsymbol{\sigma}_{ij}n_j\overline{\boldsymbol{u}}_i\mathrm{d}s-\iint_{S_p}[(\boldsymbol{\sigma}_{2j}n_j-\overline{\boldsymbol{p}}_2)\boldsymbol{u}_2+(\boldsymbol{\sigma}_{3j}n_j-\overline{\boldsymbol{p}}_3)\boldsymbol{u}_3]\mathrm{d}s \tag{36}$$

变分原理ⅡD 在满足一个平衡方程 $\boldsymbol{\sigma}_{1j,j}+\boldsymbol{F}_1=\boldsymbol{0}$ 的所有容许的 $\boldsymbol{u}_i,\boldsymbol{e}_{ij},\boldsymbol{\sigma}_{ij}$ 中，实际的 $\boldsymbol{u},\boldsymbol{e}_{ij},\boldsymbol{\sigma}_{ij}$ 必使下列广义泛函为驻值

$$\Pi_{\text{IID}}=\iiint_{\tau}[\boldsymbol{e}_{ij}\boldsymbol{\sigma}_{ij}-A(\boldsymbol{e}_{ij})+(\boldsymbol{\sigma}_{2j,j}+\boldsymbol{F}_2)\boldsymbol{u}_2+(\boldsymbol{\sigma}_{3j,j}+\boldsymbol{F}_3)\boldsymbol{u}_3]\mathrm{d}\tau-\iint_{S_u}\boldsymbol{\sigma}_{ij}n_j\overline{\boldsymbol{u}}_i\mathrm{d}s-\iint_{S_p}(\boldsymbol{\sigma}_{ij}n_j-\overline{\boldsymbol{p}}_i)\boldsymbol{u}_i\mathrm{d}s \tag{37}$$

也可以有满足任意一个或二个外力已知的边界条件，或任意一个或二个平衡方程的不完全的广义余能原理. 还有满足一部分平衡方程同时满足另一部分边界外力已知条件的不完全的广义变分原理. 变分原理ⅡB也就是卞学璜用来处理断裂力学中应力强度因子的变分原理(1971).

还有一类不完全的广义变分原理，所需满足的条件只在一个局部的区域内，在其他区域内的关系都可

以通过变分满足.

变分原理 IE(小位移线性弹性理论的分区不完全广义位能原理)　设弹性体的体积为 $\tau=\tau_1+\tau_2$,在 τ_1 的表面上有 S_{p1},S_{u1},在 τ_2 的表面上有 S_{p2},S_{u2},其中 $S_u=S_{u1}+S_{u2}$,$S_p=S_{p1}+S_{p2}$,在表面 S_{u1} 上满足位移已给的边界条件(7b)和在 τ_1 中满足位移应变关系式(1)的所有容许的 $\boldsymbol{u}_i$,$\boldsymbol{e}_{ij}$,$\boldsymbol{\sigma}_{ij}$ 中,实际的 $\boldsymbol{u}_i$,$\boldsymbol{e}_{ij}$,$\boldsymbol{\sigma}_{ij}$ 必使下列广义泛函为驻值

$$\begin{aligned}\Pi_{\mathrm{IE}}=&\iiint_{\tau}[A(\boldsymbol{e}_{ij})-\boldsymbol{F}_i\boldsymbol{u}_i]\mathrm{d}\tau-\\&\iiint_{\tau_2}\left[\boldsymbol{e}_{ij}-\frac{1}{2}(\boldsymbol{u}_{i,j}+\boldsymbol{u}_{j,i})\right]\boldsymbol{\sigma}_{ij}\mathrm{d}\tau-\\&\iint_{S_p}\bar{\boldsymbol{p}}_i\boldsymbol{u}_i\mathrm{d}s-\iint_{S_{u2}}(\boldsymbol{u}_i-\bar{\boldsymbol{u}}_i)\boldsymbol{\sigma}_{ij}n_j\mathrm{d}s\end{aligned}\tag{38}$$

变分原理 ⅡE(小位移线性弹性理论的分区不完全广义余能原理)　设 $\tau=\tau_1+\tau_2$,在 τ_1 的表面上有 S_{u1},S_{p1},在 τ_2 的表面上有 S_{u2},S_{p2},其中 $S_u=S_{u1}+S_{u2}$,$S_p=S_{p1}+S_{p2}$,在表面 S_{p1} 上满足外力已知的边界条件(7a)和在 τ_1 中满足平衡方程(5)的所有容许的 $\boldsymbol{u}_i$,$\boldsymbol{e}_{ij}$,$\boldsymbol{\sigma}_{ij}$ 中,实际的 $\boldsymbol{u}_i$,$\boldsymbol{e}_{ij}$,$\boldsymbol{\sigma}_{ij}$ 必使下列广义泛函为驻值

$$\begin{aligned}\Pi_{\mathrm{IIE}}=&\iiint_{\tau}[\boldsymbol{\sigma}_{ij}\boldsymbol{e}_{ij}-A(\boldsymbol{e}_{ij})]\mathrm{d}\tau+\iint_{\tau_2}(\boldsymbol{\sigma}_{ij,j}+\boldsymbol{F}_i)\boldsymbol{u}_i\mathrm{d}\tau-\\&\iint_{S_u}\boldsymbol{\sigma}_{ij}n_j\bar{\boldsymbol{u}}_i\mathrm{d}s-\iint_{S_{p2}}(\boldsymbol{\sigma}_{ij}n_j-\bar{p}_i)\boldsymbol{u}_i\mathrm{d}s\end{aligned}\tag{39}$$

还有只在表面 S_{p1} 上满足任一个或二个或全部外力已知边界条件(7a)的分区不完全广义余能原理,或只在 τ_1 内满足任一个或二个平衡方程或全部平衡方程(5)的分区不完全广义余能原理.

还有只在表面 S_{u1} 上满足任一个或二个或全部位移已知边界条件(7b)的分区不完全广义位能原理,或只在 τ_1 内满足任一个或二个或全部应变位移关系式(1)的分区不完全广义位能原理.

在下面将列举在各个区域内满足不同条件的混合变分原理.

变分原理Ⅴ(小位移线性弹性理论的分区混合变分原理) 设 $\tau=\tau_1+\tau_2$,在 τ_1 的表面上有 $S_{u1}+S_{p1}$,在 τ_2 的表面上有 $S_{p2}+S_{u2}$,且 $S_u=S_{u1}+S_{u2}$,$S_p=S_{p1}+S_{p2}$,在 τ_1 内满足应变位移关系式(1),在 S_{u1} 上满足边界位移已给条件(7b),在 τ_2 内满足平衡方程(5)和在 S_{p2} 上满足边界外力已给条件(7a)的所有容许的 $\boldsymbol{u}_i,\boldsymbol{e}_{ij},\boldsymbol{\sigma}_{ij}$ 中,实际的 $\boldsymbol{u}_i,\boldsymbol{e}_{ij},\boldsymbol{\sigma}_{ij}$ 必使下列泛函为驻值

$$\Pi_{\mathrm{V}}=\iiint_{\tau_1}[A(\boldsymbol{e}_{ij})-\boldsymbol{F}_i\boldsymbol{u}_i]\mathrm{d}\tau+\iiint_{\tau}[A(\boldsymbol{e}_{ij})-\boldsymbol{e}_{ij}\boldsymbol{\sigma}_{ij}]\mathrm{d}\tau-\iint_{S_{p1}}\overline{\boldsymbol{p}}_i\boldsymbol{u}_i\mathrm{d}s+\iint_{S_{u2}}\overline{\boldsymbol{u}}_i\boldsymbol{\sigma}_{ij}n_j\mathrm{d}s-\iint_{S_0}\boldsymbol{\sigma}_{ij}n_j^{(0)}\boldsymbol{u}_i\mathrm{d}s \tag{40}$$

其中 S_0 为 τ_1 和 τ_2 的界面,$n_j^{(0)}$ 为 τ_1 的界面 S_0 上的外向法线(从 τ_1 指向 τ_2 的)的方向余弦.

变分原理Ⅵ(小位移线性弹性理论的完全的分区混合广义变分原理)　满足(1),(2),(5),(7)诸关系的 $\boldsymbol{u}_i,\boldsymbol{e}_{ij},\boldsymbol{\sigma}_{ij}$ 的解,必使下述混合泛函为驻值

$$\Pi_{\text{Ⅵ}}=\iiint\limits_{\tau_1}\left\{A(\boldsymbol{e}_{ij})-\left[\boldsymbol{e}_{ij}-\frac{1}{2}(\boldsymbol{u}_{i,j}+\boldsymbol{u}_{j,i})\right]\boldsymbol{\sigma}_{ij}-\boldsymbol{F}_i\boldsymbol{u}_i\right\}\mathrm{d}\tau+$$
$$\iiint\limits_{\tau}[A(\boldsymbol{e}_{ij})-\boldsymbol{e}_{ij}\boldsymbol{\sigma}_{ij}-(\boldsymbol{\sigma}_{ij,j}+\boldsymbol{F}_i)\boldsymbol{u}_i]\mathrm{d}\tau-$$
$$\iint\limits_{S_{p1}}\overline{\boldsymbol{p}}_i\boldsymbol{u}_i\mathrm{d}s-\iint\limits_{S_{u1}}(\boldsymbol{u}_i-\overline{\boldsymbol{u}}_i)\boldsymbol{\sigma}_{ij}\boldsymbol{u}_j\mathrm{d}s+$$
$$\iint\limits_{S_{p2}}(\boldsymbol{\sigma}_{ij}n_j-\overline{\boldsymbol{p}}_i)\boldsymbol{u}_i\mathrm{d}s+\iint\limits_{S_{u2}}\boldsymbol{\sigma}_{ij}n_j\overline{\boldsymbol{u}}_i\mathrm{d}s-$$
$$\iint\limits_{S_0}\boldsymbol{\sigma}_{ij}n_j^{(0)}\boldsymbol{u}_i\mathrm{d}s \tag{41}$$

其中 S_0 为 τ_1 和 τ_2 的界面,n_j^* 为 τ_1 的界面 S_0 上从 τ_1 指向 τ_2 的法线的方向余弦,$S_{u1}+S_{p1}$ 为 τ_1 的表面,$S_{u2}+S_{p2}$ 为 τ_2 的表面,而且 $S_{u1}+S_{u2}=S_u$,$S_{p1}+S_{p2}=S_p$,$\tau_1+\tau_2=\tau$.

从变分原理Ⅴ或从变分原理Ⅵ,我们可以导出各级不完全的分区混合广义变分原理.

所有上述各种完全的和不完全的广义变分原理,在计算各种特殊几何形状的弹性体受到较复杂的边界条件作用的有限元解时,都是很有用处的.

3. 大位移非线性弹性理论的最小位能原理和最小余能原理

设取拖带坐标(Comoving Coordinates)系 $x_i(i=1,2,3)$,它是随着弹性体的变形而变形的. 在这个坐标系内,$\boldsymbol{u}_i(i=1,2,3)$为位移分量,位移应变关系式为

$$\boldsymbol{e}_{ij}=\frac{1}{2}(\boldsymbol{u}_{i,j}+\boldsymbol{u}_{j,i}+\boldsymbol{u}_{k,i}\boldsymbol{u}_{k,j})\qquad(i,j=1,2,3)\tag{42}$$

应力应变关系可以通过应变能密度 $A(\boldsymbol{e}_{ij})$来表示,亦即

$$\frac{\partial A}{\partial \boldsymbol{e}_{ij}}=\boldsymbol{\sigma}_{ij}\qquad(i,j=1,2,3)\tag{43}$$

如果称 $B(\boldsymbol{\sigma}_{ij})$为余能密度,则有

$$\frac{\partial B}{\partial \boldsymbol{\sigma}_{ij}}=\boldsymbol{e}_{ij}\qquad(i,j=1,2,3)\tag{44}$$

而且根据应变能密度和余能密度的定义,有

$$A(\boldsymbol{e}_{ij})+B(\boldsymbol{\sigma}_{ij})=\boldsymbol{e}_{ij}\boldsymbol{\sigma}_{ij}\tag{45}$$

这里必须指出,对于线性的小应变理论而言,$A(\boldsymbol{e}_{ij})$是$\boldsymbol{e}_{ij}$的二次不变量,$B(\boldsymbol{\sigma}_{ij})$是$\boldsymbol{\sigma}_{ij}$的二次不变量,对于非线性的弹性体而言,$A(\boldsymbol{e}_{ij})$是$\boldsymbol{e}_{ij}$的高次不变量,$B(\boldsymbol{\sigma}_{ij})$是$\boldsymbol{\sigma}_{ij}$的高次不变量. 不论是线性的或是非线性的,式(43),(44)都代表应力应变关系式.

平衡方程应该写成

$$[(\delta_{ij}+\boldsymbol{u}_{i,j})\boldsymbol{\sigma}_{jk}]_{,k}+\boldsymbol{F}_i=0 \quad (i=1,2,3) \tag{46}$$

其中$(\delta_{ij}+\boldsymbol{u}_{i,j})$代表面积元素在变形中的面积变化,$\delta_{ij}$为克氏符号

$$\delta_{ij}=\begin{cases}0 & (i\neq j)\\ 1 & (i=j)\end{cases} \tag{47}$$

边界条件为:

(a)位移已给的边界条件

$$\boldsymbol{u}_i=\overline{\boldsymbol{u}}_i \quad (i=1,2,3,\text{在 } S_u \text{ 上}) \tag{48a}$$

(b)外力已给的边界条件

$$(\delta_{ij}+\boldsymbol{u}_{i,j})\boldsymbol{\sigma}_{jk}n_k=\overline{\boldsymbol{p}}_i \tag{48b}$$

在大位移非线性弹性理论中,最小位能原理仍成立.

变分原理Ⅶ(大位移非线性弹性理论的最小位能原理)　在满足大位移应变关系式(42)和边界位移已知的条件(48a)的所有容许的$\boldsymbol{u}_i$(和$\boldsymbol{e}_{ij}$)中,实际的$\boldsymbol{u}_i(\boldsymbol{e}_{ij})$必使弹性体的总位能

$$\Pi_{\text{Ⅶ}}=\iiint_{\tau}[A(\boldsymbol{e}_{ij})-\boldsymbol{F}_i\boldsymbol{u}_i]\mathrm{d}\tau-\iint_{S_p}\overline{\boldsymbol{p}}_i\boldsymbol{u}_i\mathrm{d}s \tag{49}$$

为最小. 这里假定应力应变关系一般用(32)表示. 通过变分,在利用了式(43)和Green公式以后,可以证明从$\Pi_{\text{Ⅶ}}$的极值条件,得出平衡方程(46),及边界外力已给条件(48b). 证明是众所周知的.

变分原理Ⅷ(大位移非线性弹性理论的最小余能原理)　在满足大位移变形的平衡方程(46),及边界

外力已给条件(48b)的所有容许的 $\boldsymbol{\sigma}_{ij}$, $\boldsymbol{u}_i$ 中,实际的应力 $\boldsymbol{\sigma}_{ij}$和位移 $\boldsymbol{u}_i$ 必使弹性体的泛函

$$\Pi_{\text{Ⅷ}} = \iiint_{\tau} \left[B(\boldsymbol{\sigma}_{ij}) + \frac{1}{2}\boldsymbol{u}_{k,i}\boldsymbol{u}_{k,j}\boldsymbol{\sigma}_{ij} \right] \mathrm{d}\tau - \iint_{S_u} \overline{\boldsymbol{u}}_i (\delta_{ik} + \boldsymbol{u}_{i,k}) \boldsymbol{\sigma}_{kj} n_j \mathrm{d}s \tag{50}$$

为最小. 亦即使(50)的泛函 $\Pi_{\text{Ⅷ}}$ 为最小的 $\boldsymbol{\sigma}_{ij}$, $\boldsymbol{u}_i$,必满足边界位移已给的条件(48a). 在证明中我们利用了应力应变关系式(44),应变位移关系式(42). 证明如下

$$\delta\Pi_{\text{Ⅷ}} = \iiint_{\tau} \left(\frac{\partial B}{\partial \boldsymbol{\sigma}_{ij}} \delta\boldsymbol{\sigma}_{ij} + \frac{1}{2}\boldsymbol{u}_{k,i}\boldsymbol{u}_{k,j}\delta\boldsymbol{\sigma}_{ij} + \boldsymbol{u}_{k,i}\boldsymbol{\sigma}_{ij}\delta\boldsymbol{u}_{k,j} \right) \mathrm{d}\tau - \iint_{S_u} \overline{\boldsymbol{u}}_i \delta[(\delta_{ik} + \boldsymbol{u}_{i,k})\boldsymbol{\sigma}_{kj}] n_j \mathrm{d}s \tag{51}$$

在利用了(44)和(42)以后

$$\begin{aligned} &\frac{\partial B}{\partial \boldsymbol{\sigma}_{ij}} \delta\boldsymbol{\sigma}_{ij} + \frac{1}{2}\boldsymbol{u}_{k,i}\boldsymbol{u}_{k,j}\delta\boldsymbol{\sigma}_{ij} \\ &= \frac{1}{2}(\boldsymbol{u}_{i,j} + \boldsymbol{u}_{j,i} + 2\boldsymbol{u}_{k,i}\boldsymbol{u}_{k,j})\delta\boldsymbol{\sigma}_{ij} \\ &= \boldsymbol{u}_{k,i}(\delta_{kj} + \boldsymbol{u}_{k,j})\delta\boldsymbol{\sigma}_{ij} \end{aligned} \tag{52}$$

而且,因为 δ_{kj} 为一常量,所以

$$\boldsymbol{u}_{k,i}\boldsymbol{\sigma}_{ij}\delta\boldsymbol{u}_{k,j} = \boldsymbol{u}_{k,i}\boldsymbol{\sigma}_{ij}\delta(\delta_{kj} + \boldsymbol{u}_{k,j}) \tag{53}$$

在利用了 Green 公式后

$$\delta \Pi_{\mathrm{VIII}} = -\iiint_{\tau} \boldsymbol{u}_k \delta[(\delta_{kj} + \boldsymbol{u}_{k,j})\boldsymbol{\sigma}_{ij}]_{,i} \mathrm{d}\tau + \iint_{S_u} (\boldsymbol{u}_k - \bar{\boldsymbol{u}}_k)\delta[(\delta_{kj} + \boldsymbol{u}_{k,j})\boldsymbol{\sigma}_{ij}] n_i \mathrm{d}s \qquad (54)$$

因为 $\boldsymbol{u}_i, \boldsymbol{\sigma}_{ij}$ 满足平衡方程，所以 $[(\delta_{kj} + \boldsymbol{u}_{k,j})\boldsymbol{\sigma}_{ij}]_{,i} + \boldsymbol{F}_k = 0$，或 $\delta[(\delta_{kj} + \boldsymbol{u}_{k,j})\boldsymbol{\sigma}_{ij}]_{,i} = 0$；于是 $\delta \Pi_{\mathrm{VIII}} = 0$ 的条件为

$$\iint_{S_u} (\boldsymbol{u}_k - \bar{\boldsymbol{u}}_k)\delta[(\delta_{ki} + \boldsymbol{u}_{k,i})\boldsymbol{\sigma}_{ij}] n_j \mathrm{d}s = 0 \qquad (55)$$

或 $\boldsymbol{u}_k = \bar{\boldsymbol{u}}_k$，它是边界位移已知的条件(48a). 这就证明了大位移非线性弹性理论的最小余能原理. 当 $\boldsymbol{u}_k$ 为小位移时，$\boldsymbol{u}_k \ll 1$，从(50)中略去高级小量，即可还原为小位移的最小余能原理的泛函. 在上面有关变分原理Ⅶ和Ⅷ的证明中，不仅适用于大位移，而且也适用于非线性弹性体的问题.

武汉水利电力学院的谢定一教授于 1985 年，发表了论文《线弹性体的分态能量变分原理》，文中他提出了线弹性体力学的一类混合能量变分原理——分态能量变分原理.

他首先定义了对弹性体施行边界变换以后获得的基本体系状态，然后把基本体系状态分解成两个子状态，对这两个子状态分别建立位能泛函和余能泛函，并定义两个子状态之间的附加能量. 从而可以构造出一族分态混合能量泛函. 除了详细地论证

了三种典型的分态能量变分原理,他还列出了分态能量泛函的其他一些形式.

最后,他说明了分态能量变分原理在计算具有复杂边界条件薄板问题中的应用.①

1. 线弹性体的分态可能状态

设线弹性体 V 的边界 S 由位移边界 S_u 和力边界 S_σ 组成,即 $S=S_u+S_\sigma$. 弹性体边界的外法线方向余弦用 ν_i 表示,$i=1,2,3$. 如果 S_u 和 S_σ 都可以区分为两部分,即 $S_u=S_{u1}+S_{u2}$,$S_\sigma=S_{\sigma1}+S_{\sigma2}$(图1(a)). 我们在一部分位移边界(例如 S_{u1})上,去掉与位移对应的约束,使它成为新的力边界 $\widetilde{S}_{\sigma1}$;在一部分力边界(例如 $S_{\sigma1}$)上,加上与边界力相对应的约束,使它成为新的位移边界 $\widetilde{S}_{u1}$(图1(b)). 这样,原弹性体就变成了一个具有边界 $S=\widetilde{S}_u+\widetilde{S}_\sigma$ 的弹性体,其中 $\widetilde{S}_u=\widetilde{S}_{u1}+S_{u2}$,$\widetilde{S}_\sigma=\widetilde{S}_{\sigma1}+S_{\sigma2}$. 在施行了这种边界变换之后,得到的弹性体模型就称为原弹性体的基本体系.

① 摘自《力学学报》,1985年,第17卷第2期.

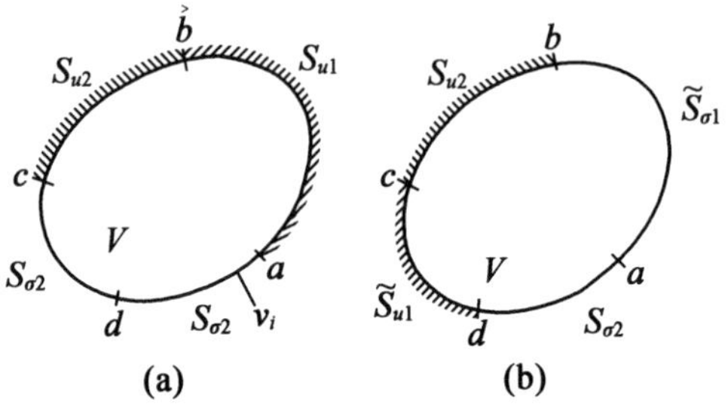

图 1

设原弹性体受体积力 $\boldsymbol{F}_i$ 作用，在 S_{u1} 和 S_{u2} 上，分别已知边界位移 $\overline{\boldsymbol{u}}'_i$ 和 $\overline{\boldsymbol{u}}''_i$，在 $S_{\sigma 1}$ 和 $S_{\sigma 2}$ 上，分别已知边界力 $\overline{\boldsymbol{T}}'_i$ 和 $\overline{\boldsymbol{T}}''_i$(图 2a).

变换 V 的边界，使 S_{u1} 成为 $\widetilde{S}_{\sigma 1}$，$S_{\sigma 1}$ 成为 $\widetilde{S}_{u1}$，得到原弹性体的基本体系. 在新的力边界 $\widetilde{S}_{\sigma 1}$，边界力设为 $\widetilde{\boldsymbol{T}}_i$，而边界位移仍需满足关系 $\boldsymbol{u}_i = \overline{\boldsymbol{u}}'_i$；在新的位移边界 $\widetilde{S}_{u1}$ 上，边界位移设为 $\widetilde{\boldsymbol{u}}_i$，而边界力仍需满足关系 $\boldsymbol{T}_i = \overline{\boldsymbol{T}}'_i$，其中 $\boldsymbol{T}_i = \boldsymbol{\sigma}_{ij}\boldsymbol{\nu}_j$(图 2b). 这种状态，称为基本体系状态.

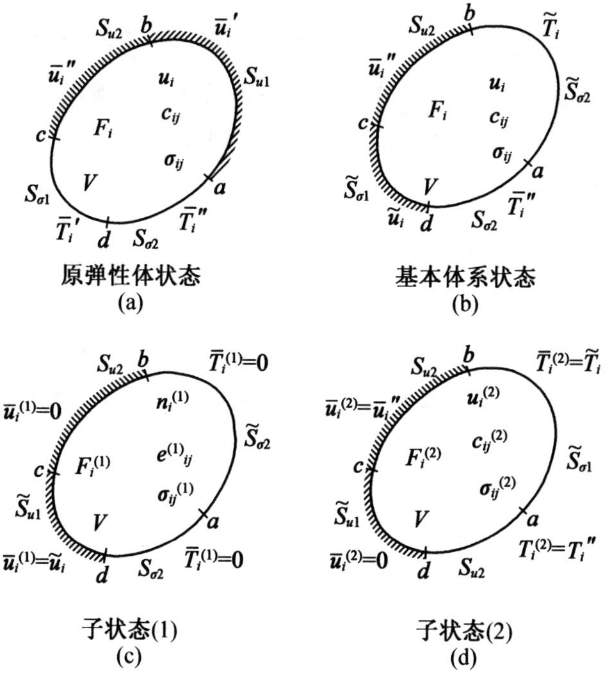

图 2

由于 V 是线弹性的，在研究基本体系状态时，可以把它分解成若干个子状态的迭加. 各子状态都由基本体系构成，仅是所受体积力及边界条件不同而已. 为了确定及叙述简单起见，本文按分解成两个子状态的情况来推证. 这两个子状态，分别称为子状态(1)和子状态(2),如图2(c)(d).

设子状态(1)受$\boldsymbol{F}_i^{(1)}$作用,$\boldsymbol{u}_i^{(1)}$,$\boldsymbol{e}_{ij}^{(1)}$,$\boldsymbol{\sigma}_{ij}^{(1)}$在$V$内需先满足的条件将在下文按不同情况分别阐述. 在边界上有下列条件,至于哪些条件需先满足,也在下文依不同情况分别阐述

$$\boldsymbol{T}_i^{(1)} = \boldsymbol{0} \qquad (在\ \tilde{S}_{\sigma 1}和\ S_{\sigma 2}上) \tag{1.1}$$

$$\boldsymbol{u}_i^{(1)} = \tilde{\boldsymbol{u}}_i \qquad (在\ \tilde{S}_{u1}上) \tag{1.2}$$

$$\boldsymbol{u}_i^{(1)} = \boldsymbol{0} \qquad (在\ S_{u2}上) \tag{1.3}$$

设子状态(2)受$\boldsymbol{F}_i^{(2)}$作用,$\boldsymbol{u}_i^{(2)}$,$\boldsymbol{e}_{ij}^{(2)}$和$\boldsymbol{\sigma}_{ij}^{(2)}$在$V$内需先满足的条件,在下文根据不同情况分别阐述. 在边界上有下列条件,至于哪些条件需先满足,仍在后文阐述

$$\boldsymbol{T}_i^{(2)} = \tilde{\boldsymbol{T}}_i \qquad (在\ \tilde{S}_{\sigma 1}上) \tag{1.4}$$

$$\boldsymbol{T}_i^{(2)} = \overline{\boldsymbol{T}}''_i \qquad (在\ S_{\sigma 2}上) \tag{1.5}$$

$$\boldsymbol{u}_i^{(2)} = \boldsymbol{0} \qquad (在\ \tilde{S}_{u1}上) \tag{1.6}$$

$$\boldsymbol{u}_i^{(2)} = \overline{\boldsymbol{u}}''_i \qquad (在\ S_{u2}上) \tag{1.7}$$

根据迭加原理,显然

$$\boldsymbol{F}_i = \boldsymbol{F}_i^{(1)} + \boldsymbol{F}_i^{(2)} \tag{1.8a}$$

$$\boldsymbol{u}_i = \boldsymbol{u}_i^{(1)} + \boldsymbol{u}_i^{(2)} \tag{1.8b}$$

$$\boldsymbol{e}_{ij} = \boldsymbol{e}_{ij}^{(1)} + \boldsymbol{e}_{ij}^{(2)} \tag{1.8c}$$

$$\boldsymbol{\sigma}_{ij} = \boldsymbol{\sigma}_{ij}^{(1)} + \boldsymbol{\sigma}_{ij}^{(2)} \tag{1.8d}$$

如果基本体系状态是一个可能状态，那么对应的子状态(1)和子状态(2)的总和，就称为原弹性体的一个分态可能态.

2. 两个子状态都以一类力学量表达的分态能量变分原理

设子状态(1)为位能子状态，在 V 内仅 $\boldsymbol{u}_i^{(1)}$ 是自变函数，它满足几何方程

$$\boldsymbol{e}_{ij}^{(1)}=\frac{1}{2}(\boldsymbol{u}_{i,j}^{(1)}+\boldsymbol{u}_{j,i}^{(1)})$$

及位移边界条件(1. 2)，(1. 3). 根据经典位能泛函[①②③]，可写出子状态(1)的位能

$$\varPi_{P1}^{(1)}=\iiint_V[A(\boldsymbol{u}_{i,j}^{(1)})-\boldsymbol{F}_i^{(1)}\boldsymbol{u}_i^{(1)}]\mathrm{d}v \qquad (2.1)$$

式中，$\varPi_{P1}^{(1)}$ 的脚标“$P1$”表示位能是单变量的，$A(\boldsymbol{u}_{i,j}^{(1)})$ 是子状态(1)的应变能密度.

设子状态(2)为余能子状态，在 V 内仅 $\boldsymbol{\sigma}_{ij}^{(2)}$ 是自变函数，它满足平衡方程 $\boldsymbol{\sigma}_{ij,j}^{(2)}+\boldsymbol{F}_i^{(2)}=\mathbf{0}$ 及力边界条件(1. 4)，(1. 5). 根据经典余能泛函，可写出子状态(2)的余能

---

① 胡海昌. 弹性力学的变分原理及其应用. 北京：科学出版社，1981.

② 鹫津久一郎. 弹性学の变分原理概论. ユンヒコ——タによる构造工学讲座Ⅱ－3－A. 日本钢构造协会编. 培风馆，昭和 47 年，1972.

③ 钱伟长. 变分法及有限元(上册). 北京：科学出版社，1980.

$$\Pi_{C1}^{(2)} = \iiint_V B(\boldsymbol{\sigma}_{ij}^{(2)})\,\mathrm{d}v - \iint_{S_{u2}} \boldsymbol{T}_i^{(2)} \overline{\boldsymbol{u}}''_i \mathrm{d}s \quad (2.2)$$

式中,$\Pi_{C1}^{(2)}$的脚标“$C1$”表示余能是单变量的,$B(\boldsymbol{\sigma}_{ij}^{(2)})$是子状态(2)的余能密度.

这样的子状态(1)和(2)的总和,就称为弹性体V的两个子状态都以一类力学量表达的分态可能状态.分态能量泛函Π_{11}由三项组成,可表示为

$$\Pi_{11} = \Pi_{P1}^{(1)} - \Pi_{C1}^{(2)} + H \quad (2.3)$$

式中,Π_{11}的脚标有两个数字“1”,我们以前后两个数字的位置顺序分别表示“子状态(1)”和“子状态(2)”,以数字的值表示该子状态中V内所含力学量的数目(下文同),Π_{11}表示两个子状态的泛函都只包含一类力学量.H是两个子状态间的交叉能量附加项,定义为

$$H = \iint_{\tilde{S}_{\sigma 1}} \widetilde{\boldsymbol{T}}_i \overline{\boldsymbol{u}}_{i'} \mathrm{d}s + \iint_{\tilde{S}_{u1}} (\boldsymbol{T}_i^{(2)} - \overline{\boldsymbol{T}}'_i)\widetilde{\boldsymbol{u}}_i \mathrm{d}s \quad (2.4)$$

分态混合能量变分原理是,在线弹性体的一切分态可能状态中,真实的分态状态使分态能量泛函为驻值

$$\delta\Pi_{11} = \delta\Pi_{P1}^{(1)} - \delta\Pi_{C1}^{(2)} + \delta H = 0 \quad (2.5)$$

证明如下:对式(2.1)变分,注意到子状态(1)满足边界条件(1.2)(1.3),可以得到

$$\delta \Pi_{P1}^{(1)} = -\iiint\limits_{V} (\boldsymbol{\sigma}_{ij,j}^{(1)} + \boldsymbol{F}_i^{(1)}) \delta \boldsymbol{u}_i^{(1)} \mathrm{d}v +$$

$$\iint\limits_{\tilde{S}_{u1}} \boldsymbol{T}_i^{(1)} \delta \tilde{\boldsymbol{u}}_i \mathrm{d}s + \iint\limits_{\tilde{S}_{\sigma1}+S_{\sigma2}} \boldsymbol{T}_i^{(1)} \delta \boldsymbol{u}_i^{(1)} \mathrm{d}s \quad (2.6)$$

对式(2.2)变分，注意到子状态(2)满足边界条件(1.4)和(1.5)，可以得到

$$\delta \Pi_{C1}^{(2)} = \iint\limits_{\tilde{S}_{\sigma1}} \boldsymbol{u}_i^{(2)} \delta \tilde{\boldsymbol{T}}_i \mathrm{d}s + \iint\limits_{\tilde{S}_{u1}} \boldsymbol{u}_i^{(2)} \delta \boldsymbol{T}_i^{(2)} \mathrm{d}s +$$

$$\iint\limits_{\tilde{S}_{u2}} (\boldsymbol{u}_i^{(2)} - \bar{\boldsymbol{u}}''_i) \delta \boldsymbol{T}_i^{(2)} \mathrm{d}s \quad (2.7)$$

对式(2.4)进行变分，有

$$\delta H = \iint\limits_{\tilde{S}_{\sigma1}} \bar{\boldsymbol{u}}'_i \delta \tilde{\boldsymbol{T}}_i \mathrm{d}s + \iint\limits_{\tilde{S}_{u1}} [(\boldsymbol{T}_i^{(2)} - \bar{\boldsymbol{T}}'_i) \delta \tilde{\boldsymbol{u}}_i + \tilde{\boldsymbol{u}}_i \delta \boldsymbol{T}_i^{(2)}] \mathrm{d}s$$

$$(2.8)$$

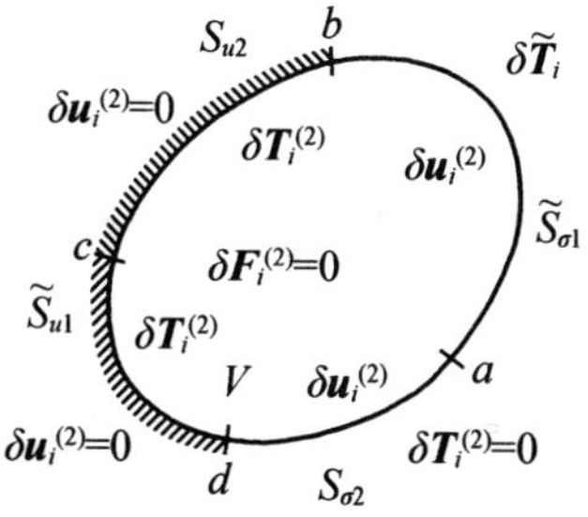

图3　子状态(2)的变分

图3是子状态(2)的变分，如果把它作为一种状态，把子状态(1)作为另一种状态，根据虚功互等定理，在这样两种状态之间存在如下关系

$$\iint_{\tilde{S}_{u1}} \tilde{\boldsymbol{u}}_i \delta \boldsymbol{T}_i^{(2)} \mathrm{d}s + \iint_{\tilde{S}_{\sigma 1}} \boldsymbol{u}_i^{(1)} \delta \widetilde{\boldsymbol{T}}_i \mathrm{d}s = 0 \tag{2.9}$$

把式(2.9)代入式(2.8)，可以得到

$$\delta H = \iint_{\tilde{S}_{u1}} (\boldsymbol{T}_i^{(2)} - \overline{\boldsymbol{T}}_i') \delta \tilde{\boldsymbol{u}}_i \mathrm{d}s - \iint_{\tilde{S}_{\sigma 1}} (\boldsymbol{u}_i^{(1)} - \overline{\boldsymbol{u}}'_i) \delta \widetilde{\boldsymbol{T}}_i \mathrm{d}s \tag{2.10}$$

把式(2.6)(2.7)和式(2.10)代入变分方程式(2.5)，可以得到与泛函 Π_{11} 变分驻值条件等价的一组条件

$$\boldsymbol{\sigma}_{ij,j}^{(1)} + \boldsymbol{F}_i^{(1)} = \boldsymbol{0} \quad (\text{在 } V \text{ 内}) \tag{2.11a}$$

$$\boldsymbol{T}_i^{(1)} = \boldsymbol{0} \quad (\text{在 } \widetilde{S}_{\sigma 1} \text{ 和 } S_{\sigma 2} \text{ 上}) \tag{2.11b}$$

$$\boldsymbol{u}_i^{(2)} = \boldsymbol{0} \quad (\text{在 } \widetilde{S}_{u1} \text{ 上}) \tag{2.12a}$$

$$\boldsymbol{u}_i^{(2)} = \tilde{\boldsymbol{u}}''_i \quad (\text{在 } S_{u2} \text{ 上}) \tag{2.12b}$$

$$\boldsymbol{T}_i^{(1)} + \boldsymbol{T}_i^{(2)} = \overline{\boldsymbol{T}}'_i \quad (\text{在 } \widetilde{S}_{u1} \text{ 上}) \tag{2.13a}$$

$$\boldsymbol{u}_i^{(1)} + \boldsymbol{u}_i^{(2)} = \overline{\boldsymbol{u}}'_i \quad (\text{在 } \widetilde{S}_{\sigma 1} \text{ 上}) \tag{2.13b}$$

式(2.11)是子状态(1)未先要求满足的平衡方程和边界条件(1.1)；式(2.12)是子状态(2)未先要求满足的边界条件(1.6)和(1.7)；条件(2.13a)是综合两

个子状态在变换边界 $\widetilde{S}_{u1}$ 上应满足的边界力平衡条件；条件(2.13b)是综合两个子状态在变换边界 $\widetilde{S}_{\sigma 1}$ 上应满足的位移协调条件. 这样，两个子状态都以一类力学量表达的分态混合能量变分原理获证.

3. 两个子状态都以二类力学量表达的分态能量变分原理

仍设子状态(1)为位能子状态，但泛函用 $\boldsymbol{u}_i^{(1)}$ 和 $\boldsymbol{\sigma}_{ij}^{(1)}$ 二类力学量表达，其中要求 $\boldsymbol{u}_i^{(1)}$ 在位移边界上先满足条件(1.2)，在 V 内先满足几何方程. 根据二类变量 Hellinger-Reissner 广义位能泛函[①②]，可以写出子状态(1)的以二类力学量表达的广义位能

$$\Pi_{P2}^{(1)} = \iiint_V \left[\frac{1}{2}\boldsymbol{\sigma}_{ij}^{(1)} (\boldsymbol{u}_{i,j}^{(1)} + \boldsymbol{u}_{j,i}^{(1)}) - B(\boldsymbol{\sigma}_{ij}^{(1)}) - \boldsymbol{F}_i^{(1)} \boldsymbol{u}_i^{(1)} \right] \mathrm{d}v - \iint_{S_{u2}} \boldsymbol{T}_i^{(1)} \boldsymbol{u}_i^{(1)} \mathrm{d}s \tag{3.1}$$

式中，$\Pi_{P2}^{(1)}$ 的脚标“$P2$”表示在 V 内的位能是二类变量形式的.

也设子状态(2)为余能子状态，但泛函也用 $\boldsymbol{u}_i^{(2)}$ 和 $\boldsymbol{\sigma}_{ij}^{(2)}$ 二类力学量表达. 其中，只要求 $\boldsymbol{\sigma}_{ij}^{(2)}$ 在边界 $\widetilde{S}_{\sigma 1}$

① 胡海昌. 弹性力学的变分原理及其应用. 北京：科学出版社，1981.

② 鷲津久一郎. 弹性学の变分原理概论. ユンヒコ——タによる构造工学讲座Ⅱ－3－A. 日本钢构造协会编. 培风馆，昭和 47 年，1972.

上先满足条件(1.4),根据二类变量 Hellinger-Reissner 广义余能泛函,可以写出

$$\Pi_{C2}^{(2)}=\iiint\limits_{V}[B(\boldsymbol{\sigma}_{ij}^{(2)})+(\boldsymbol{\sigma}_{ij,j}^{(2)}+\boldsymbol{F}_{i}^{2})\boldsymbol{u}_{i}^{(2)}]\mathrm{d}v-$$

$$\iint\limits_{S_{\sigma 2}}(\boldsymbol{T}_{i}^{(2)}-\overline{\boldsymbol{T}}''_{i})\boldsymbol{u}_{i}^{(2)}\mathrm{d}s-\iint\limits_{S_{u2}}\boldsymbol{T}_{i}^{(2)}\overline{\boldsymbol{u}}''_{i}\mathrm{d}s \quad (3.2)$$

式中,$\Pi_{C2}^{(2)}$的脚标“$C2$”表示在 V 内的余能是二类变量形式的.

两个子状态间的交叉能量附加项,仍由式(2.4)定义,相应的分态能量泛函 Π_{22}可表示为

$$\Pi_{22}=\Pi_{P2}^{(1)}-\Pi_{C2}^{(2)}+H \quad (3.3)$$

变分原理是,真实解使泛函 Π_{22}为驻值

$$\delta\Pi_{22}=\delta\Pi_{P2}^{(1)}-\delta\Pi_{C2}^{(2)}+\delta H=0 \quad (3.4)$$

对式(3.1)(3.2)变分后,与式(2.10)一起代入变分方程(3.4),可以得到与 Π_{22}变分驻值条件等价的下列条件

$$\frac{\partial B(\boldsymbol{\sigma}_{ij}^{(1)})}{\partial\boldsymbol{\sigma}_{ij}^{(1)}}=\boldsymbol{e}_{ij}^{(1)} \quad (在 V 内) \quad (3.5a)$$

$$\boldsymbol{\sigma}_{ij,j}^{(1)}+\boldsymbol{F}_{i}^{(1)}=\boldsymbol{0} \quad (在 V 内) \quad (3.5b)$$

$$\boldsymbol{T}_{i}^{(1)}=\boldsymbol{0} \quad (在 \widetilde{S}_{\sigma 1} 和 S_{\sigma 2} 上) \quad (3.5c)$$

$$\boldsymbol{u}_{i}^{(1)}=\boldsymbol{0} \quad (在 S_{u2} 上) \quad (3.5d)$$

$$\frac{\partial B(\boldsymbol{\sigma}_{ij}^{(2)})}{\partial\boldsymbol{\sigma}_{ij}^{(2)}}=\frac{1}{2}(\boldsymbol{u}_{i,j}^{(2)}+\boldsymbol{u}_{j,i}^{(2)}) \quad (在 V 内) \quad (3.6a)$$

$$\boldsymbol{\sigma}_{ij,j}^{(2)} + \boldsymbol{F}_i^{(2)} = \mathbf{0} \quad (\text{在 } V \text{ 内}) \tag{3.6b}$$

$$\boldsymbol{T}_i^{(2)} = \overline{\boldsymbol{T}}''_i \quad (\text{在 } S_{\sigma 2} \text{ 上}) \tag{3.6c}$$

$$\boldsymbol{u}_i^{(2)} = \mathbf{0} \quad (\text{在 } \widetilde{S}_{u1} \text{ 上}) \tag{3.6d}$$

$$\boldsymbol{u}_i^{(2)} = \overline{u}''_i \quad (\text{在 } S_{u2} \text{ 上}) \tag{3.6e}$$

$$\boldsymbol{T}_i^{(1)} + \boldsymbol{T}_i^{(2)} - \overline{\boldsymbol{T}}'_i = \mathbf{0} \quad (\text{在 } \widetilde{S}_{u1} \text{ 上}) \tag{3.7a}$$

$$\boldsymbol{u}_i^{(1)} + \boldsymbol{u}_i^{(2)} - \overline{\boldsymbol{u}}'_i = \mathbf{0} \quad (\text{在 } \widetilde{S}_{\sigma 1} \text{ 上}) \tag{3.7b}$$

条件(3.5)是子状态(1)在 V 内未先要求满足的平衡方程及在边界上未先满足的条件(1.1),(1.3);条件(3.6)是子状态(2)在 V 内未先要求满足的几何方程、平衡方程以及未先要求满足的边界条件(1.5)至(1.7);式(3.7a),(3.7b)是综合两个子状态在变换边界 $\widetilde{S}_{u1}$ 上应满足的边界力平衡条件及在 $\widetilde{S}_{\sigma 1}$ 上应满足的位移协调条件.

4. 两个子状态都以三类力学量表达的分态能量变分原理

设子状态(1)仍是位能子状态,位能以 $\boldsymbol{u}_i^{(1)}$,$\boldsymbol{e}_{ij}^{(1)}$,$\boldsymbol{\sigma}_{ij}^{(1)}$ 三类力学量表达,要求在边界 $\widetilde{S}_{u1}$ 上满足条件(1.2),根据三类变量的胡海昌-鹫津久一朗广义位

能泛函①②,可写出子状态(1)的广义位能

$$\Pi_{P3}^{(1)}=\iiint_V\Big\{A(\boldsymbol{e}_{ij}^{(1)})-\boldsymbol{F}_i^{(1)}\boldsymbol{u}_i^{(1)}-\boldsymbol{\sigma}_{ij}^{(1)}\big[\boldsymbol{e}_{ij}^{(1)}-\frac{1}{2}(\boldsymbol{u}_{i,j}^{(1)}+\boldsymbol{u}_{j,i}^{(1)})\big]\Big\}\mathrm{d}v-\iint_{S_{u2}}\boldsymbol{T}_i^{(1)}\boldsymbol{u}_i^{(1)}\mathrm{d}s \tag{4.1}$$

子状态(2)仍设为余能子状态,也以 $\boldsymbol{u}_i^{(2)}$, $\boldsymbol{e}_{ij}^{(2)}$, $\boldsymbol{\sigma}_{ij}^{(2)}$ 三类力学量表达,要求在力边界 $\widetilde{S}_{\sigma1}$ 上满足条件(1.4),根据三变量胡海昌余能泛函③,可写出子状态(2)的广义余能

$$\Pi_{C3}^{(2)}=\iiint_V[\boldsymbol{\sigma}_{ij}^{(2)}\boldsymbol{e}_{ij}^{(2)}-A(\boldsymbol{e}_{ij}^{(2)})+(\boldsymbol{\sigma}_{ij,j}^{(2)}+\boldsymbol{F}_i^{(2)})\boldsymbol{u}_i^{(2)}]\mathrm{d}v-\iint_{S_{u2}}\boldsymbol{T}_i^{(2)}\overline{\boldsymbol{u}}''_i\mathrm{d}s-\iint_{S_{\sigma2}}(\boldsymbol{T}_i^{(2)}-\overline{\boldsymbol{T}}''_i)\boldsymbol{u}_i^{(2)}\mathrm{d}s \tag{4.2}$$

两子状态间的能量附加项,仍由式(2.4)定义. 相应的分态能量泛函 Π_{33} 可表示为

$$\Pi_{33}=\Pi_{P3}^{(1)}-\Pi_{C3}^{(2)}+H \tag{4.3}$$

变分原理是,真实解使泛函 Π_{33} 为驻值

---

① 胡海昌. 弹性力学的变分原理及其应用. 北京:科学出版社,1981.

② 鹫津久一郎. 弹性学の变分原理概论. ユンヒコ——タによる构造工学讲座Ⅱ-3-A. 日本钢构造协会编. 培风馆,昭和47年,1972.

③ 胡海昌. 弹性力学的变分原理及其应用. 北京:科学出版社,1981.

$$\delta\Pi_{33}=\delta\Pi_{P3}^{(1)}-\delta\Pi_{C3}^{(2)}+\delta H=0 \qquad (4.4)$$

将式(4.1)(4.2)变分,与式(2.10)一起代入变分方程(4.4),可以得到与Π_{33}变分驻值条件等价的条件.篇幅有限,从略.

5. 分态能量变分原理的其他形式

在弹性体V的两个子状态中,还可以依据满足条件的不同,分别具有不同形式的位能或余能.从而可以构造出分态能量变分原理家族中的一些其他形式的泛函.当两个子状态的泛函所包含的力学量不相等时,有这样六种分态能量变分原理

$$\Pi_{32}=\Pi_{P3}^{(1)}-\Pi_{C2}^{(2)}+H \qquad (5.1)$$

$$\Pi_{31}=\Pi_{P3}^{(1)}-\Pi_{C1}^{(2)}+H \qquad (5.2)$$

$$\Pi_{23}=\Pi_{P2}^{(1)}-\Pi_{C3}^{(2)}+H \qquad (5.3)$$

$$\Pi_{21}=\Pi_{P2}^{(1)}-\Pi_{C1}^{(2)}+H \qquad (5.4)$$

$$\Pi_{13}=\Pi_{P1}^{(1)}-\Pi_{C3}^{(2)}+H \qquad (5.5)$$

$$\Pi_{12}=\Pi_{P1}^{(1)}-\Pi_{C2}^{(2)}+H \qquad (5.6)$$

以上六种再加上Π_{11},Π_{22},Π_{33}总共有九种,这九种分态混合能量变分原理,又可称为"完整的"分态能量变分原理.所谓"完整的",是指在建立泛函时,凡需要先满足的一组条件,它的所有分量都完整地得到满足;凡不要求先满足的条件,它的所有分量都完整地不满足.

与此相对应，还有“非完整的”分态混合能量变分原理. 所谓“非完整的”，是指在建立泛函时，凡要求先满足的任一组条件，它的所有分量不是完整地都被满足；凡不要求先满足的任一组条件，它有一部分分量却能先满足.

例如，在构造位能 $\Pi_{P3}^{(1)}$ 时，如果在边界 $\tilde{S}_{u1}$ 上只有两个位移分量能满足条件(1.2)：$\boldsymbol{u}_1^{(1)}=\tilde{\boldsymbol{u}}_1,\boldsymbol{u}_2^{(1)}=\tilde{\boldsymbol{u}}_2$，这时 $\Pi_{P3}^{(1)}$ 就具有相应的一种非完整的形式并与式(4.1)有区别

$$\begin{aligned}\Pi_{P3}^{(1)}=&\iiint_V\Big\{A(\boldsymbol{e}_{ij}^{(1)})-\boldsymbol{F}_i^{(1)}\boldsymbol{u}_i^{(1)}-\\&\boldsymbol{\sigma}_{ij}^{(1)}\Big[\boldsymbol{e}_{ij}^{(1)}-\frac{1}{2}(\boldsymbol{u}_{i,j}^{(1)}+\boldsymbol{u}_{j,i}^{(1)})\Big]\Big\}\mathrm{d}v-\\&\iint_{S_{u2}}\boldsymbol{T}_i^{(1)}\boldsymbol{u}_i^{(1)}\mathrm{d}s-\iint_{\tilde{S}_{u1}}\boldsymbol{T}_i^{(1)}(\boldsymbol{u}_3^{(1)}-\tilde{\boldsymbol{u}}_3)\mathrm{d}s\end{aligned}\tag{5.7}$$

显然，非完整的位能或余能有许多种形式. 如果在两个子状态中，位能和余能泛函同时或不同时是非完整的，那么在九种完整的分态变分原理的每一种中，都可派生出多种非完整的分态变分原理. 在此就不一一罗列了.

所有完整的和非完整的分态能量变分原理，构成分态能量变分原理族，九种完整的原理是这一族的主

干,各种非完整的原理是该族的支系. 族中任何两种原理都可以依据某些条件是否需要先满足而相互转化.

6. 弹性薄板的分态能量变分原理

弹性薄板的分态可能状态见图4所示,为简单起见,设板边界是光滑(无角点)的. 限于篇幅,对所用符号不作说明,图4与图2有着对应的关系.

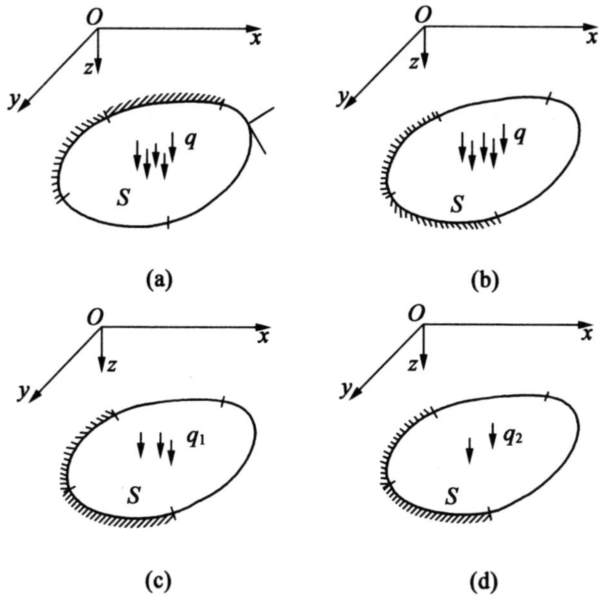

图4

薄板的分态混合能量泛函是

$$\Pi_{22} = \Pi_{P2}^{(1)} - \Pi_{C2}^{(2)} + H \tag{6.1}$$

$$\Pi_{P2}^{(1)} = \iint_S \{ (- M_X^{(1)} w_{,XX}^{(1)} - 2M_{XY}^{(1)} w_{,XY}^{(1)} - M_Y^{(1)} w_{,YY}^{(1)}) - \frac{1}{2D(1-\mu^2)}[(M_X^{(1)} + M_Y^{(1)})^2 + 2(1+\mu) \cdot (M_{XY}^{(1)} - M_X^{(1)} M_Y^{(1)})] - q_1 w^{(1)} \} \mathrm{d}x\mathrm{d}y \tag{6.2}$$

$$\Pi_{C2}^{(2)} = \iint_S \{ \frac{1}{2D(1-\mu^2)}[(M_X^{(2)} + M_Y^{(2)})^2 + 2(1+\mu) \cdot (M_{XY}^{(2)2} - M_X^{(2)} M_Y^{(2)})] + (Q_{X,X}^{(2)} + Q_{Y,Y}^{(2)} + q_2) \cdot w^{(2)} \} \mathrm{d}x\mathrm{d}y - \int_{C_{u2}} (V_v^{(2)} \overline{w}'' - M_v^{(2)} \overline{w}''_{,v}) \mathrm{d}c \tag{6.3}$$

$$H = \int_{\tilde{C}_{\sigma 1}} (\tilde{V}_v \overline{w}' - \tilde{M}_v \overline{w}'_{,v}) \mathrm{d}c + \int_{\tilde{C}_{u1}} [(V_v^{(2)} - \overline{V}'_v) \tilde{w} - (M_v^{(2)} - \overline{M}'_v) \tilde{w}_{,v}] \mathrm{d}c \tag{6.4}$$

变分原理表示为

$$\delta\Pi_{22} = \delta\Pi_{P2}^{(1)} - \delta\Pi_{C2}^{(2)} + \delta H = 0 \tag{6.5}$$

它等价于下列条件

$$\left.\begin{aligned} & w_{,XX} = -\frac{1}{D(1-\mu^2)}(M_X - \mu M_Y) && (6.6a) \\ & w_{,XY} = -\frac{1}{D(1-\mu)} M_{XY} && (6.6b) \\ & w_{,YY} = -\frac{1}{D(1-\mu^2)}(M_Y - \mu M_X) && (6.6c) \\ & Q_{X,X} + Q_{Y,X} + q = 0 && (6.6d) \end{aligned}\right\} \text{(对于子状态(1)和(2),在 } S \text{ 内)}$$

$$\left.\begin{aligned}V_v^{(1)}+V_v^{(2)}-\overline{V}_v'=0\\ M_v^{(1)}+M_v^{(2)}-\overline{M}_v'=0\end{aligned}\right\}\text{(在 }\widetilde{C}_{u1}\text{ 上)}\qquad\begin{aligned}&(6.7\text{a})\\&(6.7\text{b})\end{aligned}$$

$$\left.\begin{aligned}w^{(1)}+w^{(2)}-\overline{w}'=0\\ w_{,v}^{(1)}+w_{,v}^{(2)}-\overline{w}_{,v}'=0\end{aligned}\right\}\text{(在 }\widetilde{C}_{\sigma1}\text{ 上)}\qquad\begin{aligned}&(6.8\text{a})\\&(6.8\text{b})\end{aligned}$$

式中,μ 为材料的泊桑比,V_v 为板边界上的等效剪力. 条件(6.6)是两个子状态在板中面 S 内应满足的平衡方程和物理方程. 式(6.7)是在变换后的新边界 $\widetilde{C}_{u1}$上的力平衡条件. 式(6.8)是在新边界 $\widetilde{C}_{\sigma1}$上的位移协调条件.

薄板的分态能量变分原理,可以用来计算具有复杂边界的弹性薄板.

举一个简单的例子[①]. 图 5(a)所示的一块三边简支一边固定的文克尔弹性地基板,长宽比 $\lambda=\dfrac{b}{a}$,考虑受均布荷载 q 作用的静力问题.

① 谢定一. 分区与分项能量原理及应用. 北京:清华大学,1980.

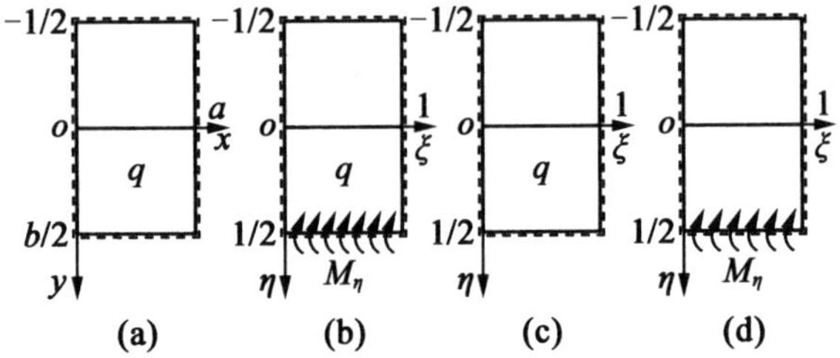

图 5

基本方程是

$$\nabla^2 \nabla^2 w + \frac{K}{D} w = \frac{q}{D} \tag{6.9}$$

设无量纲坐标及挠度为 $\xi = \frac{x}{a}, \eta = \frac{y}{b}$ 和 $\overline{w}$

$$w = \overline{w} \frac{qa^4}{D} \times 10^{-3} \tag{6.10}$$

则基本方程无量纲化为

$$\left(\frac{\partial^2}{\partial \xi^2} + \frac{1}{\lambda^2} \cdot \frac{\partial^2}{\partial \eta^2}\right)\left(\frac{\partial^2}{\partial \xi^2} + \frac{1}{\lambda^2} \cdot \frac{\partial^2}{\partial \eta^2}\right)\overline{w} + k\overline{w} = 10^3 \tag{6.11a}$$

$$k = \frac{Ka^4}{D} \tag{6.11b}$$

板的无量纲内力及内力

$$\begin{cases} M_\xi = -\left(\overline{w}_{,\xi\xi} + \dfrac{\mu}{\lambda^2}\overline{w}_{,\eta\eta}\right) & (6.12\text{a}) \\ \vdots \end{cases}$$

$$\begin{cases} M_X = M_\xi q a^2 \times 10^{-3} & (6.13\text{a}) \\ \vdots \end{cases}$$

在固定边 $\eta = \dfrac{1}{2}$ 处去掉转动约束，得到四边简支的基本体系（图 5b），在 $\eta = \dfrac{1}{2}$ 处，M_η 以力参数表示为

$$M_\eta = \sum_{m=1,3}^{\infty} M_{\eta m} \sin \gamma_m \xi \qquad (6.14)$$

分基本体系状态为两个子状态，如图 5(c)(d). 有

$$\begin{cases} \overline{w} = \overline{w}^{(1)} + \overline{w}^{(2)} & (6.15\text{a}) \\ M_\xi = M_\xi^{(1)} + M_\xi^{(2)} & (6.15\text{b}) \\ \vdots \end{cases}$$

设子状态(1)，(2)的挠度 $\overline{w}^{(1)}$，$\overline{w}^{(2)}$ 分别为

$$\overline{w}^{(1)} = \sum_{m=1,3}^{\infty} \overline{w}_m^{(1)} \sin \gamma_m \xi \qquad (6.16\text{a})$$

$$\overline{w}^{(2)} = \sum_{m=1,3}^{\infty} \overline{w}_m^{(2)} \sin \gamma_m \xi \qquad (6.16\text{b})$$

子状态(1)在 $q_1 = q$ 作用下，让 $\overline{w}^{(1)}$ 满足基本方程(6.11a)及四边简支的边界条件，可得

$$\overline{w}_m^{(1)} = \overline{w}_{0m}^{(1)} (1 + A_m \cosh \alpha_m \eta \cos \beta_m \eta + B_m \sinh \alpha_m \eta \sin \beta_m \eta) \qquad (6.17\text{a})$$

$$\overline{w}_{,\eta}^{(1)} = \sum_{m=1,3}^{\infty} \overline{w}_{0m}^{(1)} (-E_{\eta} k_{sm} + F_{m} k_{4m}) \sin \gamma m \xi \sin \gamma_{m} \xi \tag{6. 17b}$$

$$M_{\xi}^{(1)} = \sum_{m=1,3}^{\infty} M_{\xi m}^{(1)} \sin \gamma_{m} \xi \tag{6. 17c}$$

$$\vdots$$

其中 $\alpha_m, \beta_m, \gamma_m, A_m, \cdots, M_{\xi m}^{(1)}$ 都是与 m 有关的已知参数,表达式从略. $\overline{w}_{0m}^{(1)}$ 是

$$\overline{w}_{0m}^{(1)} = \frac{4 \times 10^{3}}{\gamma_{m} (\gamma_{m}^{4} + k)} \tag{6. 18}$$

子状态(2)在 $q_2 = 0$ 及 $\eta = \frac{1}{2}$ 处的弯矩 M_{η} 作用下,让 $\overline{w}^{(2)}$ 满足基本方程(6. 11a)及四边简支的边界条件,可得到含待定参数 $M_{\eta m}$ 的

$$\begin{aligned} \overline{w}_{m}^{(2)} = \frac{M_{\eta m}}{2\sqrt{k} t_{m}} (& k_{2m} \cos \text{h}\ \alpha_{m} \eta \cos \beta_{m} \eta - \\ & k_{1m} \sinh \alpha_{m} \eta \sin \beta_{m} \eta - \\ & t_{rm} k_{4m} \cosh \alpha_{m} \eta \sin \beta_{m} \eta + \\ & t_{rm} k_{3m} \sinh \alpha_{m} \eta \cos \beta_{m} \eta) \end{aligned} \tag{6. 19a}$$

$$\begin{aligned} \overline{w}_{,\eta}^{(2)} = \frac{1}{\sqrt{k}} \sum_{m=1,3}^{\infty} \frac{M_{\eta m}}{2 t_{m}} (& t_{rm} G_{m} k_{1m} - t_{rm} H_{m} k_{2m} - \\ & I_{m} k_{3m} + J_{m} k_{4m}) \sin \gamma_{m} \xi \end{aligned} \tag{6. 19b}$$

$$M_{\xi}^{(2)} = \sum_{m=1,3}^{\infty} M_{\xi m}^{(2)} \sin \gamma_{m} \xi \tag{6. 19c}$$

于是,分态能量变分原理 $\delta\Pi_{22}=0$ 可以大幅度简化,得到变换边界 $\widetilde{C}_{\sigma 1}(\eta=\frac{1}{2}$处)上的变形协调条件

$$\overline{w}_{,\eta}^{(1)}+\overline{w}_{,\eta}^{(2)}=0 \tag{6.20}$$

从而解出待定参数

$$M_{\eta m}=\frac{2\sqrt{k}t_m\overline{w}_{0m}^{(1)}(E_mk_{3m}-F_mk_{4m})}{t_{rm}(G_mk_{1m}-H_mk_{2m})-I_mk_{3m}+J_mk_{4m}} \tag{6.21}$$

将求出的 $M_{\eta m}$ 代入(6. 19a),(6. 19c)…可得子状态(2)的位移与内力. 将(6. 17a)与(6. 19a)分别代入式(6. 16a),(6. 16b),再代入式(6. 15a),可求出无量纲位移 $\overline{w}$,由式(6. 10)获板的位移. 将式(6. 17c),…和(6. 19c),…对应代入式(6. 15b),…,求出各项无量纲内力 M_ξ 等,由式(6. 13a),…获板的内力.

给定材料的泊桑比 μ,给板长宽比 λ 以不同的取值,就能编制出这种弹性地基板的位移、内力计算系数表.

7. 结语

分态能量变分原理是一族新型的混合能量变分原理,适用于线弹性体小位移问题. 这类变分原理的突出特点是,把原弹性体分成子状态,有的子状态以位能为特征,有的子状态以余能为特征. 因此,分态能量变分原理是综合应用各类位能原理和余能原理于一体的能

量变分原理.

用分态能量变分原理,可以分析具有复杂边界的线弹性体,当弹性体因边界复杂而不便计算时,可以进行边界变换,建立具有简单边界的基本体系($\widetilde{\boldsymbol{u}}_i$ 和 $\widetilde{\boldsymbol{T}}_i$ 包含待定参数),然后根据需要分成以位能为特征和以余能为特征的子状态,构造分态能量泛函,根据泛函的变分驻值条件求解.

弹性结构的分析方法,有静力法和能量法两大类.静力法中的位移法、力法和混合法相应于能量法中的位能原理、余能原理和分区混合能量原理.线弹性体的分态混合能量变分原理,则相应于联合法.

从式(2.11)至(2.13)或从式(3.5)至(3.7)可以看出,通常所用的迭加法,仅仅是分态能量变分原理在一些条件能先满足时的一种特殊情形.

本书是一位中国学者用英文写的一部数学著作,虽然现在读数学的人英文水平都很高了,但是还是不如看母语方便,所以有机会我们还是要多出中文版专著.

1995年,商务印书馆出版了印度学者高善必(D. D. Kosambi)的史学著作《印度古代文化与文明史纲》.高善必是马克思主义历史学家、数学家,哈佛大学的数学博士.译者王树英在序言里引用季羡林的两段话,概括了此书的价值所在:“印度真正用马克思主义观点研究印度历史的,高善必应该说是第一人."

"这是一部学术价值很高的著作.他一空依傍,独立创新,印度和世界学术界给以很高的评价,它完全是当之无愧的,译为汉文,实有必要."

刘培杰

2022 年 6 月 30 日

于哈工大

刘培杰数学工作室
已出版(即将出版)图书目录——原版影印

书 名	出版时间	定 价	编号
数学物理大百科全书.第1卷(英文)	2016-01	418.00	508
数学物理大百科全书.第2卷(英文)	2016-01	408.00	509
数学物理大百科全书.第3卷(英文)	2016-01	396.00	510
数学物理大百科全书.第4卷(英文)	2016-01	408.00	511
数学物理大百科全书.第5卷(英文)	2016-01	368.00	512
zeta 函数,q-zeta 函数,相伴级数与积分(英文)	2015-08	88.00	513
微分形式:理论与练习(英文)	2015-08	58.00	514
离散与微分包含的逼近和优化(英文)	2015-08	58.00	515
艾伦·图灵:他的工作与影响(英文)	2016-01	98.00	560
测度理论概率导论,第2版(英文)	2016-01	88.00	561
带有潜在故障恢复系统的半马尔柯夫模型控制(英文)	2016-01	98.00	562
数学分析原理(英文)	2016-01	88.00	563
随机偏微分方程的有效动力学(英文)	2016-01	88.00	564
图的谱半径(英文)	2016-01	58.00	565
量子机器学习中数据挖掘的量子计算方法(英文)	2016-01	98.00	566
量子物理的非常规方法(英文)	2016-01	118.00	567
运输过程的统一非局部理论:广义波尔兹曼物理动力学,第2版(英文)	2016-01	198.00	568
量子力学与经典力学之间的联系在原子、分子及电动力学系统建模中的应用(英文)	2016-01	58.00	569
算术域(英文)	2018-01	158.00	821
高等数学竞赛:1962—1991年的米洛克斯·史怀哲竞赛(英文)	2018-01	128.00	822
用数学奥林匹克精神解决数论问题(英文)	2018-01	108.00	823
代数几何(德文)	2018-04	68.00	824
丢番图逼近论(英文)	2018-01	78.00	825
代数几何学基础教程(英文)	2018-01	98.00	826
解析数论入门课程(英文)	2018-01	78.00	827
数论中的丢番图问题(英文)	2018-01	78.00	829
数论(梦幻之旅):第五届中日数论研讨会演讲集(英文)	2018-01	68.00	830
数论新应用(英文)	2018-01	68.00	831
数论(英文)	2018-01	78.00	832

刘培杰数学工作室
已出版(即将出版)图书目录——原版影印

书　　名	出版时间	定　价	编号
湍流十讲(英文)	2018－04	108.00	886
无穷维李代数:第3版(英文)	2018－04	98.00	887
等值、不变量和对称性(英文)	2018－04	78.00	888
解析数论(英文)	2018－09	78.00	889
《数学原理》的演化:伯特兰·罗素撰写第二版时的手稿与笔记(英文)	2018－04	108.00	890
哈密尔顿数学论文集(第4卷):几何学、分析学、天文学、概率和有限差分等(英文)	2019－05	108.00	891
偏微分方程全局吸引子的特性(英文)	2018－09	108.00	979
整函数与下调和函数(英文)	2018－09	118.00	980
幂等分析(英文)	2018－09	118.00	981
李群,离散子群与不变量理论(英文)	2018－09	108.00	982
动力系统与统计力学(英文)	2018－09	118.00	983
表示论与动力系统(英文)	2018－09	118.00	984
分析学练习.第1部分(英文)	2021－01	88.00	1247
分析学练习.第2部分,非线性分析(英文)	2021－01	88.00	1248
初级统计学:循序渐进的方法:第10版(英文)	2019－05	68.00	1067
工程师与科学家微分方程用书:第4版(英文)	2019－07	58.00	1068
大学代数与三角学(英文)	2019－06	78.00	1069
培养数学能力的途径(英文)	2019－07	38.00	1070
工程师与科学家统计学:第4版(英文)	2019－06	58.00	1071
贸易与经济中的应用统计学:第6版(英文)	2019－06	58.00	1072
傅立叶级数和边值问题:第8版(英文)	2019－05	48.00	1073
通往天文学的途径:第5版(英文)	2019－05	58.00	1074
拉马努金笔记.第1卷(英文)	2019－06	165.00	1078
拉马努金笔记.第2卷(英文)	2019－06	165.00	1079
拉马努金笔记.第3卷(英文)	2019－06	165.00	1080
拉马努金笔记.第4卷(英文)	2019－06	165.00	1081
拉马努金笔记.第5卷(英文)	2019－06	165.00	1082
拉马努金遗失笔记.第1卷(英文)	2019－06	109.00	1083
拉马努金遗失笔记.第2卷(英文)	2019－06	109.00	1084
拉马努金遗失笔记.第3卷(英文)	2019－06	109.00	1085
拉马努金遗失笔记.第4卷(英文)	2019－06	109.00	1086
数论:1976年纽约洛克菲勒大学数论会议记录(英文)	2020－06	68.00	1145
数论:卡本代尔1979:1979年在南伊利诺伊卡本代尔大学举行的数论会议记录(英文)	2020－06	78.00	1146
数论:诺德韦克豪特1983:1983年在诺德韦克豪特举行的Journees Arithmetiques数论大会会议记录(英文)	2020－06	68.00	1147
数论:1985－1988年在纽约城市大学研究生院和大学中心举办的研讨会(英文)	2020－06	68.00	1148

刘培杰数学工作室
已出版(即将出版)图书目录——原版影印

书　　名	出版时间	定　价	编号
数论:1987 年在乌尔姆举行的 Journees Arithmetiques 数论大会会议记录(英文)	2020—06	68.00	1149
数论:马德拉斯 1987:1987 年在马德拉斯安娜大学举行的国际拉马努金百年纪念大会会议记录(英文)	2020—06	68.00	1150
解析数论:1988 年在东京举行的日法研讨会会议记录(英文)	2020—06	68.00	1151
解析数论:2002 年在意大利切特拉罗举行的 C. I. M. E. 暑期班演讲集(英文)	2020—06	68.00	1152
量子世界中的蝴蝶:最迷人的量子分形故事(英文)	2020—06	118.00	1157
走进量子力学(英文)	2020—06	118.00	1158
计算物理学概论(英文)	2020—06	48.00	1159
物质,空间和时间的理论:量子理论(英文)	2020—10	48.00	1160
物质,空间和时间的理论:经典理论(英文)	2020—10	48.00	1161
量子场理论:解释世界的神秘背景(英文)	2020—07	38.00	1162
计算物理学概论(英文)	2020—06	48.00	1163
行星状星云(英文)	2020—10	38.00	1164
基本宇宙学:从亚里士多德的宇宙到大爆炸(英文)	2020—08	58.00	1165
数学磁流体力学(英文)	2020—07	58.00	1166
计算科学:第 1 卷,计算的科学(日文)	2020—07	88.00	1167
计算科学:第 2 卷,计算与宇宙(日文)	2020—07	88.00	1168
计算科学:第 3 卷,计算与物质(日文)	2020—07	88.00	1169
计算科学:第 4 卷,计算与生命(日文)	2020—07	88.00	1170
计算科学:第 5 卷,计算与地球环境(日文)	2020—07	88.00	1171
计算科学:第 6 卷,计算与社会(日文)	2020—07	88.00	1172
计算科学. 别卷,超级计算机(日文)	2020—07	88.00	1173
多复变函数论(日文)	2022—06	78.00	1518
复变函数入门(日文)	2022—06	78.00	1523
代数与数论:综合方法(英文)	2020—10	78.00	1185
复分析:现代函数理论第一课(英文)	2020—07	58.00	1186
斐波那契数列和卡特兰数:导论(英文)	2020—10	68.00	1187
组合推理:计数艺术介绍(英文)	2020—07	88.00	1188
二次互反律的傅里叶分析证明(英文)	2020—07	48.00	1189
旋瓦兹分布的希尔伯特变换与应用(英文)	2020—07	58.00	1190
泛函分析:巴拿赫空间理论入门(英文)	2020—07	48.00	1191
卡塔兰数入门(英文)	2019—05	68.00	1060
测度与积分(英文)	2019—04	68.00	1059
组合学手册. 第一卷(英文)	2020—06	128.00	1153
* 一代数、局部紧群和巴拿赫 * 一代数丛的表示. 第一卷,群和代数的基本表示理论(英文)	2020—05	148.00	1154
电磁理论(英文)	2020—08	48.00	1193
连续介质力学中的非线性问题(英文)	2020—09	78.00	1195
多变量数学入门(英文)	2021—05	68.00	1317
偏微分方程入门(英文)	2021—05	88.00	1318
若尔当典范性:理论与实践(英文)	2021—07	68.00	1366
伽罗瓦理论. 第 4 版(英文)	2021—08	88.00	1408

刘培杰数学工作室
已出版(即将出版)图书目录——原版影印

书　　名	出版时间	定　价	编号
典型群,错排与素数(英文)	2020—11	58.00	1204
李代数的表示:通过 gln 进行介绍(英文)	2020—10	38.00	1205
实分析演讲集(英文)	2020—10	38.00	1206
现代分析及其应用的课程(英文)	2020—10	58.00	1207
运动中的抛射物数学(英文)	2020—10	38.00	1208
2—纽结与它们的群(英文)	2020—10	38.00	1209
概率,策略和选择:博弈与选举中的数学(英文)	2020—11	58.00	1210
分析学引论(英文)	2020—11	58.00	1211
量子群:通往流代数的路径(英文)	2020—11	38.00	1212
集合论入门(英文)	2020—10	48.00	1213
酉反射群(英文)	2020—11	58.00	1214
探索数学:吸引人的证明方式(英文)	2020—11	58.00	1215
微分拓扑短期课程(英文)	2020—10	48.00	1216
抽象凸分析(英文)	2020—11	68.00	1222
费马大定理笔记(英文)	2021—03	48.00	1223
高斯与雅可比和(英文)	2021—03	78.00	1224
π 与算术几何平均:关于解析数论和计算复杂性的研究(英文)	2021—01	58.00	1225
复分析入门(英文)	2021—03	48.00	1226
爱德华·卢卡斯与素性测定(英文)	2021—03	78.00	1227
通往凸分析及其应用的简单路径(英文)	2021—01	68.00	1229
微分几何的各个方面.第一卷(英文)	2021—01	58.00	1230
微分几何的各个方面.第二卷(英文)	2020—12	58.00	1231
微分几何的各个方面.第三卷(英文)	2020—12	58.00	1232
沃克流形几何学(英文)	2020—11	58.00	1233
彷射和韦尔几何应用(英文)	2020—12	58.00	1234
双曲几何学的旋转向量空间方法(英文)	2021—02	58.00	1235
积分:分析学的关键(英文)	2020—12	48.00	1236
为有天分的新生准备的分析学基础教材(英文)	2020—11	48.00	1237
数学不等式.第一卷.对称多项式不等式(英文)	2021—03	108.00	1273
数学不等式.第二卷.对称有理不等式与对称无理不等式(英文)	2021—03	108.00	1274
数学不等式.第三卷.循环不等式与非循环不等式(英文)	2021—03	108.00	1275
数学不等式.第四卷.Jensen 不等式的扩展与加细(英文)	2021—03	108.00	1276
数学不等式.第五卷.创建不等式与解不等式的其他方法(英文)	2021—04	108.00	1277

刘培杰数学工作室
已出版(即将出版)图书目录——原版影印

书　名	出版时间	定　价	编号
冯·诺依曼代数中的谱位移函数:半有限冯·诺依曼代数中的谱位移函数与谱流(英文)	2021—06	98.00	1308
链接结构:关于嵌入完全图的直线中链接单形的组合结构(英文)	2021—05	58.00	1309
代数几何方法.第1卷(英文)	2021—06	68.00	1310
代数几何方法.第2卷(英文)	2021—06	68.00	1311
代数几何方法.第3卷(英文)	2021—06	58.00	1312

代数、生物信息和机器人技术的算法问题.第四卷,独立恒等式系统(俄文)	2020—08	118.00	1199
代数、生物信息和机器人技术的算法问题.第五卷,相对覆盖性和独立可拆分恒等式系统(俄文)	2020—08	118.00	1200
代数、生物信息和机器人技术的算法问题.第六卷,恒等式和准恒等式的相等 问题、可推导性和可实现性(俄文)	2020—08	128.00	1201
分数阶微积分的应用:非局部动态过程,分数阶导热系数(俄文)	2021—01	68.00	1241
泛函分析问题与练习:第2版(俄文)	2021—01	98.00	1242
集合论、数学逻辑和算法论问题:第5版(俄文)	2021—01	98.00	1243
微分几何和拓扑短期课程(俄文)	2021—01	98.00	1244
素数规律(俄文)	2021—01	88.00	1245
无穷边值问题解的递减:无界域中的拟线性椭圆和抛物方程(俄文)	2021—01	48.00	1246
微分几何讲义(俄文)	2020—12	98.00	1253
二次型和矩阵(俄文)	2021—01	98.00	1255
积分和级数.第2卷,特殊函数(俄文)	2021—01	168.00	1258
积分和级数.第3卷,特殊函数补充:第2版(俄文)	2021—01	178.00	1264
几何图上的微分方程(俄文)	2021—01	138.00	1259
数论教程:第2版(俄文)	2021—01	98.00	1260
非阿基米德分析及其应用(俄文)	2021—03	98.00	1261
古典群和量子群的压缩(俄文)	2021—03	98.00	1263
数学分析习题集.第3卷,多元函数:第3版(俄文)	2021—03	98.00	1266
数学习题:乌拉尔国立大学数学力学系大学生奥林匹克(俄文)	2021—03	98.00	1267
柯西定理和微分方程的特解(俄文)	2021—03	98.00	1268
组合极值问题及其应用:第3版(俄文)	2021—03	98.00	1269
数学词典(俄文)	2021—01	98.00	1271
确定性混沌分析模型(俄文)	2021—06	168.00	1307
精选初等数学习题和定理.立体几何.第3版(俄文)	2021—03	68.00	1316
微分几何习题:第3版(俄文)	2021—05	98.00	1336
精选初等数学习题和定理.平面几何.第4版(俄文)	2021—05	68.00	1335
曲面理论在欧氏空间 En 中的直接表示(俄文)	2022—01	68.00	1444
维纳—霍普夫离散算子和托普利兹算子:某些可数赋范空间中的诺特性和可逆性(俄文)	2022—03	108.00	1496
Maple 中的数论:数论中的计算机计算(俄文)	2022—03	88.00	1497
贝尔曼和克努特问题及其概括:加法运算的复杂性(俄文)	2022—03	138.00	1498

刘培杰数学工作室
已出版(即将出版)图书目录——原版影印

书　名	出版时间	定　价	编号
复分析:共形映射(俄文)	2022－07	48.00	1542
微积分代数样条和多项式及其在数值方法中的应用(俄文)	2022－08	128.00	1543
蒙特卡罗方法中的随机过程和场模型:算法和应用(俄文)	2022－08	88.00	1544

狭义相对论与广义相对论:时空与引力导论(英文)	2021－07	88.00	1319
束流物理学和粒子加速器的实践介绍:第2版(英文)	2021－07	88.00	1320
凝聚态物理中的拓扑和微分几何简介(英文)	2021－05	88.00	1321
混沌映射:动力学、分形学和快速涨落(英文)	2021－05	128.00	1322
广义相对论:黑洞、引力波和宇宙学介绍(英文)	2021－06	68.00	1323
现代分析电磁均质化(英文)	2021－06	68.00	1324
为科学家提供的基本流体动力学(英文)	2021－06	88.00	1325
视觉天文学:理解夜空的指南(英文)	2021－06	68.00	1326
物理学中的计算方法(英文)	2021－06	68.00	1327
单星的结构与演化:导论(英文)	2021－06	108.00	1328
超越居里:1903年至1963年物理界四位女性及其著名发现(英文)	2021－06	68.00	1329
范德瓦尔斯流体热力学的进展(英文)	2021－06	68.00	1330
先进的托卡马克稳定性理论(英文)	2021－06	88.00	1331
经典场论导论:基本相互作用的过程(英文)	2021－07	88.00	1332
光致电离量子动力学方法原理(英文)	2021－07	108.00	1333
经典域论和应力:能量张量(英文)	2021－05	88.00	1334
非线性太赫兹光谱的概念与应用(英文)	2021－06	68.00	1337
电磁学中的无穷空间并矢格林函数(英文)	2021－06	88.00	1338
物理科学基础数学.第1卷,齐次边值问题、傅里叶方法和特殊函数(英文)	2021－07	108.00	1339
离散量子力学(英文)	2021－07	68.00	1340
核磁共振的物理学和数学(英文)	2021－07	108.00	1341
分子水平的静电学(英文)	2021－08	68.00	1342
非线性波:理论、计算机模拟、实验(英文)	2021－06	108.00	1343
石墨烯光学:经典问题的电解解决方案(英文)	2021－06	68.00	1344
超材料多元宇宙(英文)	2021－07	68.00	1345
银河系外的天体物理学(英文)	2021－07	68.00	1346
原子物理学(英文)	2021－07	68.00	1347
将光打结:将拓扑学应用于光学(英文)	2021－07	68.00	1348
电磁学:问题与解法(英文)	2021－07	88.00	1364
海浪的原理:介绍量子力学的技巧与应用(英文)	2021－07	108.00	1365
多孔介质中的流体:输运与相变(英文)	2021－07	68.00	1372
洛伦兹群的物理学(英文)	2021－08	68.00	1373
物理导论的数学方法和解决方法手册(英文)	2021－08	68.00	1374
非线性波数学物理学入门(英文)	2021－08	88.00	1376
波:基本原理和动力学(英文)	2021－07	68.00	1377
光电子量子计量学.第1卷,基础(英文)	2021－07	88.00	1383
光电子量子计量学.第2卷,应用与进展(英文)	2021－07	68.00	1384
复杂流的格子玻尔兹曼建模的工程应用(英文)	2021－08	68.00	1393
电偶极矩挑战(英文)	2021－08	108.00	1394
电动力学:问题与解法(英文)	2021－09	68.00	1395
自由电子激光的经典理论(英文)	2021－08	68.00	1397

刘培杰数学工作室
已出版(即将出版)图书目录——原版影印

书　　名	出版时间	定　价	编号
曼哈顿计划——核武器物理学简介(英文)	2021－09	68.00	1401
粒子物理学(英文)	2021－09	68.00	1402
引力场中的量子信息(英文)	2021－09	128.00	1403
器件物理学的基本经典力学(英文)	2021－09	68.00	1404
等离子体物理及其空间应用导论. 第1卷,基本原理和初步过程(英文)	2021－09	68.00	1405
拓扑与超弦理论焦点问题(英文)	2021－07	58.00	1349
应用数学:理论、方法与实践(英文)	2021－07	78.00	1350
非线性特征值问题:牛顿型方法与非线性瑞利函数(英文)	2021－07	58.00	1351
广义膨胀和齐性:利用齐性构造齐次系统的李雅普诺夫函数和控制律(英文)	2021－06	48.00	1352
解析数论焦点问题(英文)	2021－07	58.00	1353
随机微分方程:动态系统方法(英文)	2021－07	58.00	1354
经典力学与微分几何(英文)	2021－07	58.00	1355
负定相交形式流形上的瞬子模空间几何(英文)	2021－07	68.00	1356
广义卡塔兰轨道分析:广义卡塔兰轨道计算数字的方法(英文)	2021－07	48.00	1367
洛伦兹方法的变分:二维与三维洛伦兹方法(英文)	2021－08	38.00	1378
几何、分析和数论精编(英文)	2021－08	68.00	1380
从一个新角度看数论:通过遗传方法引入现实的概念(英文)	2021－07	58.00	1387
动力系统:短期课程(英文)	2021－08	68.00	1382
几何路径:理论与实践(英文)	2021－08	48.00	1385
论天体力学中某些问题的不可积性(英文)	2021－07	88.00	1396
广义斐波那契数列及其性质(英文)	2021－08	38.00	1386
对称函数和麦克唐纳多项式:余代数结构与Kawanaka恒等式(英文)	2021－09	38.00	1400
杰弗里·英格拉姆·泰勒科学论文集:第1卷. 固体力学(英文)	2021－05	78.00	1360
杰弗里·英格拉姆·泰勒科学论文集:第2卷. 气象学、海洋学和湍流(英文)	2021－05	68.00	1361
杰弗里·英格拉姆·泰勒科学论文集:第3卷. 空气动力学以及落弹数和爆炸的力学(英文)	2021－05	68.00	1362
杰弗里·英格拉姆·泰勒科学论文集:第4卷. 有关流体力学(英文)	2021－05	58.00	1363

刘培杰数学工作室
已出版(即将出版)图书目录——原版影印

书　　名	出版时间	定　价	编号
非局域泛函演化方程:积分与分数阶(英文)	2021-08	48.00	1390
理论工作者的高等微分几何:纤维丛、射流流形和拉格朗日理论(英文)	2021-08	68.00	1391
半线性退化椭圆微分方程:局部定理与整体定理(英文)	2021-07	48.00	1392
非交换几何、规范理论和重整化:一般简介与非交换量子场论的重整化(英文)	2021-09	78.00	1406
数论论文集:拉普拉斯变换和带有数论系数的幂级数(俄文)	2021-09	48.00	1407
挠理论专题:相对极大值,单射与扩充模(英文)	2021-09	88.00	1410
强正则图与欧几里得若尔当代数:非通常关系中的启示(英文)	2021-10	48.00	1411
拉格朗日几何和哈密顿几何:力学的应用(英文)	2021-10	48.00	1412
时滞微分方程与差分方程的振动理论:二阶与三阶(英文)	2021-10	98.00	1417
卷积结构与几何函数理论:用以研究特定几何函数理论方向的分数阶微积分算子与卷积结构(英文)	2021-10	48.00	1418
经典数学物理的历史发展(英文)	2021-10	78.00	1419
扩展线性丢番图问题(英文)	2021-10	38.00	1420
一类混沌动力系统的分歧分析与控制:分歧分析与控制(英文)	2021-11	38.00	1421
伽利略空间和伪伽利略空间中一些特殊曲线的几何性质(英文)	2022-01	68.00	1422
一阶偏微分方程:哈密尔顿—雅可比理论(英文)	2021-11	48.00	1424
各向异性黎曼多面体的反问题:分段光滑的各向异性黎曼多面体反边界谱问题:唯一性(英文)	2021-11	38.00	1425
项目反应理论手册.第一卷,模型(英文)	2021-11	138.00	1431
项目反应理论手册.第二卷,统计工具(英文)	2021-11	118.00	1432
项目反应理论手册.第三卷,应用(英文)	2021-11	138.00	1433
二次无理数:经典数论入门(英文)	2022-05	138.00	1434
数,形与对称性:数论,几何和群论导论(英文)	2022-05	128.00	1435
有限域手册(英文)	2021-11	178.00	1436
计算数论(英文)	2021-11	148.00	1437
拟群与其表示简介(英文)	2021-11	88.00	1438
数论与密码学导论:第二版(英文)	2022-01	148.00	1423

刘培杰数学工作室
已出版(即将出版)图书目录——原版影印

书　　名	出版时间	定　价	编号
几何分析中的柯西变换与黎兹变换:解析调和容量和李普希兹调和容量、变化和振荡以及一致可求长性(英文)	2021—12	38.00	1465
近似不动点定理及其应用(英文)	2022—05	28.00	1466
局部域的相关内容解析:对局部域的扩展及其伽罗瓦群的研究(英文)	2022—01	38.00	1467
反问题的二进制恢复方法(英文)	2022—03	28.00	1468
对几何函数中某些类的各个方面的研究:复变量理论(英文)	2022—01	38.00	1469
覆盖、对应和非交换几何(英文)	2022—01	28.00	1470
最优控制理论中的随机线性调节器问题:随机最优线性调节器问题(英文)	2022—01	38.00	1473
正交分解法:涡流流体动力学应用的正交分解法(英文)	2022—01	38.00	1475
芬斯勒几何的某些问题(英文)	2022—03	38.00	1476
受限三体问题(英文)	2022—05	38.00	1477
利用马利亚万微积分进行 Greeks 的计算:连续过程、跳跃过程中的马利亚万微积分和金融领域中的 Greeks(英文)	2022—05	48.00	1478
经典分析和泛函分析的应用:分析学的应用(英文)	2022—03	38.00	1479
特殊芬斯勒空间的探究(英文)	2022—03	48.00	1480
某些图形的施泰纳距离的细谷多项式:细谷多项式与图的维纳指数(英文)	2022—05	38.00	1481
图论问题的遗传算法:在新鲜与模糊的环境中(英文)	2022—05	48.00	1482
多项式映射的渐近簇(英文)	2022—05	38.00	1483
一维系统中的混沌:符号动力学,映射序列,一致收敛和沙可夫斯基定理(英文)	2022—05	38.00	1509
多维边界层流动与传热分析:粘性流体流动的数学建模与分析(英文)	2022—05	38.00	1510
演绎理论物理学的原理:一种基于量子力学波函数的逐次置信估计的一般理论的提议(英文)	2022—05	38.00	1511
R^2 和 R^3 中的仿射弹性曲线:概念和方法(英文)	2022—08	38.00	1512
算术数列中除数函数的分布:基本内容、调查、方法、第二矩、新结果(英文)	2022—05	28.00	1513
抛物型狄拉克算子和薛定谔方程:不定常薛定谔方程的抛物型狄拉克算子及其应用(英文)	2022—07	28.00	1514
黎曼-希尔伯特问题与量子场论:可积重正化、戴森-施温格方程(英文)	2022—08	38.00	1515
代数结构和几何结构的形变理论(英文)	2022—08	48.00	1516
概率结构和模糊结构上的不动点:概率结构和直觉模糊度量空间的不动点定理(英文)	2022—08	38.00	1517

刘培杰数学工作室

已出版(即将出版)图书目录——原版影印

书　名	出版时间	定　价	编号
反若尔当对:简单反若尔当对的自同构	2022－07	28.00	1533
对某些黎曼一芬斯勒空间变换的研究:芬斯勒几何中的某些变换	2022－07	38.00	1534
内诣零流形映射的尼尔森数的阿诺索夫关系	即将出版		1535
与广义积分变换有关的分数次演算:对分数次演算的研究	即将出版		1536
强子的芬斯勒几何和吕拉几何(宇宙学方面):强子结构的芬斯勒几何和吕拉几何(拓扑缺陷)	即将出版		1537
一种基于混沌的非线性最优化问题:作业调度问题	即将出版		1538
广义概率论发展前景:关于趣味数学与置信函数实际应用的一些原创观点	即将出版		1539

书　名	出版时间	定　价	编号
纽结与物理学:第二版(英文)	2022－09	118.00	1547
正交多项式和 q－级数的前沿(英文)	即将出版		1548
算子理论问题集(英文)	即将出版		1549
抽象代数:群、环与域的应用导论:第二版(英文)	即将出版		1550
菲尔兹奖得主演讲集:第三版(英文)	即将出版		1551
多元实函数教程(英文)	即将出版		1552

联系地址：哈尔滨市南岗区复华四道街 10 号　哈尔滨工业大学出版社刘培杰数学工作室

网　　址：http://lpj.hit.edu.cn/

邮　　编：150006

联系电话：0451－86281378　　13904613167

E-mail:lpj1378@163.com